SPECIALIZED TREATMENT SYSTEMS

INDUSTRIAL WASTE TREATMENT PROCESS ENGINEERING

VOLUME III

Specialized Treatment Systems

Gaetano Joseph Celenza

CRC Press
Taylor & Francis Group
Boca Raton London New York

CRC Press is an imprint of the
Taylor & Francis Group, an **informa** business

First publlished 2000 by Technomic Publishing Company, Inc.

Published 2019 by CRC Press
Taylor & Francis Group
6000 Broken Sound Parkway NW, Suite 300
Boca Raton, FL 33487-2742

© 2000 by Taylor & Francis Group, LLC
CRC Press is an imprint of Taylor & Francis Group, an Informa business

First issued in paperback 2019

No claim to original U.S. Government works

ISBN 13: 978-0-367-44746-5 (pbk)
ISBN 13: 978-1-56676-769-9 (hbk)

Visit the Taylor & Francis Web site at
http://www.taylorandfrancis.com

and the CRC Press Web site at
http://www.crcpress.com

Library of Congress Catalog Card No. 99-69167

*These books are dedicated to my parents and brother
who encouraged my efforts in engineering.*

Table of Contents

Introduction

WHAT ARE THE BOOKS ABOUT?

THE title defines the scope of the books. *Industrial*, indicating problems characteristic of, and solutions aimed at, manufacturing and processing plants. *Waste Treatment*, indicating treatment of wastes, solids, and residuals generated in manufacturing. *Process Engineering*, indicating that design application and selection procedures are detailed. The tone of the technical discussions reflects industrial realities, requiring economic and site specific process engineering considerations to avoid noncompliance notoriety and fines, and to optimize capital and operating expenditures.

Waste control is treated as a single subject; sequencing process controls applied to (1) the raw materials entering the plant, (2) the generation source, (3) passage through a production plant, and (4) to a final central treatment. A facility treatment problem is addressed as two related subjects. One involving *manufacturing evaluation* to eliminate or minimize waste generation, and the other involving selection and application of appropriate *unit operations* to treat resulting effluents. The entire industrial problem is presented in a three volume set. *Volume I* details source correction, pollution prevention, development of a facility treatment system, and primary chemical treatment; *Volume II* details biological treatment unit operations; and *Volume III* details special treatment systems.

These books are written to guide experienced engineers through the various steps of industrial liquid and solid waste treatment. However, the structure of the text allows a wider application suitable to various experience levels; by beginning each chapter with a simplified explanation of applicable theory, expanding to practical design discussions, and finishing with system Flowsheets and Case Study detail calculations. The reader can "enter or leave" a chapter according to their specific needs. As a result, it can serve as a primer for students engaged in environmental engineering studies, or a "one stop" source for experienced engineers. It includes basic design principles which could be applied to municipal systems with significant industrial influents. In fact, substantial portions of the books are applicable to primary, secondary and tertiary municipal plant treatment. Conventional treatment methods are emphasized, allowing evaluation and implementation of emerging technologies akin to these methods, permitting enhancement to achieve improved performance. Innovative technologies are not downgraded, but encouraged as a choice based on understanding the basic operating principles; but not employed as a fad to replace applicable available technology.

HOW ARE THE BOOKS ORGANIZED?

Industrial Waste Treatment Process Engineering is prepared as a step-by-step implementation manual, detailing the selection and design of industrial liquid and solid waste treatment systems. Great emphasis is placed on identifying and compensating for highly variable and unpredictable manufacturing waste characteristics. The books are written as a *single source* industrial waste management text, consolidating all the process engineering principles required to evaluate a complex industrial waste problem; starting with pollution prevention and source correction, and finishing with "end-of-pipe" treatment. A complete treatment facility is discussed, not just an isolated unit process; including the ancillary as well as the major equipment. Emphasis is placed on the fate of the contaminants, warning of potential problems resulting from by-products formed; preventing a waste treatment problem from becoming an air emission or solid disposal problem.

What makes the books unique is the level of process engineering details included in both the facility evaluation and unit operations sections. The facility evaluation includes a step-by-step review of each major and support manufacturing operation; identifying probable contaminant discharges, practical prevention measures, and point source control procedures. This general plant review is followed by procedures to conduct a site specific pollution control program.

The unit operation chapters contain all the required details to complete a treatment process design; including:

(1) Basic concepts explaining the applicable design relations

(2) How adjustments are made for specific industrial waste characteristics

(3) How basic concepts are applied to equipment sizing

(4) Available equipment and system configurations

(5) Guidelines for selecting the appropriate equipment

(6) Process operating limits

(7) Reported industrial performance data

(8) Required design data

(9) Case studies with detailed calculations for the prominent applications

(10) Conceptual Engineering Flowsheets

(11) Commonly reported design and operating deficiencies

(12) A discussion of the fate of the contaminants; emphasizing total contaminant destruction

(13) General mechanical and associated engineering guidelines

PROCESS ENGINEERING PROCEDURES

The facility evaluation procedures discussed follow the process engineering considerations used in developing the manufacturing facilities. The manufacturing plant is subdivided into related production modules, which sequentially follow the production flow. This includes raw material storage, raw material preparation, the basic manufacturing process, the product separation process, product purification, the finishing operation, by-product recovery, packaging, and support facilities. The flow "in-and-out" of each of these segments is examined to identify potential pollution sources, and logically identify corrective action that would eliminate, minimize, or source correct any waste generation. Although a chemical plant is suggested, many of these separate operations are required in most product manufacturing, in one form or another. This breakdown into individual operations allows the Process Engineer to generically identify similar plant operations, and apply the suggested step-by-step analysis to implement applicable source correction actions.

The unit treatment operations discussed are those commonly applied for municipal waste or water treatment, those employed for industrial tertiary treatment, and those used for thermal destruction of hazardous wastes. Multiple references are included to identify sources of the design criteria cited, so that the interested reader can further review the concept details. Municipal and water treatment applications are frequently mentioned because they are the source of much of the available operating data. Although some of the principles described originate from operating data from these facilities,

final design must be adjusted for industrial conditions to compensate for differences in treatment reactivity, waste volume variability, and the peculiar properties of industrial wastes. Not only do the properties of industrial wastes vary from municipal or water treatment influents, they are not consistent from industry to industry, or many times from plant to plant in the same industry.

In most cases, unit operation process design methodology is described in terms of models, with specific constants defining the range of applicability. The design method is further analyzed in terms of step-by-step procedures required to complete a system design. These models are supplemented by:

(1) Commonly applied capacity limits expressed in terms of hydraulic loadings, system residence time, or organic loadings

(2) Specific waste characteristics affecting capacity limits and process performance, and therefore the limits of the unit operation

(3) Practical guidelines for selecting and applying the unit operation

Industrial application of these models and associated process engineering procedures requires that an adjustment be made for the waste characteristics and variations, as well as scale-up from theoretical to actual operating conditions. Each chapter prescribes the following:

(1) The required design data that must be obtained from site specific testing

(2) Historical industrial performance data to gauge the applicability to meet required effluent quality

(3) Operating limits for the unit operation

(4) By-products produced from the process so that the fate of the contaminants can be evaluated

(5) General engineering limits to define the mechanical, control, or environmental limits of the system

(6) In some cases, safety factors to upgrade a system design from the theoretical to the operating level

The unit process engineering discussions culminate with Case Studies (calculations) and a Preliminary Concept Flowsheet.

The format in which the unit operations are presented could mislead the reader into ignoring the design criteria verification requirements, believing that a simple parameter can be used to size a system, leaving the details to the plant operator. The most important elements of Process Engineering are not, and cannot be, included in the discussions—common sense, engineering skills, and experience. All Process Engineering requires interpretation of available design methods to suit a site specific problem. Any shortcuts can only lead to a poor operating, or in extreme cases, a failed system; leading to noncompliance, fines, prohibitive operating costs, and endless capital to "patch" the system.

Finally, in reviewing the suggested steps in developing a treatment system the Process Engineer may not agree that a certain step is necessary, or reason that it could be scaled down considerably, or that it will require more attention than indicated. If this book has forced the engineer to consider all elements of a design, adjusting some elements from that indicated in the text to suit specific requirements, *it has served its primary purpose.*

PROCESS CALCULATIONS

Following the basic concepts, each unit operation chapter contains a complete discussion of the process design variables, detailing the individual elements influencing the equipment design and operation. The use of these variables are illustrated through process calculations, referred to as Case Studies. They represent a carefully sequenced procedure to size the system equipment within a range of acceptable operating parameters. They are called Case Studies because adequate examples are included to illustrate the various processing conditions and configurations, emphasizing operating limitations.

In some examples the results indicate (for the conditions evaluated) that the process has limited, or no, applicability because of the waste characteristics, the waste variability, or practical implementation. This is done to emphasize limitations encountered in full scale operations.

Finally, the calculations included in the texts were computer generated. The results were not always rounded out to whole numbers to avoid having to manually check each operation to the sensitivity of a preceding value, and because it could increase the chance of transferring an incorrect number to the text. Therefore, the number of decimals presented are a matter of convenience, and not intended to demonstrate the precision of an evaluation.

PROCESS FLOWSHEETS

Preliminary Conceptual Flowsheets are included in most unit operations to represent typical, not all, configurations employed; and to enable the Process Engineer to *initiate* the required Process and Engineering Diagrams. They will have to be altered and enhanced to meet site specific requirements. The instrumentation indicated is basic, and in many cases minimum, and must be upgraded to meet the specific plant control philosophy.

The instrumentation symbols are those commonly used in Chemical Process Engineering. The combined symbols, represented in a "balloon," are summarized and explained in Table I.1.

TABLE I.1. Instrumentation Symbols.

Symbol	Placement of Symbol		
	Primary	Second	Final
A	Analysis—pH, turbidity, etc.	Alarm	—
C	—	—	Control (ler)
E	Voltage	Element	—
F	Flow	—	—
G	—	—	Gauge
H	—	—	High (value)
I	—	Indicating	—
K	Time	—	—
L	Level	—	Low (value)
P	Pressure	—	—
R	—	Recording	—
S	Speed	—	Switch
T	Temperature	—	Transmit(ter)
V	—	—	Valve

A typical loop would include a sequence of elements, using a combination of symbols listed in Table 1, to indicate a complete control system. As an example:

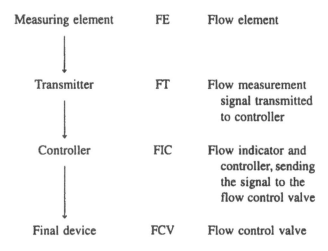

Measuring element	FE	Flow element
Transmitter	FT	Flow measurement signal transmitted to controller
Controller	FIC	Flow indicator and controller, sending the signal to the flow control valve
Final device	FCV	Flow control valve

Complete discussions of instrumentation applied to waste treatment systems can be reviewed in Environmental Engineering Texts [1].

HOW SHOULD THE BOOKS BE USED?

The three volume book set, and the chapters in the books, have been presented in the order that Process Engineering is commonly applied. Accordingly, the industrial waste treatment design can be initiated starting with Chapter I-1.

First, a Pollution Prevention Evaluation should be conducted within the plant following the procedures outlined in Chapter I-1.

Next, a preliminary central treatment evaluation should be conducted to treat the remaining wastes generated, as discussed in Chapter I-2. This includes:

- a complete waste characterization
- a regulatory review
- preliminary selection of treatment system candidates
- laboratory tests to screen viable treatment options
- pilot studies to develop design criteria

Finally, a complete treatment facility concept should be developed, as discussed in Chapter I-3.

After the preliminary concept has been completed, detail design of the specific unit processes comprising the selected treatment system can be completed. The applicable unit operations are presented in the texts as follows:

- equalization in Volume I, Chapter I-4
- chemical treatment in Volume I, Chapters I-5 to I-9
- flotation in Volume I, Chapter I-10
- aerobic biological treatment in Volume II, Chapters II-2 to II-5
- anaerobic biological treatment in Volume II, Chapter II-7
- aerobic sludge treatment in Volume II, Chapter II-6
- anaerobic sludge treatment in Volume II, Chapter II-7
- sedimentation in Volume II, Chapter II-8
- industrial and hazardous waste incineration in Volume III, Chapter III-1
- adsorption in Volume III, Chapter III-2
- ion Exchange in Volume III, Chapter III-3
- wastewater stripping in Volume III, Chapter III-4
- filtration in Volume III, Chapter III-5
- membrane technology in Volume III, Chapter III-6
- dewatering in Volume III, Chapter III-7

The unit operations should be evaluated in a step-by-step procedure involving the following:

- A preliminary estimate of the equipment size should be made, using the common loadings cited, to establish site and economic feasibility.
- Determine the proper configuration for the technology selected.
- Review the operating limitations for the technology selected.
- Review the reported performance data.
- Do the detail process calculations required to establish the size of various pieces of equipment, according to the procedures described in the Process Engineering Criteria section and Case Study calculations.
- Review the General Engineering Criteria for engineering, construction, or operating limitations.
- Review the reported deficiencies as a check list to avoid commonly encountered process, mechanical, and general engineering pitfalls.
- Develop a Preliminary Process Flowsheet, as illustrated in the individual chapters, tailored to the *specific site and process conditions*.
- Work with the Instrument Engineer, and other engineering disciplines to develop the Process Flow Diagram, the Process and Instrument Diagram, and the plant layout.

This outlines the recommended use of the books' contents in the development of an industrial waste treatment system, with further details provided in the individual chapters.

REFERENCES

1. Water Environment Federation: *WPCF Manual of Practice No OM 6: Process Instrumentation and Control—Operation and Maintenance,* 1984.

Industrial and Hazardous Waste Incineration

Incineration is employed to thermally destroy combustible organics: generated as hazardous wastes, plant residues, or secondary treatment plant sludges.

INCINERATORS are used to destroy combustible hazardous wastes, manufacturing wastes and residues, and treatment plant sludges. As illustrated in Figure 1.1, major incinerator ancillary components include waste storage, feed preparation, combustion, optional heat recovery, and exhaust emission controls. Waste storage facilities assure a continuous and consistent feed supply to the incinerator to maintain a stable combustion reaction. Storage is part of a feed preparation system, which can include pretreatment, conditioning, and blending.

An incinerator functions as a controlled reactor designed to convert combustible feed materials to the inert products indicated in Table 1.1. The required reaction is perpetuated by sustaining a combustion flame as a constant energy source and maintaining suitable *reaction oxidation* conditions by providing adequate residence time, a controlled chamber temperature, and proper (mixing) turbulence. The combustion process can be conducted in a single chamber or in two stages, with the latter common for solid waste materials. In a two-stage unit, the primary chamber promotes pyrolysis, volatilization, and initial combustion, whereas the secondary combustion chamber serves as a fume incinerator to complete the oxidation process.

Because of the high energy levels required to promote oxidation, heat recovery and minimizing supplementary fuel demand are important cost saving considerations. Pollution control is mandatory to meet the many regulatory requirements governing air emissions from combustion equipment. This could include particulate removal, acid scrubbing and neutralization, sulfur and nitrogen oxides control, and controlling products of incomplete combustion.

Combustion occurs in a controlled environment involving three primary components: (1) energy releasing compounds containing carbon, hydrogen and sulfur, (2) the other reactant oxygen, and (3) nonreactants such as nitrogen, water, and excess oxygen, acting as diluents or a heat sink in the system. Other waste constituents serve no useful purpose, complicating the combustion process by generating air emissions or ash that must be treated and disposed or by depositing on critical combustion equipment components.

Incineration is commonly associated with destruction of high organic content wastes with heating values adequate to sustain the combustion reaction. However, mandatory detoxification requirements for contaminated site remediation has resulted in incineration of toxic wastes and soils with negligible heat content to completely destroy trace amounts of hazardous components using high temperature, rapid oxidation. In such cases a fourth component contributes to the oxidation process, supplementary fuel to supply energy to support the combustion. This represents the most complex of combustion systems since most of the clean exhaust components are fuel combustion products, whereas the waste contributes all the reaction products of concern. This is because the process is driven to achieve complete destruction of low concentration feed components, resulting in a high potential for ash and slag generation as well as forming a variety of exhaust products difficult to identify or evaluate.

Generally, the combustion process is simple, occurring at low pressures and elevated temperatures. However, the combustion chemistry is complex, involving not only the simple oxidation of the feed constituents but the formation of potentially difficult to dispose ash and minute quantities of varied exhaust compounds of concern. The health risks associated with incinerator exhaust combustion products and incomplete products of combustion are constantly being evaluated and debated. Because of these and other complexities, designing an incineration system requires an understanding of the total combustion system and its related components.

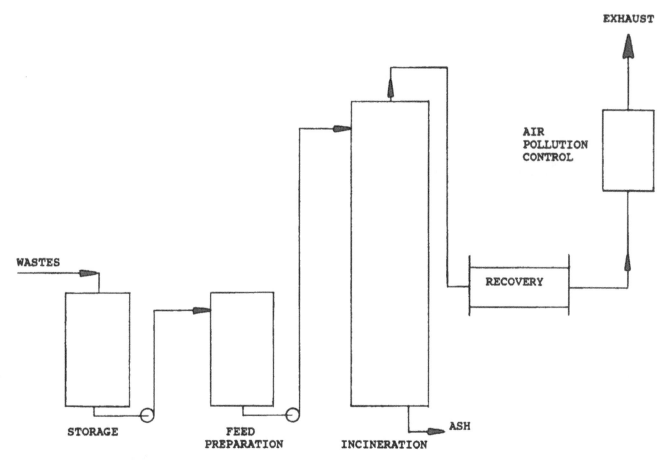

Figure 1.1 Incinerator system sketch.

BASIC CONCEPTS

Combustion involves the rapid chemical reaction of oxygen with volatile waste components and fuel. The primary function of an incinerator is reducing the volume of bulk wastes or converting regulated waste components to less hazardous secondary compounds. Oxidation of fuel and organic combustible components results in heat release, which enables the reactants to reach and sustain the ignition temperature, which with adequate retention time and reactant mixing (turbulence) results in high destruction efficiency. Although high oxidation efficiency can be easily achieved, the waste components and their partition path in the combustion process greatly affect the viability of incineration as a treatment method.

At a glance, the combustion process is simple enough. Feed material is converted to final products discharged either as an ash or exhaust products. The actual conversion is a complex one in which feed volatilization and primary combustion can occur in the first stage, followed by conversion to final products in the secondary chamber. Depending on the feed's physical state, this can occur in a single chamber (liquid and gaseous feed) or in two stages (solid wastes). Feed partition to ash and the exhaust, and their final composition, is governed by (1) the combustion temperature and in some advanced combustion systems pressure, (2) feed composition and concentration, and (3) reaction kinetics preferentially driving the products toward equilibrium concentrations.

Combustion conditions are governed by resulting *energy* and *material* transfers, whereas waste conversion is gov-

TABLE 1.1. Combustion Products.

Typical Products		
C	to	CO_2
H	to	H_2O
S	to	SO_2
Cl	to	HCl
F	to	HF
O	to	O_2
N	to	N_2
N	to	NOx
H_2O	to	Steam
Inert Volatile	to	Ash
Metals	to	Fumes
Metals	to	Ash

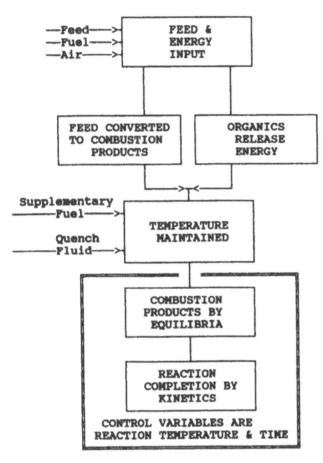

Figure 1.2 Incinerator operating functions.

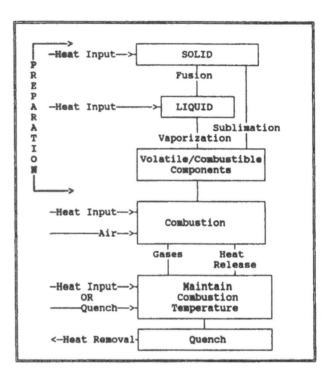

Figure 1.3 Heat transfer mechanisms.

erned by reaction kinetics and chemical equilibrium. These factors are interrelated as indicated in Figure 1.2. The process is initiated by feed and energy input, and completed by subsequent "absorption" of sensible heat by the exhaust products. The energy exchanges drives the combustion gases to an equilibrium temperature. The combustion chamber residence time and temperature govern the system thermodynamics, specifically the reaction rate and the equilibria products formed.

ENERGY TRANSFER

Combustion temperature is the basic operating control. Energy released from the waste is transferred to the combustion products to reach a combustion temperature adequate to support organic waste destruction by continuous oxidation. Heat transfer mechanisms involve *release* of heat of combustion from the oxidation process, balanced by *absorption* of the feed water heat of vaporization, heats of fusion or sublimation, and the process exhaust gases sensible heat capacity. All the energy steps are illustrated in Figure 1.3.

The process starts with heat input to volatilize solid and liquid waste constituents to gasses and suitable components

that sustain the combustion process. Heat of combustion released from the volatile waste organics and supplementary fuel are primary energy sources, the released energy transferred to the combustion products as sensible heat. The heat release and absorption mechanisms result in an equilibrium temperature that may not be within the desired combustion range. Supplementary fuel input or quench are added until the desired oxidation chamber temperature is achieved. Therefore, two dominant energy transfers occur until a suitable energy equilibrium is reached. Combustion energy released heating the gases, and the excess air or water quench limiting the chamber temperature to the desired level. The overall energy balance is indicated in Figure 1.4. The specific energy mechanisms that contribute to the combustion are detailed in Figures 1.5 to 1.7, categorized as large or small

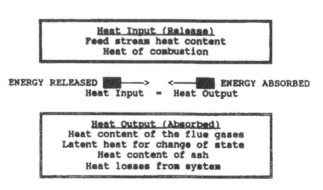

Figure 1.4 Energy balance.

```
                    Heat of Combustion

              ██──────>RELEASED

The heat released when one unit of organic is
converted to combustion products is defined as
the heat of combustion, sometimes called the heat
of reaction.  Using methane as an example:

      CH₄  +  2 O₂  =  CO₂  +  2 H₂O

In this case, the Higher Heating Value at a reference
temperature 20 degrees Centigrade is 192 Kcal per mole
of methane. The higher heating value includes the heat
of evaporation of the  water which  would be available
when  the  steam  condenses.   However,  since  most
combustion  processes operate  well above  the heat of
condensation of water, the heat of condensation  is
excluded, and the lower (or net) heating value is used
in combustion calculations. When hydrogen is not one
of the combustion products the higher and lower
heating values are identical. Although the heats of
combustion of many organic compounds are readily
available, because the composition of many wastes are
usually unknown or highly variable, the heat of
combustion is best determined by laboratory tests.
```

Figure 1.5 Large heat releases from waste to combustion chamber.

heat sources based on the heat capacity per unit weight of the reactant or product. Energy balance details are developed in the Appendix and illustrated in Case Studies 24, 25, and 26.

MATERIAL BALANCES

An incinerator material balance is based on a stoichiometric accounting of the reactants, flue gas products, and residues. The single criteria for a material balance calculation is that the weight of all materials entering (Material Input) the incinerator must equal the weight of all the materials discharged (Material Output), as indicated in Figure 1.8.

The accuracy of a material balance depends on a detailed analysis of the waste feeds, either identifying the chemical constituents or establishing a total ultimate analysis. In most cases a complete analysis of all chemical constituents is impractical, and the waste is identified according to ash, water, and volatile composition. The volatile material is further characterized according to its carbon, hydrogen, nitrogen, sulfur, and halide content. The waste is also analyzed

```
                    Sensible Heats *

              ██<──────ABSORBED

Sensible heat: The heat content of substances
(ENTHALPY) at a specific temperature is based
on the sensible heat content of the individual
components at that temperature, relative to a base
temperature.  In the case of liquids, where the
specific heat is relatively constant over the
temperature range encountered, the heat content is a
product of the weight times the specific heat constant
times the temperature difference encountered; i.e.,

Heat capacity = pounds per hour·M_cp·(T_out - T_in).

Unlike liquids, the specific heat of gases
(BTU per pound per degree change) changes
significantly with temperature, so that method chosen to
calculate the enthalpy (heat content at a specific
temperature) depends on the accuracy required. These include:

 -Use enthalpy tables for the common components such
  as air, oxygen, nitrogen, carbon dioxide, etc.
 -Use sensible heat equations such as:

    M_cp = a + bT + cT² dT³

  to establish the enthalpy value at the desired
  temperatures.

 -Approximate the value using an average specific
  heat value over the temperature range considered.

 For the accuracy required for most incineration
 evaluations use of an average specific heat is
 adequate.

              ██──────> HEAT RELEASED

* Sensible heat can also be RELEASED from preheated
feed streams.
```

Figure 1.7 Small heat sinks (absorbers), component enthalpy.

for toxic components specified by regulatory requirements. Material balance details are developed in the appendix and illustrated in Case Studies 24, 25, and 26.

REACTION THERMODYNAMIC PRINCIPLES

Combustion thermodynamic principles are similar to those described for chemical reactions in Chapter I-5, involving (1) the Arrhenius principle relating reaction kinetics, temperature, and energy, (2) reaction rates governing reactor retention time, and (3) equilibrium principles establishing the exhaust products (pollutants). In fact, reaction rate kinetics is seldom a criteria considered in incinerator design. However, the process will be controlled by thermodynamic considerations, and in the actual operation the *science* of thermody-

```
                    Latent Heats

              ██<──────ABSORBED

The heat required to change phases is termed latent
heat, the most common being that required to evaporate
water into steam.  Common phase changes in incineration
include the following:

Heat of fusion:  solid to liquid: ash to molten form.
Heat of sublimation:  solid to gas:  Release of
                             volatiles from solids
Heat of vaporization: liquid to gas:
                             evaporation of water

The heat absorbed in the phase change, must be accounted
for in the incineration energy balance
```

Figure 1.6 Large heat sinks (absorbers), phase change energy.

```
  ┌─────────────────────────┐
  │ Material Input          │
  │      Wastes             │
  │       Air               │
  │       Fuel              │
  │                         │
  │ Material Output         │
  │    Flue gases           │
  │  Ash or Blowdown        │
  └─────────────────────────┘
```

Figure 1.8 Material balance.

namics will over rule the *ART* of selecting operating conditions. Incinerator *primary* combustion chamber design is defined by selecting

(1) A reasonable heat release flux or residence time consistent with an economical vessel design
(2) A reaction temperature in the 980 to 1200°C (1800 to 2200°F) range
(3) A proper physical geometry to provide combustion chamber turbulence and mixing
(4) Mechanical criteria aimed at effective liquid feed atomization or solid waste conditioning to enhance combustion

Specific criteria are governed by the waste characteristics, which control the *primary* chamber volume and oxidation temperature. Because of the large quantities of materials processed in an incinerator, resulting in significant heat release and exhaust gases volumes, residence time is a limited incinerator variable. The wide range of reaction times available in chemical and biological reactors, for relatively smaller processing volumes, is not feasible in combustion treatment processes. The incinerator size is a balance of (high) temperature and (low) time, controlled by the limits of a reasonable and economical combustion vessel design. In addition to oxidation time, some portion of the residence time is required to convert aqueous and solid wastes into a volatile combustible state.

The *secondary combustion chamber* design is generally based on providing adequate retention time to convert the intermediate primary chamber gaseous products into acceptable exhaust products. In the final combustion phase there are a multitude of potential reactions involving volatile unreacted feed components, as well as products and intermediates, resulting in minute quantities of a variety of end products. The total combustion chemistry is too complex to be completely understood, or to fully define. In the past any reaction beyond the primary combustion was irrelevant. However, there is renewed interest in all secondary reactions. Their significance magnified in hazardous waste incinerator design because of specifically regulated exhaust chemicals that are considered potentially hazardous in minute concentrations.

Chemical equilibrium principles affords the opportunity to qualitatively evaluate and estimate quantities of specific regulated exhaust chemicals, based on a suggested reaction path, and assuming that equilibrium conditions have been achieved. In most cases equilibrium conditions are not achieved and the estimated pollutants can be considered *worst-case* results. An obvious limitation of theoretical equilibrium evaluation is that the reaction path must be known. This is complicated by the fact that all the possible gas phase reactions are not known, nor will all the exhaust constituents actually be known until the system is in operation and the exhaust monitored. Even then, you can only monitor what you suspect or expect in the exhaust.

In estimating the possible exhaust constituents, the Process Engineer must first identify the pollutants of concern, assume a reaction path, and then calculate the equilibrium value as an approximation of the maximum emission level. Combustion reactions have been presented in early Chemical Engineering texts, such as that presented by J.C. Hottel of Massachusetts Institute of Technology [11]. This information is commonly reproduced in many incinerator texts in the form represented in Figure 1.9. Reactions contributing to the formation of some commonly regulated pollutants, or potential pollutants of concern, are indicated in Table 1.2.

The data cited in Figure 1.9 and Table 1.2 are only a few of the possible reactions which could occur, but they are reactions affecting exhaust characteristics, and of general interest in combustion systems. Kinetic data for formation or destruction of *principal organic hazardous constituents* (POHCs), *products of incomplete combustion* (PICs), or *dioxins/furans* are not readily available.

The use of Figure 1.1 data can be illustrated by the simple reaction:

$$\tfrac{1}{2}N_2 + \tfrac{1}{2}O_2 = NO$$

The corresponding equilibrium expression for this mechanism is

$$K_p = [NO]/[N_2]^{\frac{1}{2}}[O_2]^{\frac{1}{2}}$$

Realizing that for any given temperature the value of K_p is a constant, the formation of NO under a variety of conditions can be evaluated. Qualitatively the following conditions will influence NO formation:

(1) An increased reaction temperature will increase K_p, increasing nitrogen oxide formation.
(2) Excess oxygen is commonly employed to drive the principle combustion reaction toward completion. However, excessive air increases the nitrogen and oxygen content, resulting in increased nitrogen oxide formation.
(3) Combined waste nitrogen increases the potential for nitrogen oxides formation.
(4) Operating the primary chamber at oxygen deficient conditions reduces the potential for nitrogen oxide formation.
(5) Recycling of exhaust gas to lower the exhaust chamber temperature results in a lowering of the K_p value, thereby lowering nitrogen oxide.

Based on Case Study 24 data, the effect of increasing nitrogen concentration, assuming the above reaction path, is illustrated in Table 1.3. Realizing that estimating exhaust concentrations from equilibrium constants may not be pre-

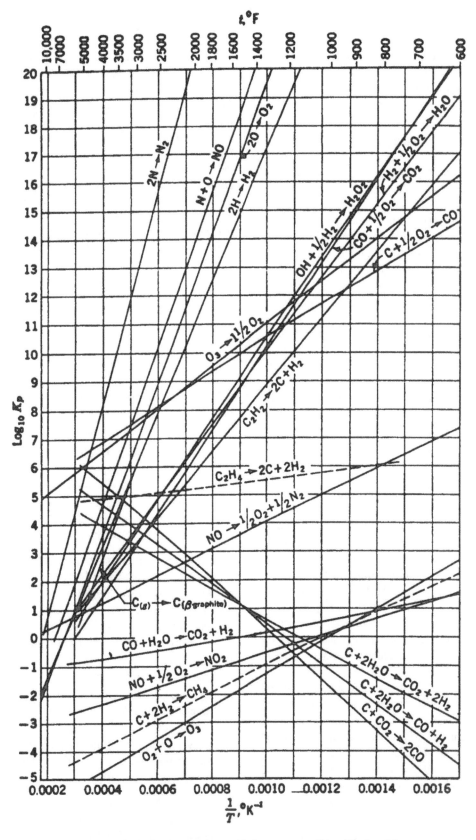

Figure 1.9 Combustion reaction equilibrium constants. [11, public domain]

6

TABLE 1.2. Common Exhaust Products.

Carbon Monoxide Mechanisms		
$CO + \frac{1}{2} O_2$	$=$	CO_2
$C + CO_2$	$=$	$2\,CO$
$C + \frac{1}{2} O_2$	$=$	CO
$C + H_2O$	$=$	$CO + H_2$
$CO + H_2O$	$=$	$CO_2 + H_2$
Nitrogen Oxide Mechanisms		
$\frac{1}{2} N_2 + \frac{1}{2} O_2$	$=$	NO
$NO + \frac{1}{2} O_2$	$=$	NO_2
$NO + \frac{1}{2} N_2$	$=$	N_2O
$N + O$	$=$	NO
Sulfur Mechanisms		
SO_2	$=$	$SO + \frac{1}{2} O_2$
$SO_2 + 3 H_2$	$=$	$H_2S + 2 H_2O$
Other Mechanisms		
$\frac{1}{2} N_2 + 3/2 H_2$	$=$	NH_3
$C + \frac{1}{2} N_2$	$=$	CN

TABLE 1.4. New Technologies.

Wet air oxidation
Molten salt
Circulating bed
Plasma arc
Advanced electric reactor
High temperature fluid wall
Vertical tube reactor
Pyrolysis
Supercritical Water
Oxygen Enriched
Infrared System
Westinghouse/O'Connor

cise, it does allow an analysis of the effects of operating parameters on final emission characteristics.

INCINERATION SYSTEMS

Incinerators can be classified into two broad categories, as either a fixed chamber or a moving bed. In turn, fixed chambers can be either single or dual chamber; single chambers are primarily employed for liquids, whereas dual chamber units can be employed for solids, sludge, refuse, or viscous liquids. Moving beds fall into a variety of types, the most common being rotary kilns, multiple hearths, moving grate and fluidized beds. Rotary kilns, fluidized beds, circulating beds, multiple hearths, and liquid injection incinerators are commonly employed for industrial waste treatment. Some emerging technologies are listed in Table 1.4.

The primary combustion chamber of an incinerator can include a wide variety of configurations, as discussed below. In the past, single stage combustion was viable, the process driven to complete the reaction using massive air supplies, and the exhaust discharged untreated. Because air pollution control is mandatory, and heat recovery a viable option, uncontrolled air incineration is neither technically or economically feasible. Instead, oxygen is introduced using a forced or induced air system, controlled to optimize system performance, and the exhaust treated to meet regulatory requirements. Pressurized incinerators are viable if the downstream pressure losses are moderate, allowing transport of the combustion chamber gases without an excessive operating pressure. Where downstream equipment losses are significant, induced air systems are used at the incinerator outlet as transport devices to downstream air pollution or recovery equipment, or as part of a combined, complex transport system to manage the process flows. Induced fans are commonly employed in hazardous waste incinerators to assure a slightly negative chamber pressure, preventing leakage into the workplace.

When applied to industrial or hazardous wastes, incinerators are part of a complete waste treatment system involving

TABLE 1.3. Effect of Excess Air to NOX Produced.

Excess Air, %	Moles/hr O_2	Moles/hr N_2	Total	ppm	NO Moles/hr	NO lb/hr	Quench lb/hr	Fuel (3) lb/hr
5	3.2	212	522	50	0.026	0.78	384	—
10	6.4	224	561	68	0.038	1.14	278.5	—
25	16.0	264	599	109	0.065	1.95	—	9
40	29.7	350	709	142	0.101	3.03	—	99
50	41.1	423	804	163	0.131	3.93	—	176
60	55.1	512	920	182	0.167	5.01	—	270
75	84.1	697	1160	209	0.242	7.26	—	465
100	174.9	1275	1910	246	0.470	14.10	—	1070

Notes:
(1) Assumed nitrogen reaction: $\frac{1}{2} \cdot N_2 + \frac{1}{2} \cdot O_2 + = NO$.
(2) Calculation based on $K = 0.001$ at $T = 1800°F$.
(3) Fuel added to maintain 1800°F temperature, compensating for quenching effect of excess air. Small quantity of excess fuel combustion air also included in calculation.
(4) Refer to Case Study No. 24 for 25% excess air condition.

considerable upstream and downstream equipment. Upstream storage is included to assure a continual and consistent feed, allowing waste feed preparation and planning of treatment campaigns, as well as provide waste storage during furnace repair or maintenance. A consistent feed composition assures a reasonably constant burn rate, extending the furnace brick and critical parts operating life, minimizing regulated incomplete combustion products releases resulting from cycling combustion conditions. Waste preparation involves sorting, sizing, and blending capabilities as part of the storage facilities, assuring that solids are reduced to combustible sizes and liquids are pumpable and concentrated to achieve acceptable heat release rates.

Downstream from the incinerator, flue gas quenching, air pollution control, and optional heat recovery may be required. Air pollution control generally includes secondary combustion to assure destruction of all organic volatiles; followed by particulate removal, NO_x reduction, and removal of all acid derivatives.

LIQUID INJECTION INCINERATOR

The major components of a liquid injector incinerator include a fuel burner to produce a combustion flame, and waste nozzles to inject liquids, all of which are enclosed in a horizontal or vertical lined combustion chamber. Waste feed nozzles can be top, side-center, or side-tangential entering, the nozzle action greatly influencing incinerator configuration and the internal process. Combustion effectiveness for any nozzle orientation depends on an intimate mixing of the vaporized waste components and oxygen throughout the chamber volume, maintained under highly oxidizing conditions. Waste is sprayed from nozzles forming a fine mist, enhancing vaporization and oxidation.

The incinerator process starts with fuel burners igniting, maintaining a flame using supplementary fuel or clean high caloric value waste, and a proportional amount of air. Poorly combustible liquid wastes and supplementary air are fed through separate nozzles, atomized into an area influenced by the flame. Additional (secondary) air is fed directly to the combustion chamber to sustain combustion conditions. Excess air will be required to drive the reaction to completion, or to temper the gases to the combustion temperature.

Combustion products exit the combustion chamber to a quench section. There they are cooled to a suitable temperature before entering the air pollution control equipment, after which they are discharged through a stack. Heat recovery may be employed to preheat feed streams or generate steam. High-nitrogen-containing waste compounds could generate excessive nitrogen oxide emissions, requiring appropriate controls such as ammonia injection into the combustion chamber or employing two-stage combustion. In a two-stage combustion system primary chamber reduction conditions are maintained by operating at a reduced temperature and deficient oxygen levels, utilizing recycled cooler gases, and

the combined cooled gases are fed to a secondary chamber to complete the combustion.

Common Design Criteria

Typical design conditions are indicated in Figure 1.10, generally representing a basic nonswirling type unit. Specific conditions depend on the selected combustion configuration. Burn rates can be increased to 890,000 kcal/hr/cubic meter (100,000 BTU/hr/cubic foot) and higher with increased combustion chamber turbulence. Excess oxygen conditions can vary from 10 to 60% of the theoretical stoichiometric value, depending on the quantity required to (1) suit quench requirements to maintain a desired reaction temperature, (2) induce turbulence, at excess air values ranging from 25 to 60%, or (3) treat hazardous wastes, with excess air values ranging from 10 to 60% and as high as 200% for extreme conditions [4,9,17].

The average reported hazardous waste incinerator capacity is 7 million kcal/hr (28 MMBTU/hr), with unit capacities ranging from 18 to 25 MMkcal/hr (70 to 100 MMBTU per hour) [16,17,19]. Physical dimensions are governed by construction and transport restrictions. In addition, the unit must be configured to restrain flame impingement to the walls, and assure geometric considerations to optimize turbu-

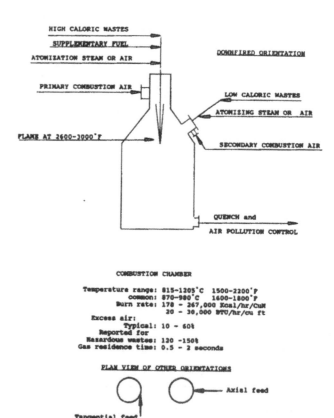

Figure 1.10 Liquid incinerator operating data (adapted from References [16,17]).

TABLE 1.5. Theoretical Combustion Chamber Physical Characteristics.

MMkcal/hr (MMBTU/hr)		cu meter (cu feet)		Diameter		Length		
				M	ft	M	ft	L/D
1.3	(5)	7	(250)	1.4	(4.7)	4.3	14	3:1
2.5	(10)	14	(500)	1.8	(6.0)	5.5	18	3:1
5.0	(20)	28	(1000)	2.3	(7.5)	7.0	23	3:1
7.1	(28)	40	(1400)	2.6	(8.4)	7.6	25	3:1
10.1	(40)	57	(2000)	2.9	(9.5)	8.5	28	3:1
20.2	(80)	113	(4000)	4.2	(13.7)	8.2	27	2:1
25.2	(100)	142	(5000)	4.5	(14.9)	8.8	29	2:1

lence conditions. Theoretical combustion volumes and corresponding physical dimensions based on nonswirl units with a burn rate of 178,000 kcal/hr/cubic meter (20,000 BTU/hr/ft^3) are illustrated in Table 1.5. Actual dimensions will be adjusted to suit economical construction criteria. If high swirl vortex configurations are employed, with burn rates up to 890,000 kcal/hr/cubic meter (100,000 BTU/hr/ft^3), the combustion chamber volumes are proportionally smaller. The total height of liquid incinerators will be extended with the inclusion of a stack, either self supporting or supported by the incinerator structure.

Configurations

Liquid incinerator configuration is primarily governed by the effectiveness of the waste atomization, to minimize furnace wall impingement by the flame or waste and maximize internal distribution of the reactants. Waste impingement is a major concern contributing to refractory erosion, and a primary consideration in establishing the chamber geometry. Waste impingement is minimized by assuring proper waste atomization and vaporization before any droplets can reach the walls, a condition highly dependent on the waste nozzle design.

Chamber volume depends on the required residence time. Ideally, the combustion chamber physical dimensions should be selected to achieve a length to equivalent diameter ratio ranging from 2 to 3:1, which is conducive to good mixing. However, any inlet flow provisions, operating conditions, or internal mechanical devices which promote turbulence eases the geometric requirements. In practice, the incinerator dimensions may have to be adjusted to conform with available space, operating access to major components, and optimizing construction costs. Organic destruction efficiency is governed by physical characteristics ensuring intimate contact between air and vaporized waste at an effective reaction temperature. These include the burner design, mixing capabilities, and waste nozzle design.

Exhaust gases are discharged to a quench tank to moderate the exhaust temperature. Tank water is recirculated and replenished as needed. In any of the configurations, the quantity of solids removed in the quench tank depends on the

particulate distribution and the retention time. Solids concentration in the tank must be controlled, with provisions made for chamber blowdown and water makeup.

Liquid waste incinerators are horizontal or vertically refractory lined cylindrical chambers, classified according to mixing effectiveness as nonswirling, down-fired quench, or vortex type units.

Nonswirling Type

A nonswirling configuration produces minimum turbulence, which when combined with an axial orientation of the burner, results in a nonuniform chamber temperature. These basic units can be horizontally or vertically mounted, the orientation primarily a space limitation or construction cost consideration. Because they provide minimum turbulent conditions, low burn rates ranging from 180,000 to 270,000 kcal/hr/cubic meter (20,000 to 30,000 BTU/hr/ft^3) are employed. Quench vessels are commonly separate from the combustion chamber. Figure 1.10 illustrates a downflow configuration, specifying burn rates for nonswirling conditions.

Down-Fired Quench Unit

These systems, also referred to as *submerged quench reactors*, are similar to the nonswirling type, except that increased turbulence results from the downward gas movement. The vertical combustion chamber is commonly mounted on top of a water-filled quench tank, cooling the entering combustion gases to temperatures suitable for direct discharge or downstream air pollution control equipment. The flame is emitted from a down-fired burner, mounted on the top-center of the combustion chamber. The waste feed nozzles are commonly located at the upper level, at the vessel side, injecting atomized waste into the flame "envelope." These units have a special advantage when used for wastes containing high salt content, with low melting points and high slag potential. The quench is employed as a salt bath to capture the separated solids at temperatures below their melt point. The increased gas movement and resulting chamber turbulence results in allowable burn rates at the higher end of the nonswirl range. Their effectiveness depends on the chamber physical characteristics, which are based on proprietary designs. Figure 1.10 illustrates an axial feed orientation.

Vortex Type

These units are designed for maximum mixing in a highly turbulent environment, a result of axial swirl or tangential burner entries creating a cyclonic or vortex flow pattern. The chamber turbulence is further increased by introducing secondary air in a similar tangential pattern. The high mixing patterns results in heat release rates ranging from 356,000 to 890,000 kcal/hr/cubic meter (40,000 to 100,000 BTU/hr/ft^3) and corresponding smaller combustion chambers than axial fed incinerators [4]. Figure 1.10 illustrates a tangential feed orientation.

Feed Injection

Where liquid wastes are injected into the combustion chamber depends on their heating value. Clean, high caloric wastes with heating values of 4000 kcal/kg (8000 BTU/lb) or greater are commonly used as fuel, injected into the primary fuel burners. Low (or no) heating value wastes are injected through separate nozzles into a high temperature area surrounding the flame "envelope" to prevent flame quenching (flameout), assuring flame stability and waste destruction. Like waste impingement, burner location and design to minimize flame impingement is a prime concern in maximizing refractory life.

Atomizing Nozzles [4,15,19]

There are a variety of atomizing nozzles, categorized as internal or external and as mechanical, pressure, or gas-fluid devices. The nozzle selection depends on waste characteristics, its fluidization capabilities, and solids concentration. Relatively clean waste streams, and most fuel nozzles, can utilize mechanical atomization, with nozzle complexity increasing when used for wastes with poor fluid flow properties and high solid concentration.

Mechanical Atomization Rotary Cup

Atomization is achieved in a rotary cup when waste flowing from a hollow shaft enters into a spinning cup, forming a cone of fine mist droplets, converging with a tangential air stream simultaneously exiting the cone. Rotary cups have a capacity ranging from 40 to 1000 liters/hr (10 to 250 gph), and a turndown ratio of 5 to 1. Because atomization is a result of mechanical energy expended by the rotary cup, required liquid operating pressures are low. Factors governing nozzle atomization are the waste viscosity, which must be less than 300 SSU, and its sensitivity to the tangential air flow, which could result in flameout if too high and

droplet wall impingement if too low. These nozzles are not recommended for wastes with high solids content, because of plugging potential from suspended solids, or dissolved salts deposited after liquid evaporation.

Mechanical Atomization Pressure

Atomization is achieved by applying waste pressures from 520 to 1040 kPa (75 to 150 psig), forcing the liquid through fixed nozzles, applying tangential air to improve the spray pattern. Turndown ratios ranging from 2.5 to 3.0 to 1 can be obtained, with improved turndowns up to 10:1 if provision is made for external fluid recirculation to reduce internal nozzle flow variation. Nozzle capacities range from 40 to 400 liters per hour (10 to 100 gph), typically applicable to fluids with up to 150 SSU viscosity. Nozzle erosion and plugging prohibits the use of these nozzle for wastes with high dissolved or suspended solids concentration. Upstream filters are commonly employed to protect the nozzles from suspended solids plugging.

External Two-Fluid Flow Nozzles

These nozzles employ a shearing effect to the waste as a result of air or steam impinging the liquid stream external to the nozzle, the resulting turbulence atomizing the feed. Low pressure air nozzles have an applicable waste turndown of 3 to 6:1, utilizing air at 3 to 7.5 cubic meters per liter (400 to 1000 scf per gallon) of waste, at air pressures ranging from 3.5 to 35 kPa (0.5 to 5 psig). These nozzles can be employed for wastes with viscosities from 200 to 1500 SSU.

High-pressure external nozzles utilize air or steam to generate atomization external to the nozzle, at pressures of 200 to 1000 kPa (30 to 150 psig) or greater. Air use ranges from 0.6 to 1.6 cubic meters per liter of waste (80 to 215 scf per gallon), the equivalent steam ranging from 0.25 to 0.5 kg per liter of waste (2 to 4 pounds per gallon). They can operate at turndown ratios ranging from 3 to 4:1, with waste viscosities at 150 to 5000 SSU. When used as primary fuel nozzles, the nozzle flames can be relatively long and must be carefully located in the combustion chamber design. Because flow restriction is minimal, low or high pressure units are capable of operating within the pumpable range of wastes and at relatively high solids content.

Internal Two-Fluid Flow Nozzles

Internal mix nozzles disperse the waste within the nozzle, using steam or air as the dispersion fluid. They generally operate at pressures similar to high pressure nozzles, operating at turndown ratios ranging from 3 to 4:1. Because of internal flow restrictions their application is limited to fuel or clean, low-viscosity (less 100 SSU), low-solid content wastes. Although relatively inexpensive, they are easily plugged and subjected to intensive wear, especially when used for the wrong service.

Sonic Nozzles

Sonic nozzles can be considered a special case of internal mixing nozzles, where atomizing energy is created from high frequency sound waves. Because they are constructed with almost no waste flow restriction, they are especially effective with slurries and sludges, as well as with viscous or solids containing wastes. They are difficult to adjust for waste variations, operating at very low turndown ratios, and must be carefully tailored to expected design conditions.

Nozzle Construction

Nozzles are constructed of a variety of materials, with 316 stainless steel and Hastelloy C suitable for exposure to corrosive conditions at moderate temperature. Increased high temperature and abrasion protection can be achieved using special alloys or ceramics such as alumina, silicon carbide, or tungsten carbide. Selecting the correct nozzle material is critical for industrial incinerators, which often operate at severe corrosive conditions and high temperatures.

Nozzle protection against plugging and burn-out is critical, with the ability to remove, replace, and clean nozzles a prime consideration. Nozzles should never remain in the burn area if waste feed is terminated. In such cases, provisions should be made for cooling to prevent excessive tip temperatures. In addition, thermal shock damage can be prevented by providing for slow nozzle injection and withdrawal. Waste leaks into the atomizing air piping must be prevented, especially during on-line inspection, requiring careful design of the connecting pipe system. Feed line solids settling and plugging can be prevented by providing a waste circulating loop, with a very short takeoff to avoid dead end piping, as well as heat tracing and insulation to prevent solids precipitation. Secondary air should be injected in the waste nozzle annuli to improve combustion and keep nozzles cool. Incinerator viewing ports should be provided for on-line nozzle inspection.

ROTARY KILN

Rotary kilns are commonly employed for solid (or near solid) wastes, requiring water evaporation and volatilization in preparation for combustion. These initial steps, along with some partial oxidation, occurs in the primary combustion chamber. The gases produced are completely oxidized in the secondary combustion chamber, which is principally a gas incinerator. Although most inert feed solids are deposited as ash in the primary chamber, some flyash is transported with the product gases to the secondary chamber, where they pass through to the air pollution control equipment.

A combustion flame is maintained in both chambers with individual fuel burners, requiring supplementary fuel and primary air to help maintain each chamber temperature. Solid wastes are fed to a primary chamber through a handling system consisting of a conveyor and feed hopper, a ram

feed, or a pack feed system. Slurries are commonly fed through a lance. Drums are batch fed. Adequate fuel and air are injected in the secondary chamber burners to complete oxidation of the primary exhaust products.

Solids waste handling includes storage, removing metallic and large inorganic solids, size reduction of oversized combustible materials, safe preparation of highly reactive materials, and incinerator charging. Reducing oversized feed material improves treatment efficiency and can include granulators, dicing machines, grinders, hammer mills, or attrition mills.

Common Design Criteria

Typical design conditions are indicated in Figure 1.11, with conditions adjusted to suit specific process conditions and combustion configurations, as discussed below. Excess oxygen requirements are based on balancing the quench requirements, driving the primary reactions, and minimizing NO_x formation.

Common Physical Dimensions

Rotary kiln components include a feed system, a primary combustion chamber (kiln), a secondary combustion chamber, an ash removal system, air pollution control equipment, and a stack. The primary combustion chamber consists of a refractory lined cylindrical kiln rotating at peripheral speeds of 1 to 5 fpm [4,17]. Solids residence time and

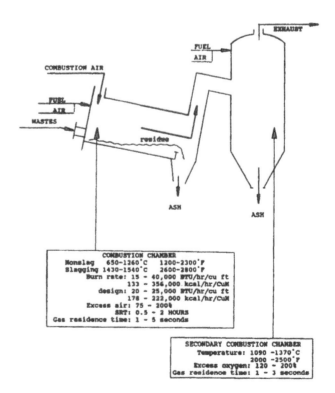

Figure 1.11 Rotary kiln operating data (adapted from References [16,17]).

burn rate establishes the kiln volume, with the physical dimensions affected by other considerations. First, the kiln length must be adequate to allow sufficient time for drying, volatilization, ignition, and burnout. In turn, the diameter is selected to minimize kiln entrainment to the secondary combustion chamber. Common practice is for a length to diameter ratio of less than 10, in the range of 3 to 4. Kilns are frequently designed to operate at a slight vacuum of 1.5 to 5 cm of water (0.5 to 2 inches of water) so that leakage is into the system and not to the working area. This necessitates inclusion of an induced draft fan. Refractory selection is primarily governed by feed composition, temperature, and the presence of specific waste components such as sodium. The temperature maintained depends on whether the kiln is operating in the slagging or ash mode. Service conditions which aggravate brick destruction or erosion, such as drum feed or thermal shock, results in more extensive and expensive brick selection.

The average hazardous waste rotary kiln is rated at 15 million kcal per hour (60 MMBTU/hr), with release rates as high as 23 (90) reported [16]. A kiln diameter can exceed 5 meters (15 feet); the corresponding lengths for cocurrent systems are constructed for a L/D ratio between 4 and 10, with countercurrent kilns much longer [25]. A low ratio, resulting in a relatively large diameter and lower velocity, is frequently employed to minimize particulate carryover. A high ratio produces a longer chamber, frequently requiring a slower rotational speed, resulting in a longer required solids residence time. The kiln outer chamber is commonly steel constructed and refractory protected. Afterburners are constructed as vertical or horizontal chambers similar to gas incinerators, steel constructed and refractory protected, with physical geometry and dimensions adjusted to meet site restrictions.

Primary Chamber

Kiln design is optimized by adjusting combustion chamber time, temperature, and turbulence. Basic to controlling these variables are the solids residence time, physical kiln dimensions (length to depth ratio), the burn rate, gaseous velocity, and excess air. The heat release is controlled at 223,000 to 356,000 kcal/hr/cubic meter (25,000 to 40,000 BTU/hr/ft³), and the physical dimensions are set to maintain the gaseous velocity at approximately 5 meters per second (15 fps).

The kiln temperature and oxygen level are major design factors that must be carefully chosen to meet performance requirements through the expected feed range. The kiln temperature is a result of the heat release of the combustible feed and the quench capacity of the exhaust products; and if required, adjusted with supplementary fuel or quench. The selected operating temperature is governed by regulatory mandates, the required destruction efficiency, and refractory limits. Generally, high kiln temperatures allow greater

throughput and produce more efficient oxidation conditions but increase slagging, volatile metal release, and entrainment problems.

The excess air used, generally in the range of 75 to 200% of the stoichiometric quantity, significantly affects the flue gas temperature. As an example, 1% excess air can result in approximately a 5°C (10°F) decrease in flue gas temperature, requiring supplementary fuel to maintain the combustion temperature. Any oxygen above that required to affect the reaction is generally driven by the system quench requirements. Control options available to minimize excess oxygen quantities include

(1) Minimize incinerator quench air by injecting water or steam.
(2) Where NO_x is a significant problem (pure) oxygen enrichment should be evaluated to maintain the kinetic advantages of oxygen and reduce the associated nitrogen volume and nitrogen oxide formation. Inherent system leakage limits the economic quantities of pure oxygen that can be used.

A significant performance parameter is the solids residence time (Rs), defined by the relationship: [30]

$$Rs = (1.77 \cdot A \cdot L \cdot F)/(S \cdot D \cdot N) \text{ (minutes)}$$

where A is the angle of repose, typically 35° for solids; L is the kiln length; F is a dimensionless factor to adjust for the presence of feed lifters, flights, or dams in the kiln, varying from 1 for smooth kilns and 0.5 for the presence of lifters; S is the slope in degrees; D is the kiln diameter; and N is the kiln speed in rpm.

Solids retention time can be adjusted to optimize the heat transfer characteristics between the kiln and secondary chamber. Because heat is more effectively transferred through solids than gases, it is more fuel efficient to push the kiln than the afterburner. This is accomplished by minimizing the temperature difference between the kiln and afterburner (temperature approach), by controlling the solids retention time. Long retention times (60 to 120 minutes) result in maximum solids temperature, the temperature approach being minimal, and the operation pushed toward a slagging kiln. A short retention time (30 to 60 minutes) results in a minimum solids temperature, a temperature approach of 170 to 280°C (300 to 500°F), operating as an ashing kiln.

Solids residence time also affects kiln gas residence time as a result of the volume occupied by the ash. The effective kiln volume being equal to the total volume adjusted for ash capacity, which for an ashing kiln can range from 7.5 to 15%, and a slagging kiln 4 to 6% of the total volume. The corresponding kiln gas residence time (Kt) is defined as:

Kt = Effective kiln volume, cubic feet/kiln gas flow, cfm

A significant primary chamber operating parameter is its capacity, which can be increased by increasing the charge, kiln temperature, rotating speed, feed temperature, preheating combustion air or decreasing excess air, feed moisture content, and kiln leakage. Capacity can be optimized by instrumenting key operating variables to monitor system performance.

Another significant operating parameter is the feed method; which can be batch, semicontinuous, or continuous. Batch systems produce peak heat and pollution releases (puffs), resulting in unstable conditions affecting both the combustion chamber and the downstream exhaust control equipment. Cycling operations, common with varying or batch feed conditions, result in a system sensitive to the process turn-down ratio, and the response time. Most burners have a turndown ratio of no more than four to one, so that outside this range the kiln utilizes excessive fuel at low capacity, and cannot develop an effective temperature at peak loads. Waste storage, blending to balance waste feed characteristics, and reducing excessive heat load variations is essential for effective incinerator performance.

Afterburner

Secondary Combustion Chambers (SCC) can be configured in the vertical-downflow, vertical-upflow, or horizontal position. Specific configuration considerations include structural limitations, locating and positioning of downstream pollution control or recovery equipment, and costs. Generally, configuration has little effect on the system performance, relative to other design considerations such as temperature, time, and turbulence.

SCC temperature is the most significant design and operating variable, generally ranging from 980 to 1375°C (1800 to 2500°F), many times mandated by the regulatory requirements. Regulators may allow lower temperatures for nonhazardous waste, increased to about 1095°C (2000°F) for hazardous wastes, and push to higher temperatures for PCB wastes. Secondary combustion chamber volumes depend on the volume of kiln gas released and the required treatment efficiency, operating at residence times from 1 to 3 seconds. Turbulence is achieved by maintaining approximately 5 meters per second (15 fps) gas velocity, and greatly affected by geometric dimensions. The SCC supplementary fuel requirements depends on the kiln combustion temperature, which governs its exhaust temperature. Fuel or high caloric clean liquid wastes are used to maintain the secondary chamber temperature. Unless heat recovery is employed the SCC gases are tempered as a preconditioning step to protect and optimize downstream air pollution equipment performance.

Rotary Kiln Configurations

Rotary kiln operating configurations can be described as cocurrent or countercurrent, conventional or rocking system, and slagging or ash.

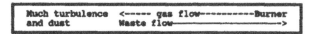

Figure 1.12 Rotary kiln: countercurrent configuration.

Cocurrent or Countercurrent

Countercurrent designs employ auxiliary fuel at the end opposite to the waste feed, as illustrated in Figure 1.12, with a distinct hot and cold end, maximizing energy transfer efficiency.

A more complex material feed and discharge system is required because gaseous products are removed at the feed end, resulting in increased particulate and partially reacted waste carryover. The increased entrainment complicates combustion control, especially solids residence time. This configuration is preferred for low caloric value wastes, allowing evaporation to occur in the lower temperature end, preventing flame cooling at the other end.

A cocurrent unit has the auxiliary burner and waste feed at the same end, as illustrated in Figure 1.13, and does not have a distinct hot or cold end. The entire unit operates near the reaction temperature range, stabilizing the combustion process. Residence times higher than those in countercurrent units are required because of the kiln temperature profiles.

Cocurrent systems temperature allow the kiln to be designed with an explicit temperature approach, allowing exit temperature control. When the temperature approach is minimized the kiln gas temperature is near that of the secondary combustion chamber, thereby minimizing SCC supplementary fuel requirements.

Rocking or Conventional

A kiln is defined as a rocking or conventional rotary kiln based on the feed method. Rocking kilns are generally batch fed with turbulence achieved by the kiln motion. A conventional rotary kiln, the most common incinerator type, is fed continuously or semicontinuously.

Slagging or Ash

The kiln operating temperature establishes whether it is a slagging or ash incinerator. An ash kiln usually operates between 650 to 985°C (1200 to 1800°F) and is more economically applied to low value caloric wastes requiring high supplementary fuel. Slagging kilns usually operate at tem-

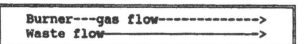

Figure 1.13 Rotary kiln: cocurrent configuration.

peratures ranging from 1430 to 1540°C (2600 to 2800°F) and by design can operate in either the ash or slagging mode. Higher kiln temperature allows increased waste throughput but can result in higher particulate and volatile metal carryover. A slagging kiln operation requires a more complex design, especially in selecting the refractory. In fact, the ash slagging results in increased refractory repair and maintenance, greatly reducing refractory life. Slagging wastes are more economically applied to high caloric wastes with moderate moisture content or where high temperature combustion is required, such as in the treatment of drummed waste.

FLUIDIZED BED INCINERATOR

Like the other incinerators mentioned, the major component of a fluidized bed is a refractory lined vertical shell serving as a combined two-stage primary and secondary combustion chamber. The chamber internals consist of a bottom distribution plate, an air inlet under the plate, and an inert bed (usually sand) supported by the plate. Bed suspension (fluidization) results from air flowing through the bottom plate, into the bed, and mixing with the bed prior to exiting. The volume above the bed acts as the secondary chamber, in place of an afterburner, oxidizing volatilized primary combustion products.

Feed to the system includes (1) make-up sand above the bed, (2) waste injected on the bed, (3) fuel oil injected on (or in) the bed, or gas injected at the bottom, and (4) fluidized air into distribution plates under the bed. Waste is fed above the bed, in the general vicinity where the auxiliary fuel nozzles are located. The fuel nozzles heat the bed, the heat transferred to waste flowing through the bed. The system is pressurized to move flue gas through the air pollution control equipment. Combustion effectiveness can be improved by preheating the air. Physical improvements are generally directed at increased turbulence, such as employing internal recirculation. Recirculation improves mixing, while allowing removal of some heavier particles, minimizes agglomeration, and improves destruction efficiency.

Fluidized beds readily adjust to wide influent waste variations, can be applied to solids or slurries (and some liquids), require low excess air and minimum fuel, and can effectively dry high water content wastes. However, fluidized beds are not suitable for wastes which form eutectics, they must be maintained relatively free of residual materials, and they could require considerable bed preparation and maintenance. Wastes having aggressive physical properties or undergoing chemical reactions could reduce bed life.

Common Design Criteria

Typical design conditions are cited in Figure 1.14, specific conditions depend on waste characteristics and treatment requirements. Fluidized bed temperatures are maintained

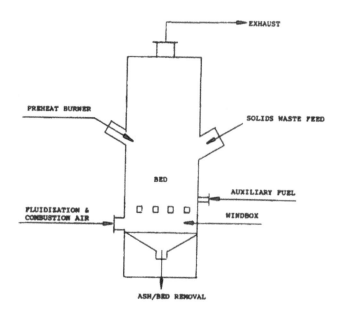

Figure 1.14 Fluidized bed operating data (adapted from References [16,17]).

well below 1093°C (2000°F) to minimize sand deformation and slagging conditions, with the free space conditions adjusted to complete oxidation. Oxygen requirements must be compatible with fluidizing air flow conditions to assure a stable bed condition, and minimize particulate carryover.

Common Physical Dimensions

Fluidized beds can be up to 15 meter (50 feet) in diameter, although 3 to 8 meter (9 to 25 foot) chambers are commonly employed; at heights from 6 to 15 meters (20 to 50 feet). The outer chamber is usually constructed of steel, and refractory protected. Provisions are made for adequate freespace to act as an afterburner, and allow particulate separation.

Operating Characteristics

The major fluidized bed operating parameters include (1) maintaining a fluidized bed and (2) ensuring a suitable temperature profile to control treatment efficiency. Air is supplied through windboxes at 21 to 35 kPa (3.0 to 5 psig) and at 16°C (60°F) or preheated by recovering some of the exit flue gas heat content. A fluidized condition will be occur at gas velocities of 1.5 to 2 meters per second (5 to 7 fps), higher velocities result in excessive particle carryover and media lost [15,17,19,24–26]. Besides impacting gas velocity, air rate affects both bed temperature profile and flue gas temperature. Typically, fluidized beds operate at 20 to 40% excess air, resulting in flue oxygen concentrations ranging from 4 to 8 vol % (based on humid gas). Single air flow systems are seldom employed because of the difficulty in balancing fluidizing and combustion air requirements. When the volume permits, air flow is partitioned to the bed and the overhead to effectively control bed fluidization and secondary chamber oxidation requirements.

Bed temperatures can range from 705 to 870°C (1300 to 1600°F), with a minimum 650°C (1200°F) required to assure a satisfactory ignition temperature. Generally, industrial wastes cannot be incinerated if the required treatment temperature exceeds 870°C (1600°F) because of the potential for excessive slagging problems. Municipal wastes treated at 760°C (1400°F) and at an adequate retention time can achieve a carbon oxidation efficiency in the 90 percent range, if its characteristics are conducive to thermal treatment.

Operating temperatures are a direct consequence of the supplementary fuel used, excess air employed, and the waste caloric value. Fuel is used to start-up and reheat the bed, as well as maintain the incineration temperature. At commonly employed temperatures the bed heat capacity is from 142,000 to 310,000 kcal/cubic meter (16,000 to 35,000 BTU/ft^3) [17,19]. The sand bed acts as a heat sink, losing approximately 6 to 12°C (10 to 20°F) per hour when inactive, requiring supplementary fuel to maintain the process temperature. Fuel can be injected at the top of the bed, or into the bed itself, once the bed is at the ignition temperature. Natural gas will not be retained by the bed and must be fired through a burner beneath the bed or as low into the bed as possible.

The incinerator second stage is designed to complete the combustion process, reducing odors and emissions, and contribute to the overall treatment efficiency. This can usually be accomplished by maintaining the conditions above the bed at 760 to 870°C (1400 to 1600°F) for 5 to 10 seconds, a maximum 55 to 140°C (100 to 250°F) above the bed temperature. However, because of bed temperature limitations, freeboard temperatures may have to be elevated to the 980 to 1370°C (1800 to 2500°F) range for 12 to 16 seconds to effectively destroy VOC components generated from hazardous wastes. In some cases a separate after-burner may be required. Freeboard temperatures are maintained using supplementary fuel, using a quench when necessary to avoid excess temperatures. SCC residence time is based on effective volume, which is close to the actual volume if adequate turbulent mixing conditions are maintained. Exhaust gases should be controlled to protect downstream equipment, avoiding temperatures less than 150°C (300°F) which promote acid deposition, and elevated temperatures which accelerate acid gas corrosion. The second stage combustion volume significantly affects process efficiency, acting not only as a final combustion chamber but as an entrainment separator for particulate removal. Physically, the chamber must have adequate volume for the required residence time, dimensioned to allow exit velocities of less than one meter per second (3 fps) for effective particulate removal, and provide adequate mixing and turbulence. Particulate entrainment, unless controlled, could result in high loadings to air pollution control equipment.

Waste Characteristics

Fluidized bed incinerators are employed for homogeneous slurry or solids, most often wastewater sludges. Waste homogeneity is important, as is sludge preparation, to protect bed properties. This includes reducing total water content to minimize bed cooling or fuel consumption, preventing bed plugging by viscous materials such as tars, and averting ash slagging. Sludges must have heat values greater than 4500 kcal/kg (8000 BTU/lb) to be autogenous. Many sludges can be fed directly onto the bed, with most of the moisture evaporated instantaneously before reaching any significant bed depth. However, dewatering to reduce the water content and increase combustibility should be considered.

Bed Characteristics

Fluid bed incinerators utilize sand as the bed media, with other substances such as the waste ash, limestone, alumina, or ceramic material sometimes used. Bed integrity is a major concern, requiring prevention of media agglomeration and plugging from captured ash or low melting point waste salts, resulting in restricted air flow. Captured bed ash is removed by continuously withdrawing and replacing a bed portion.

Bed depths range from 0.3 to 0.6 meters (1 to 2 feet), expanding 30 to 60 percent from cold conditions with a 0.6 to 0.9 m/s (2 to 3 fps) minimum fluidized air velocity, based on empty bed cross-sectional area. At operating conditions air must be adjusted to minimize bed carryover, requiring maintaining gas velocities at less than 2 meter per second (7 fps), the specific operating range dependent on the bed characteristics. Primary cyclones are included as part of many fluidized bed systems to capture and return entrained media. Approximately 5% of the bed volume is recharged every 100 to 300 hours [24,25].

Control and Instrumentation

Controls applied for a fluidized bed are specific to its configuration, although some of the control sequences discussed in the General Engineering Design Criteria section apply. The focus of fluidized bed incinerator control is obtaining a representative bed temperature, which because of bed turbulence, poor lateral mixing, and slagging temperature elements requires midpoint readings. Provision should be made to monitor, record, and control the following:

(1) Waste feed rate

(2) Waste feed temperature

(3) Windbox temperature

(4) Bed temperature

(5) Secondary combustion chamber temperature

(6) Air flow to monitor the fluidization requirements and an override control to maintain the required excess oxygen content

(7) Fuel flows to maintain the bed and SCC control temperatures

(8) Alarm to alert of excessive bed expansion

(9) Interlock controls to monitor the fuel feed if bed temperatures are too low or terminate feed if bed or refractory temperatures are too high

(10) Freeboard temperature controls system to activate the gas burner or add quench to maintain operating and safe temperatures

(11) Draft control to maintain the system pressure

(12) The combustion products should be monitored for carbon dioxide and oxygen content as well as monitoring for specific regulated pollutants

Emergency conditions that should be addressed include

(1) A purge system to control the burning sludge in event of power failure

(2) A high-temperature control to stop the fuel and sludge feed

(3) A high flue gas temperature response, activating water or an air quench, and shut-off sludge and fuel feeds

CIRCULATING FLUIDIZED BED COMBUSTOR

The Circulating Fluidized Bed Combustor is a modification of the fluidized bed, utilizing a variety of bed materials ranging from sand, limestone, or contaminated solids. Bed circulation increases bed turbulence, resulting in more uniform and lower bed temperatures, allowing greater throughput per bed area. An external cyclone separates the exhausted gas–solid mixture, recycling solids and some exhaust gases back to the bed.

This configuration results in a system more responsive to upset conditions, allowing lower combustion temperature, minimizing slagging and pollution generation, and less residual buildup. These systems can be designed for heat release rates ranging from 890,000 to 1,800,000 kcal/hr/cubic meter (100,000 to 200,000 BTU/hr/ft^3) of bed volume, at bed conditions similar to those described for fluidized bed incinerators.

MULTIPLE HEARTH

Multiple hearth furnaces consist of sequential "layers" which are vertically distanced to allow waste plowing from hearth to hearth. Solid waste is fed through the furnace roof to the top hearth, a vertical shaft containing rabble arms slowly rotates a plow moving the waste circularly to drop holes, consecutively from the top hearth to the bottom ash disposal hearth. The plow is configured like a rake, allowing an even waste distribution on each hearth. Air is distributed through a hollow center shaft supporting the rabble arms, starting from the bottom and collecting hotter gases when traveling through the shaft. Air exiting from the center shaft is sometimes recycled as preheated combustion air. Supplementary fuel is fed into the combustion zone to supply energy for combustion and replenish system radiation losses.

These furnaces can contain 4 to 14 hearths, each hearth containing 2 to 4 rabbles, the top hearth is numbered 0, and bottom (or the last) hearth 14. The system functions as four operating zones, plus a zero zone afterburner. Zone 1 consists of the drying hearths in which waste moisture is volatilized by evaporative cooling using zone 2 exiting flue gases. Zone 2 contains the combustion hearths, zone 3 the fixed carbon burning hearths, and zone 4 the cooling hearths. Primary air is fed through fuel nozzles, and combustion air flows from the furnace bottom cooling the ash and through a vertical shaft for distribution to the hearths. Exhaust gases are sometimes recycled for heat recovery in the bottom hearths. If complete combustion is not accomplished, odors or destruction efficiency is not achieved, and a secondary burner will be required. Hearths range from 2 to 8 meters (6 to 25 feet) in diameter, 4 to 20 meters (12 to 65 feet) in height, and hearth areas from 9 to 560 square meters (100 to 6000 ft^2) [24,26]. Sludge loadings range from 34 to 59 kg/square meter/hr (7 to 12 lbs/ft^2/hr) of total area for minicipal sludge, commonly operated at 40 to 50 (8 to 10).

Use of multiple hearth furnaces for destroying hazardous or controlled wastes is ineffectual because they (1) are inherently unstable, (2) are not designed for finite control as required for toxic waste destruction, (3) operate at temperature no greater than 930°C (1700°F), which may not meet regulatory approval, (4) are difficult to operate at conditions required to minimize emissions, and (5) are difficult to design with stable downstream exhaust emission control equipment. They are employed to destroy large sludge quantities, and should be considered in such cases where the facility does not have to meet stringent burn or emission control. Typical design criteria are cited in Figure 1.15.

Waste Characteristics

Multiple hearths are designed to burn low caloric value materials containing 20 to 30% combustible solids. In balance, high water content wastes are favored because of the hearth countercurrent configuration, allowing moisture re-

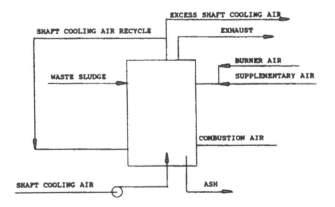

COMBUSTION CHAMBER

Temperature:
After burner 815°C (1500°F)
Drying Zone 315 - 540°C
600 - 1000°F
Combustion Range, 760 - 930°C
1400 - 1700°F
Common, 760 - 815°C
1400 - 1500°F
[930°C (1700°F) ash fusion]

Excess Air: 75 - 100%

Burn Rate: based on the entire
hearth area
34 - 59 kg/hr/SqM
7 - 12 Lbs/hr/SF

Figure 1.15 Multiple hearth operating data (adapted from References [19,26,31]).

duction by heat exchange with hot exiting flue gases. In fact, high caloric wastes could result in excessive heat releases, beyond the quenching capacity of the furnace, resulting in damage to the shaft or other critical furnace components. Low volatile combustion materials, such as oils or fats, should be fed at the lower combustion hearths to allow adequate residence time and eliminate smoking.

Operating Conditions

Critical operating concerns include start-up, maintaining an adequate temperature profile, and controlling organic emissions. Process start-up requires a gradual incinerator temperature increase from ambient conditions to approximately 705°C (1300°F) at the combustion or highest temperature zone, usually over an extended period. All moisture in the fire brick must be gradually evaporated to prevent damage. Monitoring the resulting temperature profile is critical during this period.

Once the incinerator reaches the operating temperature, feed is slowly injected at the top hearth and the process controlled to establish the required temperature profile, governed by the following criteria: [31]

(1) The top drying zone hearths will contain sludge at approximately 70°C (160°F), with gas temperatures ranging from 315 to 480°C (600 to 900°F).

(2) The combustion zone temperature will range from 760 to 815°C (1400 to 1500°F), possibly reaching 925°C (1700°F).

(3) The fixed carbon burning zone temperatures will be in the range cited for the combustion zone, with a maximum 980°C (1800°F) possible.

(4) The ash cooling zone gas temperature will be 150°C (300°F) or more, while that of the ash will range from 38°C to 200°C (100 to 400°F).

The maximum incinerator temperature should not exceed 927°C (1700 °F) to prevent damage to metal incinerator components, increased potential for ash fusion, and excessive slagging and clinker formation.

Furnace operational controls include managing sludge throughput, furnace temperature, fuel usage, and turbulence of the hot gases. Center shaft speed controls sludge throughput, which must be optimized to control the retention time and minimize sludge agglomeration. The speed must be adjusted to maximize contact exposure between the solids and hot gases, evaporation and combustion rates, and produce effective burn zones. Although reduced shaft speed should allow for higher feed destruction, too low a shaft speed could result in excessive solids agglomeration reducing the contact effectiveness.

Effective processing requires promoting conditions to complete the combustion by maintaining an adequate oxygen supply on each hearth. A minimum of 75% excess air should be utilized to enhance the combustion environment and assure adequate quenching to avoid slagging, clinker formation, or any conditions that could damage the furnace.

Combustion chamber turbulence helps maintain the temperature approach between the hearth sludge and gases, the increased mixing increasing reaction effectiveness. This can be accomplished by introducing the air at various strategically located injection points, ideally at each hearth rather than one bottom entrance point. Another important consideration is to control the volatile solids feed content to avoid excessive burn rates so that heat release is consistent with the hearth quench capacity. At the same time, complete waste volatile destruction is critical, insuring minimum release of Products of Incomplete Combustion (PIC) formation and Volatile Organic Compounds (VOC). This can be accomplished by either utilizing the zero hearth or a separate afterburner. Better control can be achieved with an afterburner because temperature, oxygen content, and turbulence control can be independent from the hearth operation.

Oxygen for Combustion Process

Excess oxygen requirements range from 75 to 100% for sludge and 15 to 25% for fuel. Air is injected into the individual hearths by damper control, with the air moving upward to the next hearths. Because air cooling requirements are related to the shaft speed, the shaft speed should be set at process conditions representing 50% inlet air dampers to assure an available operating range. On any hearth, equal air distribution is obtained by opening all dampers to the same flow level.

The individual damper controls should be adjusted to achieve the following:

(1) In the bottom hearths adequate air to cool the ash
(2) Adequate accumulated air volumes at the combustion hearths to meet excess oxygen requirements
(3) Control final hearth gas volumes (excess air) to minimize fuel requirements

Obtaining this balance is more an art than a science, based on operator experience.

Controls and Instrumentation

The controls required for a multiple hearth are specific to its configuration, although some of the basic sequences discussed in the General Engineering Design Criteria section apply. The general control philosophy involves managing throughput (retention time) by adjusting waste feed and center shaft speed. The process is stabilized when sludge leaving the evaporative hearths has lost enough moisture so that it is autogenous (if possible) when entering the combustion chamber. This requires controlling top hearth total flue gas flow and temperature by controlling hearth air flow and sludge combustion energy release. If the moisture content cannot be reduced to a level assuring autogenous conditions, then supplementary fuel must be supplied. Hearth controls include indicators to alert for excess temperature, too low a temperature, excessive air flow, and too low an air flow.

A stabilized profile can be established by providing for individual hearth supplementary fuel and air monitoring, and being able to control the furnace draft. The system draft should be set at a slight vacuum to avoid exhaust leakage into the surrounding area. If the draft is too close to zero, excessive emergency damper openings result in high ash entrainment.

Monitoring requirements include the following:

(1) Sludge feed rate
(2) Shaft rotation speed
(3) Oxygen feed rate (if possible)
(4) Fuel feed rate
(5) Temperature profiles of the hearths
(6) Flue gas flow leaving the incinerator
(7) Flue gas oxygen content
(8) Carbon monoxide concentration of the flue gas
(9) Exhaust gas analysis consistent with regulatory requirements.

The system must be equipped to control the following emergency conditions:

(1) Slow cool down during power failure to avoid damage to the incinerator
(2) High afterburner off-gas temperature indication, activating the quench system
(3) High incinerator temperature controls terminating sludge feed, monitoring residual sludge burnout, and assuring adequate air supply during the final burnout
(4) Center shaft stoppage indication, stopping the feed, and incorporating the same slow cool down controls as with a power failure

NEW TECHNOLOGIES

Specific technologies are being developed and applied for special wastes or process conditions. These include *wet air oxidation, oxygen-enriched incinerators, infrared systems, molten salt destruction, plasma arc pyrolysis,* and *supercritical fluidization*. They are not an automatic replacement of the applied technologies discussed, but an alternative applied to achieved required performance criteria for difficult to destroy wastes. Their application is based on improved performance, cost savings, energy alternatives, or a combination of these factors. Design criteria must be developed with the equipment representatives based on prototype evaluations. Because the operating experience for these emerging technologies is limited, mechanical reliability becomes a more difficult criteria to evaluate, which is difficult to establish in limited prototype studies.

PROCESS ENGINEERING DESIGN

Waste incineration is a relatively simple combustion reaction, complicated by the mandatory combustion conditions imposed for some wastes, and regulatory requirements governing incinerator exhausts and ash disposal. Performance projections must not only be based on the combustion processes but include the emission and ash control equipment.

Incinerator system effectiveness depends on many interacting operating variables, summarized in Table 1.6. The operator controls the system by adjusting the waste feed rate, oxygen feed rate, and combustion chamber temperature. Waste feed rate is a limited control rate, since in the long run it must equal generation rate. Waste properties are an important process factor, but seldom an operating parameter. The most significant properties are its physical state, ash, noncombustible content, and heating value. Temperature can be controlled by adjusting supplementary fuel input, although excessive supplementary fuel could severely impact operating costs. Oxygen can be controlled by adjusting air input; the control of which affects gas volume, gas temperature, and overall system performance. Other operating char-

TABLE 1.6. Operating Characteristics.

Variable	Operator Controllable	Critical
Waste Characteristics		
Waste generated	No	Yes
Composition	No	Yes
Water content	No	Yes
Ash content	No	Yes
Combustibility	No	Yes
Heat of combustion	No	Yes
Operating Characteristics		
Feed rate	Minimal	Yes
Feed preparation	Yes	Yes
Atomization	Yes	Yes
PC temperature	Yes	Yes
PC residence time	No	No
SCC temperature	Yes	Yes
SCC residence time	No	No
Excess oxygen	Yes	Yes
PC fuel	Yes	Yes
SCC fuel	Yes	Yes
Destruction efficiency	No	Yes
Slagging	No	Yes
Air pollution eqpt	Yes	Yes
Ash control	Yes	Yes
Emission level		
Particulate	Yes	Yes
NO_x	Yes	Yes
SO_x	Yes	Yes
Halides	Yes	Yes
CO	Yes	Yes
PICs	Limited	Yes
VOC	Limited	Yes
Metals	No	Yes

TABLE 1.7. Reported Performance Data (adapted from Reference [17]).

Range of Results	
DRE, %:	99.994–99.999
CO, ppm:	1–794.5
Total unburned hydrocarbons, ppm:	0.5–61.7
Particulates, mg/cu meter:	4–404
HCl removal, %:	98.3–99.9
Stack oxygen level, %:	3.1–14.1

REQUIRED PROCESS DESIGN DATA

Incineration can be evaluated in two phases—at the laboratory and pilot scale. Laboratory studies include analytical tests to determine the waste heating value and composition, as indicated in Table 1.9. Pilot-scale incineration tests are suggested to determine the following:

(1) A suitable feed method
(2) Optimum air requirements
(3) Fuel requirements
(4) Optimum combustion temperatures
(5) Residue content
(6) Air emissions and appropriate air control equipment
(7) Proper atomization
(8) Refractory stability
(9) Potential for slagging
(10) Slag characteristics

Regardless of how they are obtained, the data listed in Table 1.8 are required for a process design.

PROCESS DESIGN VARIABLES

The basic concepts discussed directly affect incinerator performance, and are critical in establishing individual system component design criteria. Initial design considerations

acteristics listed in Table 1.6 are significant, but not easily controlled. Treatment performance is generally focused on the combustion efficiency, although process effectiveness is a result of the performance of the individual system components between the waste stream and the atmospheric exhaust. In fact, treatment performance is commonly reported in terms of the emissions content and not the combustion efficiency. The exhaust content is a direct result of the inorganic waste composition.

REPORTED PERFORMANCE DATA

Based on historical operating experience, it can be readily concluded that combustion efficiencies exceeding 99% can be achieved, with destruction efficiencies for individual contaminants exceeding 99.99%. However, incinerator effectiveness is generally measured by the trace quantities of exhaust products, as discussed in the Fate of Contaminants and Pollution Control sections.

Reported incinerator RCRA performance data for rotary kiln, fixed hearth, liquid injection, and fluidized bed incinerators are summarized in Table 1.7 [17].

TABLE 1.8. Required Design Data.

Critical waste characteristics
  1. Listed in Table 1.9

Selected operating characteristics
  2. Combustion temperature
  3. Excess combustion oxygen

Operating characteristics that should be obtained from pilot studies, but can be estimated from treatability data
  4. Feed method
  5. Type atomization
  6. Refractory
  7. Required air pollution control
  8. Flue gas characteristics
  9. Auxiliary fuel requirements
  10. Ash quantity and characteristics

require a heat release estimate, establishing an appropriate enclosure to utilize the energy, and determining the combustion products formed. These in turn impact heat recovery considerations and air pollution control device selection. A high destruction efficiency is readily obtainable; the basic *process design* challenge is assuring a stable process, with minimum operating shutdowns or flameouts, limiting atmospheric emissions and generating a disposable ash. Although the primary combustion chamber design is a prime concern for the Process Engineer, controlling the chamber exhausts requires extensive process attention to complete the system design. In fact, the system downstream of the primary chamber could represent the major capital and operating expenditures. Therefore, in optimizing combustion parameters, the design engineer must be sensitive to the effect the selected design parameter will have on the downstream equipment. Fundamental process design considerations include

(1) Waste evaluation
(2) Feed preparation
(3) System configuration
(4) Oxidation
(5) Combustion temperature
(6) Combustion chamber turbulence
(7) Residence time
(8) Energy conservation and recovery
(9) Air pollution control
(10) Plume suppression
(11) Fate of contaminants
(12) Regulatory requirements

Waste Evaluation

Waste composition is the single most significant design consideration, subordinating the other design factors. This is because it affects incinerator selection, feed preparation, and emission characteristics. Waste characteristics can be divided into (1) flow, (2) waste composition, and (3) combustion properties. Collectively these characteristics impact the entire incineration system design.

Flow Characteristics

Primary waste flow properties include its physical state, temperature and specific gravity. The most critical design property being its physical state, liquid or solid, which establishes whether the waste will be pumped or mechanically transported as well as the applicable incinerator configuration. A liquid feed is pumped and atomized for effective combustion, whereas a solid is mechanically transported after "size" separation, reduction, and blending. A waste's physical state is not always clearly delineated, but generally a fluid with a viscosity less than 10,000 SSU is considered

a pumpable liquid and readily atomized at viscosities less than 750 SSU.

Waste feed temperature affects the system in many ways. Liquid fluid flow and atomization properties can be improved at increased temperature, thereby improving equipment performance. For any waste, solid or liquid, its feed temperature represents energy into the system, and therefore reduced supplementary fuel requirements. For that reason, preheating the feed streams is a viable heat recovery consideration. In addition, the feed temperature could affect the transport equipment materials of construction.

Specific gravity affects required feed transfer energy because for liquid or solids expended energy is directly proportional to the material weight per unit volume (specific gravity), increasing with increasing specific gravity.

Composition

Waste composition can affect the achievable destruction efficiency because of its combustibility properties, or the type of the products formed. Some compounds appear to have low destruction efficiencies because they are similar to by-products formed, as a result of incomplete combustion or by the reaction. Common among these are benzene, methylene chloride, napthalene and chloroform. When they are in the waste feed, formation of similar compounds in the combustion chamber could result in "calculated" low (or negative) destruction efficiencies.

Other significant performance characteristics related to waste content include the quantity of combustible components, volatile inorganics, inerts and total solids, alkalinity or acidity, and water.

(1) Combustible components include organics (carbon), hydrogen and oxygen, which are released as carbon dioxide, water, and molecular oxygen. Complex organic wastes can contain a variety of inorganic constituents such as nitrogen and sulfur, converted to nitrogen oxides and sulfur oxides, representing a potential air emission problem.

(2) Volatile inorganic halides are released as the corresponding acid if adequate hydrogen atoms are available, or as the molecular halogens. These compounds, along with sulfur dioxide, are common acidic air pollutants which must be controlled. Arsenic, mercury, cadmium, and chromium are normally inert inorganics but at combustion temperatures can volatilize as basic compounds, salts, or oxides.

(3) Inerts measured as the waste ash content are inorganics that in the elemental or oxidized state can represent a significant combustion process problem. If they do not collect as slag in the combustion and ash chambers, they are discharged either as exhaust particulate pollutants or as ash. Volatile inert compounds further complicate the system if they are difficult to control or remove,

such as metallic fumes. The solids in liquid wastes can result in problems to the atomization equipment. Settleable solids must be removed prior to injection because they can clog most atomization nozzles. Dissolved solids could salt out if the waste is near saturation conditions, potentially depositing at the nozzle tip and clog the nozzle. In all these cases, pretreatment should be seriously evaluated, and where practical implemented.

(4) Waste alkalinity or acidity can affect selection of waste handling and incineration equipment materials of construction. For the most part, near neutral wastes are desired, alkaline wastes can be tolerated, but acid wastes affect process stability and equipment durability.

(5) Waste water content is a direct measure of its combustibility, with increase water content requiring supplementary fuel to sustain the combustion. In essence, higher water content could increase operating costs or reduce operating capacity (if greater than originally anticipated in the design). In addition, high water content wastes could result in rapid brick deterioration from thermal shock if injected on hot services. Water in a "free" state is readily removable, so that pretreatment equipment should be included in the design. However, in some cases feed water could provide an effective quench media for wastes with high heat releases, reducing the need for excess air, and reducing nitrogen oxide formation; and therefore beneficial if controlled.

Combustion Characteristics

Combustibility can be defined as a material's oxidation potential, measured by any one of a number of parameters indicating possible combustion supporting properties. Combustion potential being defined as the ability to release energy, supporting oxidation at a "reasonable" low temperature, and able to maintain that temperature. Generally, the lower the autoignition temperature, ignition temperature, flash point, flammability limit, and volatility, the greater a material's combustibility potential in an effective oxidation environment. Critical to overall waste detoxification efficiency is organic concentration, which for low waste concentration could require more extensive combustion conditions to achieve high (regulatory) treatment efficiencies. This is commonly encountered in the site remediation cleanup of contaminated soils with low organic concentrations.

Design Waste Characteristics

Specific waste characteristics relevant to industrial incineration design has been discussed. The attempt to design an incineration system without a "defining" and complete waste analysis can only result in operating problems, and future expensive upgrading. Specific requirements are summarized

TABLE 1.9. Critical Waste Design Characteristics.

General Properties	Chemical Properties
Physical state	Specific compounds
Density	Ash content
Viscosity (liquids)	Water content
Heating value	Carbon
Flash point	Hydrogen
Solids size and shape	Sulfur
Containment (liquids)	Oxygen
Toxicity	Nitrogen
Corrosivity	Halogens
Explosivity	Metals
Biological stability (medical)	Suspended solids
Radioactivity	Salt content

in Table 1.9. An important design consideration is optimizing waste thermal reactivity, governed by two criteria, its heating value and physical state. First, any waste with a low fuel value will require rapid dispersion in a high temperature combustion chamber to assure oxidation. Its fuel heating value determines how much supplementary fuel will be required to maintain highly oxidized conditions and the combustion temperature. Fuel or high heating value organics maintains the combustion temperature, low heating value wastes reduces the temperature and under extreme conditions extinguishes the flame. Next, a waste's physical state determines its reactivity and necessary feed preparation to optimize combustion efficiency.

Feed Preparation

A *primary* consideration in developing storage and transfer facilities, and establishing safe combustion conditions, is the waste's chemical *reactivity* and *stability*. The waste must be stored, transported and delivered to the incinerator combustion chamber in a *safe* manner, consistent with the waste characteristics. Unstable or extremely reactive (explosives, toxic, volatile, etc.) wastes must be stabilized by chemical additions, chemical alteration, or diluted to a low risk state. The stabilized waste must be carefully analyzed for three important criteria affecting combustion efficiency:

Components Detrimental to Combustion
Variability
Consistency

Practical pretreatment steps must be evaluated to assure a consistent, combustible, and uniform incinerator feed. A gas is readily injected into a combustion chamber and easily oxidized. A liquid waste will require atomization to a fine mist to vaporize and achieve reasonably similar oxidation effectiveness. Sludge, solids, or residues should be shredded and reduced to particles of high area, small volume to enhance sublimation, fusion, vaporization and destruction.

Some materials impose an extremely difficult burden on incinerators, examples being destruction of drummed wastes and the detoxification of contaminated soils.

Drums and similar contained wastes are difficult to incinerate because the container must first be destroyed before the contents can be released into the combustion environment. Stable combustion conditions are impossible since large heat quantities are needed before the container is destroyed, after which five to six times the average incinerator heat capacity could be released, and oxygen demands up to 20% the average values required. Such cyclical heat releases result in excessive process demands, systematic refractory and chamber damage, and high maintenance requirements.

Destroying contaminated soils and solids containing hazardous organic constituents is extremely difficult, and a relatively inefficient oxidation process, although detoxification is frequently very effective. The major difficulty is maintaining a temperature to promote detoxification while at the same time minimize slagging, avoid releasing of toxic metals, and prevent overloading the air pollution system. Destroying organics in adsorptive materials such as clay is formidable because of poor volatilization and high moisture content, affecting not only removal efficiency but energy requirements. In such cases, feed preparation and steam stripping are important pretreatment processes in meeting remediation goals.

Requirements will be different for liquid or solids wastes, but some general considerations are applicable to all wastes:

(1) Feed preparation can improve combustibility characteristics, allowing a consistent and rapid incineration rate. Because the basic combustion reaction occurs in the gaseous state, it is essential that the feed be injected as a dispersed homogeneous liquid mist or fine particles or that solids are conditioned for maximum volatilization. For liquids this is accomplished with atomizers, while crushers and grinders are used for solids. Figure 1.16 illustrates feed preparation for various wastes.

(2) Remove large separable solid constituents that will interfere with atomization, the feed mechanism, or will not burn.

(3) Remove combustion chamber slag forming or fouling inorganic solids.

(4) Remove inorganic acids and acid-forming compounds.

(5) Control waste compounds with high nitrogen content by segregating and blending with bulk waste to avoid peak loadings.

(6) Remove, segregate, blend or dilute toxic metal wastes.

(7) Dewater high water content wastes unless the water is needed for quenching.

(8) A stable operating, economical, and low maintenance incineration process results from relatively constant combustion conditions, assuring a reasonably consistent combustion temperature. Combustion temperature is controlled by a sequence of automated steps initiated

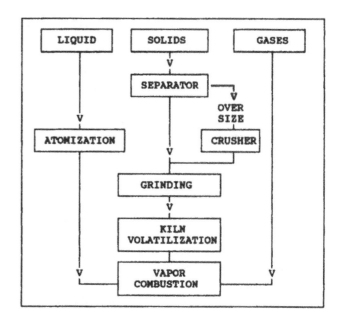

Figure 1.16 Waste preparation requirements.

by monitoring the temperature, comparing it to the set (required) temperature, and adjusting fuel input accordingly—all in a matter of seconds. If the waste feed is constantly changing, system adjustments will be frequent and sometimes large, and in extreme cases combustion temperature control is lost because of the inability to adequately react. For that reason, a constant feed composition is necessary. Because consistent plant waste generation is not reasonable, feed uniformity is best maintained with an adequately sized feed tank to provide blending and a consistent inventory for at least eight hours. This could require multiple tanks to accommodate full, preparation, testing, and feed cycles.

System Configuration

The complexity of an incineration system depends on the waste composition and quantity, its oxidation potential, the required destruction level, and allowable exhaust emissions. Waste constituents can contribute to the configuration complexity by their participation in the combustion reaction; or passing through the system, in its original or an altered form. Table 1.10 summarizes an evaluation checklist to be used to evaluate waste related design criteria. After the waste impact has been assessed, the process requirements can be met by a proper incinerator selection and incorporating required upstream and downstream support equipment.

Incinerator Selection

Before any selection is made a complete analysis should be made of the waste, based on tests indicated in Table 1.9, to identify the limiting and controlling *waste characteristics.*

TABLE 1.10. Waste Characterization Checklist.

1. Identify potential exhaust pollutants and emission levels
2. Compare emission levels with regulatory limitations
3. Estimate the ash quantity
4. Assess the potential for excessive ash slagging
5. Assess potential ash disposal problems
6. Estimated fly ash generation
7. Evaluate possible methods of minimizing emission and ash generation by temperature control, controlling other process and design parameters, or pretreatment
8. Determine a suitable combustion temperature range to balance high efficiency organic reduction, minimize air emissions, produce satisfactory ash, and reduce slagging

The extent of the characterization is site specific, but in all cases their *impact* should be established. Chemical balances should be completed to determine the probable fate of significant components from the waste to the exhaust or ash.

Next, a preliminary evaluation should be made of the basic incinerator requirements, as summarized in Table 1.11.

Proven technology should be the primary basis for selecting a suitable incinerator. This is essential to assure an effective treatment and a reasonable implementation schedule, minimizing start-up time and field revisions or adjustments. Table 1.12 offers some guides to incinerator application based on dominant waste characteristics. These guidelines are suggested as a starting point, not *absolute* choices, based on the following criteria:

(1) Liquid waste: A high organic strength aqueous or an organic liquid waste is best incinerated in a liquid injection incinerator. When injected into a rotary kiln as part of combined wastes, high water content wastes should be either mixed with the solid wastes or atomized. They should be injected into the hottest chamber area, enveloping the flame, to assure immediate and

TABLE 1.11. Preliminary Incinerator Selection Checklist.

1. Can the waste be treated in a single chamber combustion system, or is a secondary combustion chamber required?
2. Can standard technologies be applied?
3. Is a new emerging technology required? Why?
4. Are full-scale operating data available for incinerators treating similar wastes?
5. What technical level personnel will be required to implement, operate, and maintain the proposed technologies?
6. How long will it require to develop, design, build, start-up, and commission the applicable technologies?
7. What regulatory requirements are applicable to this system? How long will it require to permit, test, and get approval for each of the technologies?
8. Develop "ball-park" capital and operating costs for a complete incineration system.

complete vaporization. They should never be directly injected onto a hot refractory, which could cause thermal shock and refractory damage. An adequate supplementary fuel supply assures a constant flame, adequate heat generation, and avoids a "flame-out."

(2) Liquid waste viscosity: If the waste viscosity is greater than 10,000 SSU, with or without heating, a liquid injection incinerator, or any type that requires atomization cannot be utilized. A solid waste type incinerator is best suited.

(3) Liquid waste solids content and quality: A liquid waste with a high solids content could be difficult to handle in a liquid injection incinerator because of nozzle plugging. This is particularly true if the solids are salts near the saturation point, and "salt out" near the nozzles. If the solids are salts which become molten but do not volatilize at the combustion chamber temperatures, slag formation at critical areas of the unit, including the nozzle holes, could be a major problem. Modifications should be made to typical liquid incinerators incorporating the following:

External or Sonic Nozzles
On-Line Cleaning of the Nozzles
Recirculation of Critical Quench Fluids
Maintain Low Quench Fluid Concentrations
Provision for Clean-Out of the Critical Chambers
Increased Preventative Maintenance

As an alternative, a fluidized bed may be better able to handle situations where minimum combustion temperature and minimum temperature chamber variation is desirable. Fluidized beds tend to maintain a uniform temperature throughout, with a minimum of cold spots and resulting slag formation. However, to minimize the effect of solids collection in the bed a circulating bed may be more suitable.

(4) Liquid wastes with high volatile metal concentration: A liquid waste with high concentrations of volatile metals could present a problem for liquid incinerators. They could be driven off as volatile fumes at the vicinity of the flame, and the toxics difficult to remove in the air stream. In such cases reduction of the primary combustion temperature is desirable, or a circulating or fluidized bed considered.

(5) Solids, sludges, and slurries: Dry solid, sludge, or slurry liquid wastes are not suitable for injection incinerators. This includes bulky solids, and aqueous organic sludges or slurries. Dual chamber units are required to assure liquid evaporation, volatilization of volatile compounds, and complete combustion of released gases. Rotary kilns are commonly used for this service, fluidized beds may be suitable for some sludges or slurries.

TABLE 1.12. Impact of Waste Characteristics on Incinerator Configuration.

	Liquid	Rotary Kiln	Fixed Hearth	Fluidized Bed	Multiple Hearth	Innovative Technology
Liquid Wastes						
1. Organic containing aqueous wastes	X					
2. Solvents	X					
3. Liquid waste, high viscosity		X		X		
4. High solid composition and content	M	X		X		
5. High volatile metal content solids	M					
Solid wastes						
6. Irregular bulk solids		X	X			
7. Granular bulk solids		X	X	X		
8. Large bulk volumes					X	
9. High ash and high boiling organic content		M		X		
10. Low melting point solids	M	X	X			
11. High volatile metal content		M		X	?	
12. High combined nitrogen content		M		X	?	
Special wastes						X

X, frequently applied; M, configuration modified for application.

(6) Large solid waste quantities to be destroyed: When large quantities of sludge or relatively clean solids are to be destroyed multiple hearth furnaces can be considered to substitute vertical height for horizontal chamber length, especially if available site area is limited.

(7) Solids with high ash and high boiling organic content: Solid wastes with high ash and high boiling organic content present a special operating problem. A high operating temperature will assure a high destruction efficiency and a low organic ash content waste, but a high air emission release and refractory slagging problems could result. Too low a kiln temperature reduces the air emission and slagging problems, but reduces the destruction efficiency and increases the ash organic content. A high organic content reduces ash disposal alternatives, increasing disposal costs. Therefore, the kiln temperature will have to be carefully selected, after some testing, to optimize the design.

(8) Low melting point solids, tars, etc.: Solids with low melt points, whose viscosity can be readily reduced by heating, can be treated as a liquid waste. They can be prepared and pumped to a liquid incinerator or included as part of a feed in a solids waste rotary kiln or fixed hearth.

(9) Solids with high volatile metal content: Solid wastes with a high volatile metal content could present a special problem at high kiln temperatures, releasing high quantities of volatile metals and imposing a difficult duty on the air pollution control train. Therefore, a system should be designed to optimize the kiln temperature profile, avoiding hot spots, and minimizing volatile metal release. This may require staging the process so that the waste temperature is gradually increased, releasing the volatiles at a minimum temperature, partitioning the metals to the ash. The released organic volatiles being totally oxidized in a secondary chamber. This may also be accomplished with a fluidized bed, which may be more conveniently operated at a desired "average" combustion temperature, minimizing extreme hot spots.

(10) Solids with high combined nitrogen wastes: Solid wastes with a high combined nitrogen content incinerated at high kiln temperatures, or with high excess air, will produce conditions favoring excessive nitrogen oxides formation. Nitrogen oxide thermodynamic considerations favor reduced temperature and oxygen conditions to shift the equilibrium toward nitrogen. In this case, the advantages of gradual temperature increases encountered in a controlled air hearth, a circulating or fluidized bed, or a rotary kiln operating at reduced temperatures may be advantageous.

(11) Application of innovative technology: The variety of industrial wastes encountered with unique properties, potentially imposing high treatment costs, may necessitate evaluation of "nonstandard" technology. Some emerging technologies are listed in the Incineration Systems section.

One word of *caution*, innovative technology is a very viable alternative when incineration is selected, but available technology effectiveness is questionable. However, the operating difficulties in implementing new technology should be considered. Those that have designed and operated incinerators know that problems occur with any waste, using any technology, and that start-up problems and adjustments to any system are expensive and time consuming. Innovative technology must be reviewed on the basis of many practical considerations for an on-site plant treatment system:

(1) Can a conventional system meet the plant's waste incineration requirements?

(2) Are the problems which are requiring consideration of innovative technology insurmountable with available technology?

(3) Is the potential higher capital and operating costs acceptable?

(4) Is the probable longer development and start-up time an available luxury?

(5) Is a highly trained professional required to operate and maintain the units available?

(6) Is the plant willing to be the first, and possibly the last, to adopt innovative technology to the industry waste?

Preliminary incinerator evaluation and selection should be concluded with pilot studies to obtain the design data indicated in Table 1.8.

Support Systems

Upstream and downstream support equipment must be included to meet the process requirements and optimize incinerator efficiency. Upstream support equipment includes facilities to store, equalize, and pretreat the waste. Downstream considerations include heat recovery and pollution control. Table 1.13 summarizes a support system checklist to evaluate process requirements. Required support equipment can be included on the basis of the Table 1.13 guidelines, and discussions in the Waste Evaluation, Pretreatment Requirements, Air Pollution, and Plume Suppression sections.

Oxygen Requirements

Incinerators have been classified into two broad categories based on the air source, as controlled and uncontrolled combustion units. Historically, units were designed as open natural draft systems to assure an excess (uncontrolled) air supply, drive the oxidation process to completion, and exhaust combustion products. As emission control became a driving force, secondary combustion and particulate control was included in the system. Current pollution control require-

TABLE 1.13. Support System Selection Checklist.

1. Does high flow or composition variation require upstream equalization and storage?
2. Is waste pretreatment beneficial or practical?
3. Is waste pH adjustment required?
4. What type of feed preparation equipment is required to assure a highly atomized or fine solid feed?
5. Does the incinerator exhaust have to be quenched?
6. Is heat recovery a viable option?
7. Is particulate air emission control required?
8. Is an acid control system required?
9. Is a sulfur dioxide control system required?
10. Is nitrogen oxide exhaust control required?
11. Is a plume suppression control system required?
12. What type of ash control system is required?

ments call for sophisticated devices making uncontrolled air systems impractical and uneconomical. Excessive air increases supplementary fuel requirements and downstream heat recovery and air pollution control equipment sizes. System performance requirements mandate strict control of air flows. Indeed, stringent emission limits result in some systems being designed with a deficit air supply to the initial stage to minimize flyash carryover, or create reduction conditions minimizing NO_x formation. Excess air requirements must be tailored to process conditions and not arbitrarily selected as adhering to combustion "guidelines."

Excess Air Requirements

Although excess oxygen requirements are primarily aimed at driving the oxidation reaction to completion, the air fed into a system is a balance of the following considerations:

(1) The minimum oxygen based on a stoichiometric balance.

(2) Low oxygen will adversely affect combustion efficiency, with starved conditions resulting in incomplete combustion, high carbon monoxide concentrations, and sooting conditions.

(3) Excessive air quenches chamber temperatures, and in extreme conditions flame-out can result. Within the incinerator capacity the system controls will compensate for excess air input with increased supplementary fuel demand, increasing operating costs.

(4) In many incinerators the air supply serves as the quenching medium to maintain and limit combustion chamber temperature. High caloric value feed will require high oxygen (air) demand to serve as a quench and prevent the flame and combustion temperature from reaching critically high temperatures. In such cases excess air requirements can reach the 200 to 250% level. Air demand can be reduced by utilizing water, stream, or aqueous waste as a quench media.

(5) Some incinerator configurations (such as multihearths) promote high air leakage into the system. In such cases oxygen is not easily controlled or integrated into an automated control scheme, making finite oxygen level control difficult, if not impossible.

(6) Combustion chamber air flow significantly impacts aerodynamically induced turbulence characteristics, and with some chamber designs influence required air supply. Some incinerators are designed to utilize high air and resulting exhaust flow to create turbulent conditions to enhance the combustion process.

A review of reported incinerator performance data reveals a wide range of excess air employed, with those quoted for hazardous waste units somewhat higher than standard treatments [4,16,17]. Since hazardous waste incinerator experience is limited, and available experience many times from remediation projects where fine tuning is not always

TABLE 1.14. Incinerator Excess Air Data.

	Common	Hazardous Wastes
Liquid incinerator	10–60	120–150
Rotary kiln	75–200	75–200
Fluidized bed	40–60	
Multiple hearth	75–100	100–150

practiced, the indiscriminate use of reported high excessive oxygen levels as a "fail-safe" solution is not justified. Common excess oxygen levels are cited for specific incinerators in the incinerator systems section, and in Table 1.14. A word of caution, excess oxygen requirements often cited for operating systems are never clearly identified as to whether they are the result of combustion, quench, or configuration demands; or using maximum fan capacity because "it was there." In addition, they are commonly quoted as a wide range, and should therefore be adjusted to specific design conditions.

An indiscriminate selection of excess air will effect both capital and operating costs. First, heating and exhausting excess air results in a significant energy cost resulting in increased operating costs. In addition, the consequences of excessive exhaust volumes is larger chamber volumes, and increased incinerator costs; and increased capital and operating costs because of larger downstream air pollution and heat recovery equipment, and stacks.

Forced or Induced Draft Systems

A forced air system "pushes" air through the unit, resulting in a slight pressure; while an induced draft "pulls" gases from the system, resulting in a slight vacuum. Two criteria impact whether forced air, induced air, or a combination of both are applied. First, a forced air system simplifies controlling combustion chamber conditions, and is consistent with commonly employed automation systems. These systems introduce air at the clean side of the incinerator, employing a fan at ambient conditions and relatively low volumes. As a result, a pressure is created in the combustion chamber, potentially allowing leakage of fugitive emissions in the work place or environment. In addition, if installed alone its discharge pressure must be adequate to overcome the pressure drop through the entire downstream system. However, these fans can only be economically designed for reasonable pressures, limiting the downstream equipment that can be installed.

A second consideration is that most hazardous waste incinerators require fugitive emission control, regulated as part of work place requirements, or as part of the facility air pollution control plan. In such cases a downstream fan may be needed to induce a slight vacuum to assure leakage into the incinerator, minimizing (or eliminating) fugitive emissions. The disadvantage of these fans are that they are in-

stalled downstream from the incinerator, which could be subjected to dirty, humidified, corrosive, hot exhausts. Mechanically these fans are always exposed to more severe conditions than clean force air fans. Total system requirements may necessitate a combined "push-pull" system using a forced system to control the incinerator air requirements and a downstream system to impose a slight incinerator vacuum and to provide a driving force to move the exhaust gases through the ancillary equipment. The strategic location of the devices depends on the specific system configuration, apparent when a system hydraulic profile is completed.

Combustion Temperature

Although time, turbulence, and temperature are often cited as critical operating variables, temperature remains the most practical operating variable to control the *oxidation reaction*. As with excess air requirements, combustion temperatures cited for conventional wastes are sometimes lower than those cited for hazardous wastes, as indicated in the Incineration Systems section and Table 1.15. Increased incinerator temperatures may be valid for some hazardous wastes, but not all. In addition to promoting NO_x formation, excessive combustion temperatures can significantly effect incinerator operations by promoting slag and reducing refractory life. As with excess air, a combustion temperature operating range should be established through pilot trial runs tailored to the specific wastes.

A chamber temperature is a result of dissipating the combustion heat released by the feeds, which is absorbed as sensible heat by the product gases. The maximum attainable temperature is the adiabatic temperature reached by the product gases, assuming no losses to the surroundings. Since some loses will result, the temperature reached will always be less than this value. In fact, the chamber temperature resulting from waste combustion release will almost never be the desired control temperature, and further adjustment will be required. If the temperature is excessive, additional quenching media must be provided, and supplementary fuel if the temperature is too low. In some cases heat recovery is employed to reduce supplementary fuel requirements by preheating the inlet air, waste, or both.

Actually, modern incinerators are automatically controlled to maintain required combustion chamber conditions by

TABLE 1.15. Incinerator Temperature Ranges.

	°F	°C
Liquid incinerator	1600–1800	870–980
Rotary kiln, nonslag	1200–2300	650–1260
slag	2600–2800*	1430–1540
Fluidized bed	1400–2000	760–1095
Multiple hearth	1400–1700	760–930

*Range includes bed and freespace.

monitoring exhaust excess oxygen concentration and chamber temperature, and accordingly adjusting supplementary fuel input or influent air. In some industrial systems this could result in a "confused" system, if the automatically adjusted exhaust oxygen concentration results in excess quench capacity, reducing the required combustion temperature. In such cases the incinerator has a mind of its own, defaulting to either temperature or excess oxygen requirement, depending on the design control logic. When both excess air and combustion temperature are automatically controlled, the system must be designed with "override" provisions defaulting to some predetermined "safe" logic, or alerting the operator to make the decision.

Selecting a design temperature involves the following considerations:

(1) The reactants must be heated above the ignition temperature to the required combustion temperature.
(2) The adiabatic temperature is the maximum achievable temperature.
(3) Temperatures exceeding 1300°C (2400°F) may require special evaluation of the combustion chamber refractory, especially if the waste contains components that can attack the lining, slagging conditions occur, or aggressive process conditions are expected.
(4) A design range of 980 to 1100°C (1800 to 2000°F) is commonly employed.
(5) If solid waste is to be burned without smoke, a minimum combustion temperature of 760°C (1400°F) must be maintained.
(6) Combustion reaction rates are temperature sensitive so that a reasonable operating range must be provided for the operator to respond to the process conditions.

Combustion Chamber Turbulence

Combustion chamber turbulence is critical to combustion effectiveness, but an uncontrollable operating variable. Generally, turbulence is a direct result of combustion gas volume, quantitatively related to the mass of excess air injected and combustion products formed and volumetrically related to the chamber temperature. It is generally assumed that an adequate excess air level assures adequate combustion gas volume and appropriate turbulent conditions. This may not always be the case if the waste is highly oxygenated, and the combined oxygen radically reduces the air fed into the system.

Mixing in the combustion chamber is initiated aerodynamically by turbulent forces resulting from the combined effects of air forced into the system and the thermal activity of the flame. When this is not adequate the manufacturer may design mechanical means to accelerate turbulence, using such devices as tangential feed, nozzles, vanes, bed depth, or any other proprietary method. Combustion chamber turbulence is such an inherent part of the manufacturer's incinerator design and construction, that it is often neglected by the Process Engineer, particularly since there are very few criteria that have been acknowledged as effective for all the available configurations. As a minimum, the design engineer should review with the manufacturer the following:

(1) What incinerator features assure adequate turbulence?
(2) Is the process air requirement sufficient to assure enough combustion gas volume for adequate mixing?
(3) What is the minimum gas volume required to aerodynamically achieve the required chamber turbulence?
(4) Is turbulence adequate at the low turndown ratio?

Residence Time

Residence time is an important operating variable controlling volatilization and oxidation effectiveness. Treatment efficiency is a result of the combined effects of residence time, temperature, and turbulence. Residence times are defined in terms of the combustion stage, i.e., primary chamber, secondary chamber, or afterburner performance.

Primary Chamber

Primary chamber residence times are commonly expressed in terms of allowable burn rate, which encompasses all the chamber functions. Such criteria cannot be theoretically determined, rates are selected by experience or through specific pilot testing. Burn rates are frequently cited as heat flux (heat release per unit time per unit volume), ranging from 178,000 to 1,780,000 kcal/hr^{-1}/m^{-3} (20,000 to 200,000 BTU/hr^{-1}/ft^3), specific to the incinerator stage and configuration. Although primary chamber design is defined in terms of volume based on retention time or burn rate, its dimensions are critical to controlling particulate transport. The chamber geometry, especially the cross-sectional area, affects the combustion gas exit velocity and particulate transport capacity. Generally, the chamber length-to-diameter should be in the range of 4 to 10 for cocurrent kilns and greater for countercurrent configurations [25].

Secondary Chamber

Secondary combustion chambers or after-burners are commonly designed as gas incinerators, with residence times ranging from 1 to 3 seconds, at temperatures ranging from 980 to 1093°C (1800 to 2000°F). The basic design is frequently based on a convenient vessel size as opposed to reaction kinetics. In many cases these final chambers are overdesigned and therefore adequate if design flows are not exceeded. Operating performance is "fine tuned" by adjusting the temperature.

Energy Conservation and Recovery

Because incineration systems involve exothermic reactions with high-energy release, energy conservation and recovery are viable options. The Process Engineer must examine all opportunities for energy conservation to reduce fuel use, and resulting operating costs. In the final analysis, the benefits of energy conservation depend on the availability and cost of fuel, and recovery economics. These benefits can be readily evaluated and easily demonstrated by assigning a monetary value to conservation and recovery. Cost benefits will be directly proportional to the fuel cost and the incinerator heat recovered. The larger the incinerator system the more attractive the economics. Unfortunately, Environmental Engineers often focus on regulatory compliance and minimum capital costs, with little concern for operating costs. In fact, incinerator operating costs are usually of paramount importance, sensitive to inflationary forces, and at some point in the incineration system life span exceeds the original capital expenditures. Operating costs can be reduced by implementing energy conservation in the design, incorporating steps to minimize fuel use and recover the process energy generated.

AIR POLLUTION CONTROL

Combustion chamber exhaust gases will contain a variety of organic and inorganic constituents, all of which are derivatives of waste feed components. These include trace amounts of the organics escaping the combustion chamber(s), compounds that are formed in the combustion chamber, or products of incomplete combustion (PICs). Waste organics escaping the combustion chamber are a result of droplets transported by the combustion gases out of the flame front or gasified droplets quenched and condensed in cooler sections away from the flame front, also transported with the flue gases. Because incineration is very effective in destroying organics the quantities of original organics in the exhaust are minute, including by-products formed. However, incinerator emissions are highly regulated and allowable emission levels low.

Inorganics are commonly discharged as the carbonates or oxides of various elements present such as sodium, calcium, barium or potassium; or heavy metal oxides of antimony, arsenic, barium, beryllium, cadmium, lead, chromium (VI), mercury, silver, and thallium. Acid fumes such as hydrogen chloride, hydrogen bromide, hydrogen fluoride, and sulfur dioxide can be emitted if the halogen is present in the feed. Particulate emissions will consist of any of the inorganic materials mentioned, or any inactive feed components. Another inorganic pollutant of concern is nitrogen oxide, formed from combustion air nitrogen (usually insignificant), or combined waste nitrogen, which can be effectively converted to nitrogen oxides.

Although Process Engineers concentrate on the combustion aspects of an incineration system, it would be almost impossible to imagine installing an incineration system without air pollution control equipment. In most cases, where a treatment system does not meet specification, it does so because of noncompliance with emission standards rather than combustion inefficiency. Where emission requirements are not met, the problem can usually be attributed to poor emission control equipment performance based on arbitrary selection. Air pollution control should be developed as an integral part of the whole treatment plant rather than as a separate add-on system. The elements of an air pollution control system include a series of corrective actions that *collectively* achieve air emission requirements. Figure 1.17 illustrates the elements of an *air pollution system*. As illustrated in Figure 1.17, the major elements of an air emission control system include

(1) Waste feed inventory blending to eliminate peak contaminant concentration spikes
(2) Pretreatment of waste feed streams to remove objectionable components
(3) Selection of a supplementary fuel with low sulfur, ash and other constituents that could contribute to air pollution emissions
(4) Selecting design criteria, an incinerator configuration, and operating conditions which minimize pollution emissions
(5) Quenching the incinerator exhaust to temper the temperature to that required by *some* pollution control devices
(6) Selection of air pollution control equipment for maximum emission removal
(7) Evaluation of potential ground level concentrations over a variety of local meteorological and climatic conditions to estimate average and maximum concentrations, and if necessary conduct a risk assessment analysis to assure that potential environmental effects are minimal

Effective emission control requires employing basic design fundamentals at the feed, combustion, and emission control points.

Removal Concept

Poor air pollution control equipment performance is the result of indiscriminate selection of control devices, based on arbitrary performance, not identifying or considering specific pollutant characteristics, or not understanding the control device limitations. Removal efficiency is dependent on many factors which collectively result in separation of the contaminant from the transporting media (exhaust gases). Basically, they depend on mass transfer principles governing pollutant transport from the gas, and specific equipment removal characteristics. Relevant factors include

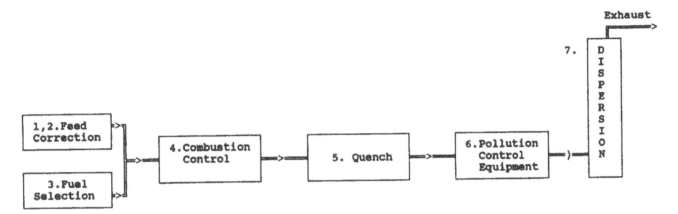

Figure 1.17 Air pollution control system.

(1) The removal driving force increases with increasing inlet concentration, resulting in higher removal rates and efficiency, although not necessarily a lower exhaust concentration, which is proportionally related to the inlet concentration.

(2) The pollutant chemical and/or physical properties.

(3) The mechanism governing pollutant removal, i.e., absorption or chemical reaction for scrubbers, impact force for particulate removal, etc.

(4) Required contact time, which could be as simple as residence time and particle velocity, or as complex as packing type, height, and size.

Often equipment is selected on the basis that it has achieved a 99% removal, or a specific emission level, for a similar incineration control system. Unless specific pollutant properties and feed concentrations have been measured, related to the control device design characteristic and performance, comparison of equipment efficiency or exhaust concentration is meaningless. Indiscriminate use of treatment efficiencies or exhaust characteristics is the most common mistake made in air pollution control, contributing to poor operating results.

Specific Incinerator Emissions

The major pollutants emitted from an incinerator include products of incomplete combustion (PICs), carbon monoxide, dioxins/furans, nitrogen oxides, sulfur dioxides, particulates, acid fumes, and heavy metal fumes. It is not within the scope of this text to detail all applicable EPA regulations and applicable wastes. Applicable regulations, proposed "draft" regulations, and guidelines are offered as a background, especially where they affect the incinerator design. Regulations are constantly changing, implementation complex, and many times legal interpretation is required to define applicable limits. The cited regulatory references, and subsequent revisions, should be reviewed for process design relevance.

Principal Organic Hazardous Constituents (POHCs)

Primary EPA incinerator regulations involve control of wastes containing any of the 400 Appendix VIII listed hazardous compounds (CFR 261, May 19, 1980), updated semiannually. In general terms, a compound can be listed as hazardous based on

(1) Ignitability, Hazard Code I

(2) Corrosivity, Hazard Code C

(3) Reactivity, Hazard Code R

(4) EP (extract procedure) toxicity, Hazard Code E

(5) Acute hazardous waste, Hazard Code H

(6) Toxic waste, defined as a carcinogenic, mutagenic, or teratogenic, Hazard Code T

The design significance is that organic wastes so designated require 99.99% removal of each principal organic hazardous constituent (POHCs), the allowable maximum emission rate for each POHC being 0.01% of the feed (EPA:CFR 264.343). A more stringent requirement of 99.9999% removal is required for wastes containing PCBs (polychlorinated biphenyls classified as F020-F028), as proposed by the U.S. EPA, Dioxin Rule (*Federal Register,* January 14, 1985).

The best treatment for POHC compliance is high combustion efficiency. This can be achieved by increasing the combustion chamber(s) temperature (most effective method), increasing the excess oxygen fed to the combustion chamber to drive the reaction kinetics toward completion, or as a final consideration increased residence time. The Process Engineer must design the combustion system to allow the operator a reasonable operating range to vary combustion temperature, excess oxygen, or both. This involves the burner design and selection, combustion air fan capacity, residence time, and downstream *air pollution control system* capacity. Where high variability or uncertainties are inherent in the waste characteristics the engineer will have to include considerable afterburner flexibility to assure maximum hydrocarbon reduction.

Products of Incomplete Combustion (PICs)

Products of incomplete combustion are primarily unreacted, partially reacted, or byproducts formed in the combustion chamber, passing through the pollution control devices. EPA "draft" guidelines suggest a 20 ppmV cap on total hydrocarbon emissions (THC) as "good operating practice," if the incinerator cannot meet a carbon monoxide hourly rolling average of 100 ppmv, corrected to a dry basis and 7% oxygen [31]. The Process Engineer should be aware of the potential for stringent PIC control, and the latest revisions to applicable regulations reviewed. Where required, control techniques for these constituents are similar to those suggested for POHCs, with considerable afterburner flexibility required to assure maximum reduction of total hydrocarbon emissions.

Carbon Monoxide

Theoretically, high carbon monoxide emission levels are an indication of ineffective combustion, probably more evident in increased PIC formation than in reduced treatment efficiencies for a specific component. Based on the assumption that good combustion conditions limits the total exhaust hydrocarbon content, the EPA has suggested that carbon monoxide control may limit PIC formation [31]. Incinerator carbon monoxide exhaust levels can generally be maintained below 100 ppmV at properly controlled combustion conditions. EPA guidelines suggest a maximum hourly rolling average of 100 ppmv, on a dry basis corrected to 7% oxygen. Whether formal limits are established, regulators will usually impose some limits in issuing an operating permit. Based on high carbon monoxide levels being indicative of poor combustion efficiency, the same control measures suggested for PICs are applicable to carbon monoxide, and should be considered as part of the incinerator design.

Dioxins/Furans

Dioxin is the name given to the family of polychlorinated dibenzo-para-dioxins (PCDD), whose primary derivative is the basic structure shown in Figure 1.18 and which can be chlorinated in multipositions resulting in 75 possible compounds. Similarly, furans are derivatives of polychlorinated dibenzofurans (PCDF) with the structure shown in Figure 1.19, which can also be chlorinated in multipositions resulting in 135 possible compounds. Most data has concentrated on what is considered the most toxic of the possible

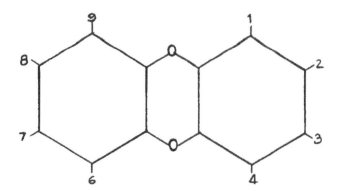

Figure 1.18 Dioxin general formula.

compounds, 2,3,7,8-tetrachlorodibenzo-*p*-dioxin (TCDD), equating related compounds to TCDD equivalence (TEQ), using toxic equivalent factors (TEF) [1].

The mechanisms by which dioxin and furans are formed has been evaluated by many investigators and summarized into three broad categories [1,2,14,23]:

(1) The compounds are present in the feed and are not destroyed in the incineration process.
(2) The compounds are formed in the combustion process vapor phase as products of incomplete combustion involving chlorinated hydrocarbons, or unchlorinated hydrocarbon and a chlorine donor.
(3) The compounds are formed in equipment downstream from the incinerator, with particulates promoting their formation from chlorinated hydrocarbons or unchlorinated hydrocarbon and a chlorine donor.

A common theory is that their formation is enhanced by hydrocarbon vaporizing and escaping from the combustion chamber, and reacting on the surface of particles with a possible chlorine donor; the potential for forming increasing in the 250 to 350°C (482 to 662°F) range. The indication being that a high carbon and/or chlorine content in the fly ash probably increases dioxins/furans formation. Other investigators have observed that oxygen has an influence on

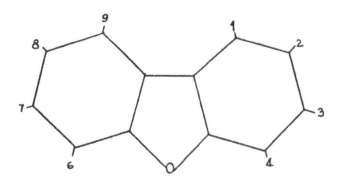

Figure 1.19 Furan general formula.

the formation of PCDD/PCDF on fly ash, with formation increasing with increased oxygen concentration.

The uncertainty of the mechanism by which dioxins/furans are formed is probably best acknowledged by the uncertainty as to how to regulate them from hazardous waste incinerators. The European Economic Community (EEC) proposed restricting the emission level of 2,3,7,8-TCDD toxic equivalent (TEQ) to less than 0.1 nanogram per normal cubic meter [1]. The U.S. EPA is regulating removal efficiency by requiring 99.99% removal of individual hazardous feed materials, and 99.9999% removal of PCB material. In addition, a high combustion efficiency must be maintained based on a carbon monoxide level less than 100 ppmV, or a waiver procedure requested substituting a total hydrocarbon (THC) emissions of less than 20 ppmV if the carbon monoxide level exceeds 100 ppmV. The EPA proposed TCDD/TCDF criteria for municipal waste combustors and hazardous waste incinerators are constantly being reviewed and revised. The Process Engineer should be aware that control could involve direct restriction of specific emission concentrations, limiting the 2,3,7,8-TCDD toxic equivalent (TEQ) emission rate, or both. The regulations cited are for background information only, and the latest regulatory requirements should be reviewed for specific design limitations.

Available data suggest that dioxin/furan formation is more a problem in municipal than industrial incinerators, probably due to the larger incinerators employed, inherently poorer combustion control, and resulting incomplete combustion. Suggested industrial waste incinerator control procedures can be summarized as follows:

(1) Eliminate or minimize PCBs and chlorine in the waste.

(2) Employ good combustion practices to minimize PICs formation by maintaining the combustion temperature at above 870°C (1600°F), maintain good air/fuel patterns, and maximizing combustion effectiveness [10]. The ability to vary the chamber oxygen content should be designed into the system to implement study results which imply that PCDF formation from PCB destruction increases with increased oxygen content [1].

(3) Reduce the possibility of PCDD/PCDF forming at temperatures between 238 to 349°C (460 to 660°F) by reducing flyash holdup in this critical temperature range [14]. In fact, the entire incinerator train should be investigated to establish what portions of the system could be at this temperature range and what probable PCDD/PCDF forming contaminants could be present. The air pollution control strategy should be to maintain the air pollution control devices well below 150°C (300°F) to eliminate the possibility of captured dioxins/furans volatilizing.

(4) Reduce particulate entrainment and promote high combustion of carbonaceous compounds in the secondary chamber [14]. In that respect, a hot cyclone used to treat rotary kiln high particulate emissions could be effective.

(5) Upgrade the air pollution control system to promote dioxin/furan removal, with special emphasis on higher flyash reductions. Some data suggest that dry reactor/spray dryer and baghouse performance could be improved using additive injections such as activated carbon upstream of the equipment [1]. Similarly, a wet air pollution control system employing a rapid quench and two-stage scrubbing, reducing the flue gases to the 180°F range, may improve exhaust quality.

Nitrogen Oxides

Nitrogen oxides can be produced from combined nitrogen components in the waste and combustion air nitrogen. At very high temperatures NO_x can be produced from nitrogen in the exhaust gases, with fuel nitrogen conversion possibly dominating at lower temperatures [15]. Both mechanisms could be significant if combustion conditions favoring NO_x formation are present. Generally, nitrogen oxide formation is a result of a high flame temperature and high oxygen concentration, the two variables that drive the equilibrium conversion to nitrogen oxides. The resulting nitric oxide equilibrium concentrations can be estimated utilizing the procedures discussed in the Basics section.

Available control technology includes

(1) Reducing flame temperatures by flue gas recirculation or steam or water injection

(2) Minimizing excess air concentration

(3) Staged combustion to selectively control the temperature and oxygen level to promote overall high combustion efficiency but minimize nitrogen oxide formation

(4) Low NO_x burners are constantly being evaluated to control combustion of high nitrogen containing wastes [20].

Proprietary systems to convert nitrogen oxides to acceptable nitrogen constituents are generally classified into two categories: selective catalytic reduction (SCR) and selective non-catalytic reduction (SNCR) methods. SCR methods involve the injection of ammonia into the flue gas prior to passing through a catalyst bed, converting nitrogen oxide to nitrogen and water. Most available systems involve Japanese technology developed for specific combustion systems and may not be applicable to industrial or hazardous wastes where constituents may be present that "poison" the catalyst. Also, standard flue gas systems, such as those used for acid removal, can be upgraded using appropriate chemical additives to enhance nitrogen oxide removal.

SNCR processes all involve injection of ammonia or urea solutions in the flue gases to convert nitrogen oxides to nitrogen and water.

Under conditions normally encountered in industrial hazardous waste incinerators, nitrogen oxides exhaust concentrations should be easily maintained at levels of less than 100 ppm by controlling combustion conditions. For industrial or hazardous incineration systems it would be expected that because of relatively small incinerator size, unlike power plants or municipal incinerators, that nitrogen oxide control would be achieved primary by combustion control, followed by enhancement of commonly employed air pollution controls. Special nitrogen control equipment would probably only be considered when wastes being treated have an unusually high nitrogen content. Industrial nitrogen oxide control would include one of a combination of the following:

(1) Storage and blending of high nitrogen content wastes
(2) Manage nitrogen containing waste loading to the incinerator
(3) Low excess air firing
(4) Two-stage combustion
(5) Flue gas recirculation to reduce combustion air
(6) Decreased flame temperature

Sulfur Oxides

Sulfur oxides are generated as a direct result of waste sulfur components oxidizing to sulfur dioxide, and some sulfur trioxide. The distribution is governed by reaction kinetics. For the most part, sulfur oxide emission control is based on wet or dry alkaline flue gas scrubbing, the selection dependent on the sulfur dioxide concentration and chemical costs. Lime is commonly used in a dry scrubber, whereas either caustic or a lime solution can be used in a wet scrubber.

The Process Engineer should consider the following criteria in developing sulfur oxide removal systems.

(1) Are there available waste water treatment facilities to dispose liquid blow down streams? If wastewater discharges cannot be tolerated then dry lime scrubbing systems are applicable; in which case solids disposal alternatives must be established.
(2) Are large quantities of alkali materials required? If large alkali chemical use is projected then lime or limestone is generally considered.
(3) Lime systems require considerable handling, preparation, and transfer equipment design evaluation because of low solubility, and transport difficulties resulting from settling and scaling.
(4) Because of its low solubility lime reactivity is slower than caustic, although it is a cheaper reagent.
(5) In small facilities, where alkali requirements are low, sodium hydroxide may be advantageous because of the lower equipment capital costs, handling convenience, and high reactivity.
(6) Where extremely high sulfur waste concentrations are expected, resulting in potentially high alkaline consumption, specialized processes involving lime injection into the combustion chamber may be worth considering.

Generally, sulfur dioxide is not a major problem in hazardous waste incinerators because waste sulfur content is normally relatively low, and low sulfur supplementary fuel is purposely selected.

Particulates

Waste solids, unless organic, will not be destroyed in the combustion process. Most will be discharged as primary combustion chamber ash, the rest as flyash. Particulate characteristics are critical in selecting control equipment. If the solid melt temperature is less than the 815 to 1090°C (1500 to 2000°F) range encountered in combustion chambers the inorganics will be vaporized and discharged as a fume, as with some metals and metal salts. In such cases, fume collection is difficult because the mixture dewpoint is suppressed by the large quantity of inerts (nitrogen and oxygen) and the small quantities of the pollutant (ppm or ppb). As a result, fume crystallization (as a solid) is minimal, and particulate collection devices ineffective.

Generally, incinerator exhaust flyash can be removed by employing a bag filter, an electrostatic precipitator, or a high energy scrubber. Flue gases containing extremely high particulate loadings are initially passed through a cyclone to economically reduce the load to the primary dust cleaning equipment, prolonging its life. Particulates are always a major problem in hazardous and industrial waste treatment systems. Current EPA regulations for hazardous waste incinerators require exhaust levels to less than 0.08 grains per dry standard cubic foot (180 mg per dry standard cubic meter), corrected to 7% excess oxygen. Local regulations may be more restrictive.

Acid Fumes

Acid fumes are formed by halogen waste components combining with available hydrogen ions to form the corresponding hydrochloric or hydrofluoric acid. These acids are removed in dry or wet scrubbing systems similar to sulfur dioxide. The EPA hazardous waste guidelines limit the allowable emissions to the *larger* of 4 pounds per hour or one percent of the feed concentration [32]. Generally, local regulations will be more restrictive.

Heavy Metal Emissions

EPA incinerator guidelines have been proposed for regulating arsenic, beryllium, cadmium, and hexavalent chromium as carcinogenic metals; and antimony, barium, lead, mercury, silver, and thallium as noncarcinogenic metals [32]. The proposed regulatory control methodology involves the following:

(1) Health standards are suggested for allowable exposure for a general population.

(2) Allowable ground level concentrations are established based on the allowable exposure value.

(3) Based on allowable ground level conditions and an effective stack height, allowable emission and feed rates are estimated for each of the carcinogenic and noncarcinogenic defined metals.

(4) The effective stack height is based on worst dispersion conditions; and the results adjusted for terrain, as well as for urban or rural conditions.

The latest revision of the cited references should be reviewed for specific details, and the allowable feed and emission rates. Table 1.16 lists the critical temperatures of the metals discussed, and corresponding chloride salt. The ability to remove the contaminant with customary air pollution equipment depends on the physical state of the component. In the solid state, the pollutants can be removed as particulates with a variety of dust control equipment. If the contaminant is in a (molten) liquid phase, the air stream must be quenched below the critical temperature for effective removal. Some vaporized metal or metal salt contaminants are difficult to condense because of high dilution, which

TABLE 1.16. Critical Metal Temperatures
(data from Reference [7]).

	Melt Point, °C	Boil Point, °C
Arsenic	817t	614
AsCl$_3$	−16	130
Cadmium	321	767
CdCl$_2$	568	961
Chromium	1857	2682
CrCl$_2$	815	1300
CrCl$_3$	877	947s
Berryllium	1277	2484
BeCl$_2$	415	532
Antimony	631	1640
SbCl$_3$	73	220
SbCl$_5$	3	140
Barium	725	1849
BaCl$_2$	962	2029
Lead	328	1751
PbCl$_2$	501	953
PbCl$_4$	−15	105e
Mercury	−39	357
Hg$_2$Cl$_2$	382s	d
HgCl$_2$	277	304
Nickel	1455	2920
NiCl$_2$	1030	970s
Silver	962	2164
AgCl	455	1564
Selenium	221	685
Se$_2$Cl$_2$	−85	127d
SeCl$_4$	—	196s
Thallium	303	1487

t = triple point; d = decomposes; e = explosive; s = sublimes.

greatly depresses the metal or metal salt dewpoint to well below the 65 to 95°C (150 to 200°F) range. Since most quench systems cool the exhaust gases to no lower than the 65 to 95°C (150 to 200°F) range, separation may be difficult for some contaminants, and they will pass through the air pollution control equipment.

Two factors further complicate the situation:

(1) The metal could be exhausted as a complex salt whose properties are not readily apparent, making control equipment selection difficult.

(2) If the metal can be removed, the by-products may be difficult to dispose, because of the components removed. Their characteristics will usually result in the liquid or solid disposal product also classified as hazardous, and subject to strict disposal regulations.

Conservatively estimated air pollution control device efficiencies for single stage treatment, as cited by the EPA are listed in Table 1.17. When the suggested control equipment efficiencies are applied to individual contaminants, it is obvious that large quantities of these contaminants cannot be tolerated in the waste stream and meet emission guidelines. Indeed, the guidelines suggest multicombinations of control equipment to achieve reasonable control limits.

The difficulty of removing volatile metallic compounds was evaluated by Barton et al. in which the scrubber capture efficiency of metals was related to the volatility [27]. Results indicated removal efficiencies of less than 50% for arsenic, lead, copper, cadmium, and beryllium with volatility temperatures less than 700°C and removal efficiencies greater than 70% for barium, strontium, magnesium, and chromium with volatility temperatures greater than 950°C.

PLUME SUPPRESSION

A steam plume results when the stack exhaust gas temperature falls below its dewpoint before it can disperse as a result of mixing with dryer, cooler ambient air. The potential for a steam plume forming can be determined on the basis of a heat transfer balance between the stack gas and ambient air. A conservative estimate of plume formation can be made from energy transfer relations, expressed by the equation [13]:

$$\frac{Hs - Ha}{Hm - Ha} = \frac{Ts - Ta}{Tm - Ta} \tag{1.1}$$

where Hs is the absolute humidity of the stack gas, weight of water vapor per weight of dry gas, Ha is the absolute humidity of the ambient air, weight of water vapor per weight of dry gas, Hm is the absolute humidity of the mixture, weight of water vapor per weight of dry gas, Ts is the stack gas temperature, Ta is the ambient air temperature, and Ts is the mixture temperature.

TABLE 1.17. EPA Heavy Metal Removal Estimates (adapted from Reference [32]).

	Ba, BE	Ag	Cr	As, Sb, Cd Pb, Ti	Hg
Wet scrubber	50	50	50	40	30
Venturi scrubber, 20–30 in WG	90	90	90	20	20
Venturi scrubber, >60 in WG	98	98	98	40	40
Electrostatic, precipitator, 4 stages	99	99	99	90	0
Wet electrostatic precipitator	97	97	96	95	60
Fabric filter	95	95	95	90	50

Note: From EPA Draft Hazardous Waste Incinerator Regulations issued January 16, 1989.

Equation (1.1) can be represented as a straight line defined by the coordinates (Ts, Hs), representing the stack gas conditions, and (Ta, Ha), representing the ambient air conditions. When this straight line is plotted on a psychometric chart a plume will probably occur if it is tangent or intersects (lies left) of the saturation humidity curve. The procedures to predict and prevent plume formation are illustrated in Figures 1.20 and 1.21.

The Process Engineer should be aware that total plume prevention is economically impossible; plume formation can only be minimized. Methods to condition an exhaust to control plume formation are illustrated in Figure 1.22.

As illustrated in Figure 1.22, gas conditioning to minimize plume formation can be accomplished by (1) reducing the exiting gas moisture, (2) reheating the gas, (3) diluting the stack gas with hot air, or (4) a combination of these techniques. The selected method is primarily dependent on costs and the stack gas conditions.

Stack gas moisture can be reduced by using a cooling tower or by an indirect heat exchanger. In both cases the stack gas is dehumidified by reduction of the stack gas to

below its dew point, to a new saturated condition at a lower temperature. This is illustrated in Figure 1.23.

Because the saturated air stream can still produce a plume, although at a lower frequency of occurrence, a higher degree of plume suppression can be obtained by reheating the gases. This is illustrated in Figure 1.24.

This can be accomplished by a direct exit burner or by heating the gases with a heat exchanger. A direct exit burner involves the least capital expenditure, but has one distinct disadvantage, namely, it adds trace quantities of water vapor, unreacted products, and products of incomplete combustion to the exhaust. This is a real problem when the incinerator must be subjected to compliance testing, since agreement must be obtained as to where to conduct the exit sampling—before or after the burner.

Finally, plume suppression can be accomplished by adding hot dilution air to the exit stack gases, if a satisfactory

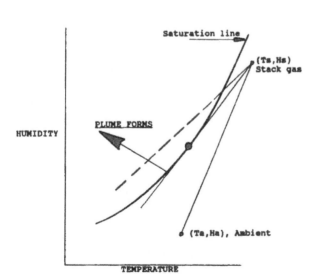

Figure 1.20 General plume formation criteria (adapted from concepts in Reference [13]).

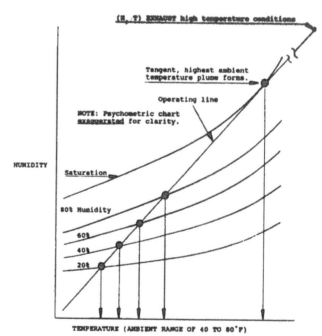

Figure 1.21 Plume formation operating line (adapted from concepts in Reference [13]).

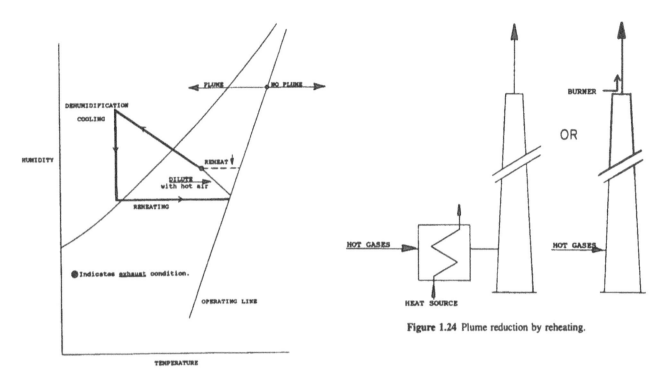

Figure 1.22 Plume prevention (adapted from concepts in Reference [13]).

Figure 1.24 Plume reduction by reheating.

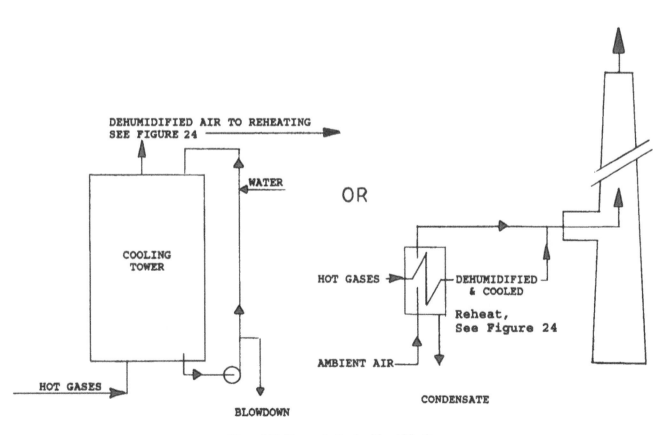

Figure 1.23 Plume reduction by dehumidification.

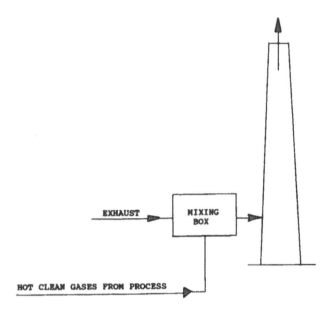

Figure 1.25 Plume reduction by dilution.

(uncontaminated) dilution stream is available, or can be economically produced. This is illustrated in Figure 1.25.

Actually, plume suppression is almost always considered in the early stages of an incineration project, with the criteria being minimum, or no (impossible) visible plume, but seldom installed, and if installed seldom operated because of high installation and operating costs. High costs for the purpose of achieving a "deception" but no environmental benefit. If the stack exhaust contains steam, no benefit is achieved; if it contains pollutants, they should be removed.

FATE OF CONTAMINANTS

The final products of an incineration process include the treated exhaust gases, ash, and dry dust or a liquid stream from pollution control systems. Collected ash and dust is commonly shipped off-site for disposal. However, waste organic or inorganic components that could concentrate in the ash may result in a hazardous waste classification, requiring stabilization techniques to insure safe disposal. Because the dry material collected is a combustion by-product, the ash and flyash collected could contain trace amounts of unreacted waste or organic combustion by-products. The ash should be analyzed for possible metal reclaim, as an alternative to stabilization, or as a first step in reclassifying as nonhazardous.

Scrubbing system blowdowns normally contain neutralized acids, dissolved solids, suspended solids, and trace amounts of unreacted waste or combustion by-products. These wastes are commonly sent to on-site industrial waste treatment systems or discharged to a municipal system. The composition of these waste streams rarely are a problem. The major problem, if any, is that the metal content of these

streams may be significant, making direct discharge difficult because of water quality permit limits.

The exhaust from incineration systems are commonly subjected to air pollution control limits for nonhazardous incinerators, and very strict emission limits for hazardous waste incinerators

REGULATORY REQUIREMENTS

Designing and operating an incinerator requires start-up permits and continuous monitoring to meet regulatory requirements. The procedures involve a sequence of regulatory steps required to assure compliance: impacting design, construction, start-up, and operation. Critical design and operating criteria result when the waste components "trigger" an EPA defined hazardous waste classification as detailed in EPA CFR part 261, discussed in the Air Pollution section. Hazardous waste compliance could vary slightly from location to location, but generally involve the following:

(1) The design must take into account RCRA Trial Burn considerations as well as the monitoring requirements imposed in the final permit.

(2) Documentation must be prepared in the design phase for the incinerator permit application, requiring plans to satisfy hazardous waste generation, storage, and treatment regulations and a proposed trial burn plan.

(3) Public notice, and public response to the notification, indicating the intent to install a hazardous waste device.

(4) Performance of a hazardous waste trial burn to assure that the device will meet regulatory compliance.

(5) Monitoring critical operating parameters during the life of the operation to assure that the conditions successfully tested in the trial burn are not exceeded or violated.

(6) Reporting any violation of permit conditions to the regulatory authorities any time during the operating life of the incinerator system.

Design considerations related to meeting the regulatory requirements discussed include the following:

(1) Assuring that the combustion residence time and temperature are adequate to meet the destruction efficiency required for the RCRA Trial Burn.

(2) Assure that the incinerator exhaust organic levels, either unreacted organic or products of incomplete combustion, are below specified limits.

(3) Assure that hydrogen chloride, chlorine, sulfur dioxide, nitrogen oxides, and particulate exhaust emissions meet regulatory requirements.

(4) Assure that toxic metals such as arsenic, mercury, cadmium, silver, chromium, and lead are not emitted in quantities in excess of the regulatory limits.

(5) Provide for adequate air pollution control equipment to assure that the pollutants mentioned are controlled.

(6) Provide adequate instrumentation to control and monitor the operation.

(7) Provide adequate monitoring equipment and sample ports to complete the RCRA Trial Burn.

(8) Provide adequate monitoring equipment to implement final permit conditions.

GENERAL ENGINEERING CRITERIA

Figures 1.26 and 1.27 illustrate typical liquid and rotary kiln Incinerator Preliminary Concept Flowsheets. A complete incineration system could involve the following process components:

(1) Waste storage facilities
(2) Waste preparation facilities
(3) Waste feed system
(4) Fuel storage facilities
(5) Incinerator system package, including
    a. Primary combustion
    b. Secondary combustion
    c. Primary and secondary air systems
    d. Heat recovery
    e. Air pollution control
    f. Ash handling system
    g. Induced fan system
    h. Stack
    i. Plume suppression
(6) Scrubber blowdown treatment facilities
(7) Ash treatment and disposal facilities

Design criteria for these components depends on the waste and incinerator selected. Some engineering considerations are discussed to specify some general design guidelines.

INCINERATOR CAPACITY

A common incinerator design problem is balancing waste generation with furnace capacity, a meaningful problem when waste is generated over an intermittent period. Under such cases, designing a facility for peak generating loads is not economical because the system will be normally operating at less than optimum conditions. In addition, frequent incinerator start-up and shut-down is not practical because

(1) Incinerator deterioration related with thermal shocks (especially refractory) is accelerated.

(2) Treatment and maintenance costs could be prohibitive. In such cases, *operating costs should be thoroughly investigated.*

(3) Emissions are maximum during system upsets, especially start-up and shut-down. Permits normally exclude

a minimum amount of start-up time from noncompliance reporting, but not as normal operating practice.

Where waste is generated intermittently, the following alternatives should be considered:

(1) Off-site contracted incineration for relatively small loads. Available off-site incinerator(s) capacity should be investigated.

(2) Larger storage facilities and a smaller incinerator to convert to a continuous treatment basis.

A similar circumstance results from the desire to operate an incinerator for one shift daily, preferably the day shift where maximum support and supervision is available. The balance here is an offset of operating costs against the cost of a bigger incinerator, which may or may not be economically feasible. In an effort to reduce costs, operating plans are sometimes proposed to shut down the incinerator on the shifts when it is not used. This is not practical! What should be done is to maintain the combustion chamber at a low burn, near the combustion temperature, allowing the system to be operated with a minimum preparation. In such cases, the incinerator design should include a tight system, minimizing heat losses, operating at minimum fuel during the "off-shifts."

WASTE TRANSPORT

When designing an incinerator the method of continuously transferring waste from the generating source to the incinerator should be carefully planned. Wastes are never directly linked from the generating source to the incinerator, but to intermediate storage facilities. Liquid wastes are usually conveyed through industrial sewers or trucked in bulk, slurries either pumped or transported in bulk, and solids are always trucked in drum or bulk loads. The storage and treatment facilities are generally designed as part of the overall manufacturing facilities and must be afforded the same safety and health protection measures governing these facilities. The design must include a "break" in the transport path from the manufacturing and employee occupied facilities to the storage facilities, and the storage facilities to the combustion equipment. This is to ensure that the combustion ignition source does not cause a major catastrophe, or that fumes cannot reach occupied facilities.

STORAGE

As discussed above, incineration systems require a balance between waste generation and treatment rates. Often wastes are collected over a fraction of the work week or day but processed over a continuous 7-day, 24-hour period. The combustion process is an instantaneous reaction, involving a fraction of time, and generally not practical for intermittent operation. Interruption in combustion operations many

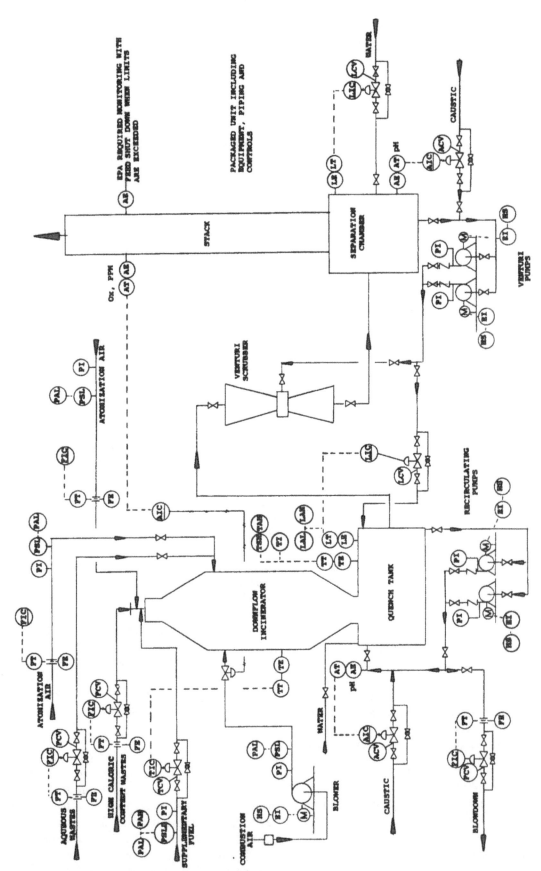

Figure 1.26 Liquid incinerator preliminary concept flowsheet.

38

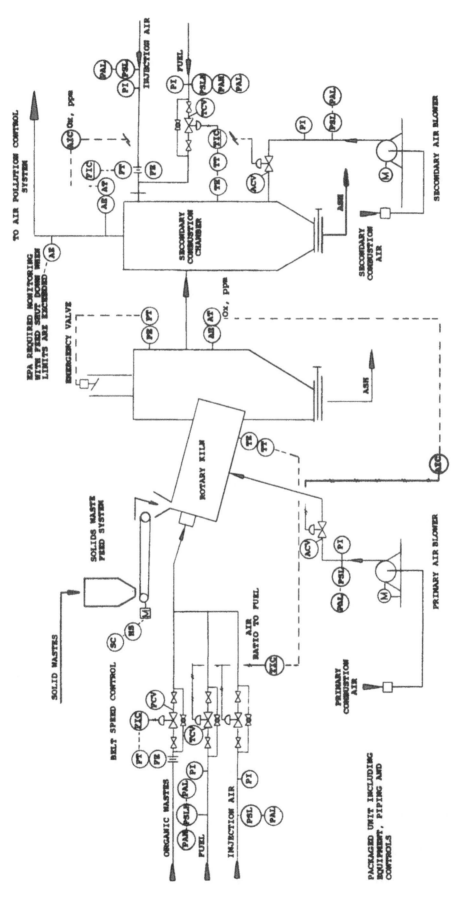

Figure 1.27 Rotary kiln preliminary concept flowsheet.

39

times result in "puff" exhaust emissions violating regulatory limits. For this reason wastes feeds are usually accumulated and stored so that a process inventory can be maintained. Storage and continuous operation is desirable because (1) the incinerator size can be reduced to accommodate an average flow, rather than peak instantaneous loading, (2) uniform feed compositions can be prepared, stabilizing the combustion process and simplifying the control instrumentation, and (3) constant incinerator heat-up and cooling is avoided.

Waste storage and feed systems must be carefully designed to assure waste compatibility with the variety of wastes encountered. Chemical compatibility depends on individual and combined waste characteristics such as chemical and physical properties, corrosivity, and reactivity. In complex manufacturing facilities a wide variety of wastes may be incinerated, the precise composition or potential reaction with other stored wastes can not always be determined. Therefore, the worst conditions must be considered in designing storage facilities.

Storage area allocation must include design criteria emphasizing prevention and a positive response to potential safety and emergency incidents. This could include

(1) Waste segregation according to compatibility, avoiding storage of compounds which could react upon release and contact.
(2) Waste isolation and containment of wastes which if released could cause ecological, safety, or health problems.
(3) Specially designed containment areas to assure that any liquid release can be restrained, controlled and cleaned.
(4) Fire protection and spill control capabilities to respond to any emergencies.
(5) Proper storage and ventilation of enclosed facilities, and protection of facilities subjected to severe winter conditions.
(6) Controlled access to storage areas by limiting vehicle traffic, and personnel access restricted to trained and authorized personnel.

Corrosivity

Waste alkalinity or acidity can range the entire scale, anywhere from 1 to 14, in some cases changing with storage time. The Process Engineer defines a systems's storage limits when selecting materials of construction for tanks and transfer systems. Therefore, the selection should be carefully evaluated for current and anticipated future requirements. A good indication of suitable materials of construction is to determine what is used in the manufacturing storage facilities. If the manufacturing storage facilities are designed for corrosive materials, in all probability corrosive wastes will be generated. Any change in the manufacturing storage facility will also require a new evaluation of the waste storage facilities. Transfer and storage equipment limits should be

clearly specified in the plant operating procedures and clearly labeled on the equipment.

Reactivity

Waste stability of individual or combined wastes is a major consideration in developing storage and transfer facilities. Some main concerns include

(1) Wastes that could undergo exothermic or pyrophoric reactions, resulting in heat or fire that could cause extreme damage to the facilities or at a minimum release a boiling liquid or gases.
(2) Waste physical changes resulting from polymerization or precipitating reactions, causing large solids deposits or slag formation, complicating transfer or pumping.
(3) Gas forming reactions resulting in vapor releases or displacement of storage tank contents, both of which could present an extremely dangerous environmental problem.

Physical Properties

Waste physical properties are an important consideration in evaluating storage requirements. Specific concerns include:

(1) Solubility of the waste components, resulting in phase separation and settling. In such cases mixing may be required to prevent phase separation and assure a uniform feed.
(2) Melt, freezing, or boiling temperatures could result in waste phase changes.
(3) Specific gravity and viscosity are important waste storage, pumping, and transfer properties.
(4) Abrasive waste solids can cause rapid erosion of transfer equipment.
(5) Safe storage of volatile waste is critical in controlling small or extreme releases from the facilities.

From a health and safety concern, containment and controlled venting of storage facilities must be evaluated—with special attention to the explosive and toxic characteristics of any volatilized materials. Materials must be classified and stored in strict compliance with federal and local regulations. Most liquid wastes will have to be stored with containment provisions. This must also include measures to prevent volatilization or direct gaseous releases. Potentially explosive wastes will require extreme care in storage, preparation, and transferring to incineration facilities. Wastes transported to storage facilities must be accurately analyzed and classified to assure storage of compatible wastes, segregation of incompatible wastes, and isolation and proper preparation of hazardous wastes.

All storage facilities should be designed considering the range of ambient conditions affecting waste temperature, potential pressure buildup, and damage to the containment vessel, such as direct sunlight, temperature, rain, snow, etc. They should be isolated from heavy traffic areas, highly populated areas, manufacturing facilities, or *other chemical storage facilities* that could accidently release chemicals incompatible with the stored waste products.

TRANSFER SYSTEMS

Chemical *reactivity* and *stability* are major concerns in transferring waste from the storage facilities to the combustion facilities. In cases where the wastes are unstable or extremely reactive they must be stabilized to assure safe transport to, and injection into, the combustion system. Stabilization methods will depend on the waste chemistry and could include adding stabilizing chemicals, converting the compound to a more stable state, or dilution to a low-risk level.

Direct wastes transfer from a central storage facility into an incinerator system is seldom feasible. Feed is commonly transferred from a central storage system to smaller day storage tanks, where batch blends can be prepared suitable for a consistent incinerator feed. In many situations more than one day or feed tank is considered to segregate wastes according to physical state (solid or liquid), caloric value, feed method, and required pretreatment.

Liquid and Slurry Wastes

Blending, phase separation, homogeneity, and ease of handling are major concerns in handling liquids or slurry wastes. Wastes should be adequately mixed to prevent stratification and separation within large storage vessels and fed from smaller day or feed tanks to assure a homogeneous feed. Day tanks are particularly suitable to meet these criteria because limited volumes can be prepared in enclosed facilities, assuring that all preparation steps can be easily monitored.

Liquid transfer systems commonly include recycle loops, with very short takeoff lengths to the injection nozzles. Velocities within the loop are maintained at a minimum of 0.6 to 0.9 meters per second (2 to 3 fps), consistent with the system hydraulics. Provision should be made for hot water or steam flushing and proper line draining when the system is not operational. Small line sizes should be avoided to minimize pluggage or cloggage, and in-line filters included at nozzle inlets. Transfer lines should be heat traced and insulated to assure fluidization of viscous liquids, prevent freezing (during flow interruption), and to prevent crystallization.

Solid Wastes

Solid wastes are commonly transported by vehicle in bulk bins or drums to a central storage facility for further processing. The most common plant wastes include biological treatment sludges, production residues, rejected production materials, and off-spec product. In some cases, building trash can be transported for disposal at the same locations. All wastes should be separated and stored in compatible facilities. The degree of manual labor employed in transporting and preparing solid wastes depends on the quantity processed, with automated systems common for large facilities. From a central storage facility the solids will be transported to local storage bins where blending, preparation, and further dewatering may be employed. Transfer method from central to local facilities depends on waste characteristics, and whether transport is bulk or in containers. Trucks are usually employed, although large bulk volumes could be mechanically or air conveyed.

INCINERATOR SYSTEM COMPONENTS

Feed Method

The feed method employed depends on the physical state of the waste and the incinerator configuration. In all cases, the waste must be fed in a form favoring rapid combustion, which may require some waste preparation and conditioning.

Solid Wastes

Solid wastes usually undergo various size reduction and sorting steps prior to incineration. In large installations this is done on a continuous basis, with solids being fed from a storage hopper, onto some type of conveying process system. The solids are screened to remove excessively large materials (usually crushed and returned), metal detectors used to remove materials which could damage the system, and completed with size reduction to produce fine and uniform material. Size reduction could include the use of grinders, hammer mills, attrition mills, dicing machines, granulators, or knife grinders. The material could also be subjected to a blending and preparation step, before or after the reduction step, using a ribbon blender, augur blender, or a cement type mixer. The prepared material is then fed to the incinerator at a controlled rate.

The feed and preparation system details depends on the specific waste. Some special considerations include

(1) Sludge stored for days may require "cake breakers" to prevent caking of feed system surfaces or transporting large sludge pieces.

(2) Small waste incineration systems can be manually fed, using a feed chamber with an air lock and a ram rod. Feed is prepared as in a continuous feed system, remov-

ing large solid or metallic components and reducing the bulk solid to an acceptable size. The resulting feed is batch fed to the feed chamber at regular intervals.

(3) High water containing solid wastes require special handling. Wastes containing significant water concentrations are usually dewatered, and the dewatered sludge either pumped or screw conveyed. The lower water content not only reduces fuel requirements, but increases refractory life by minimizing thermal shocks.

(4) Slurries are sometimes fed to the kiln using lances to preselected combustion chamber sections.

(5) Wastes with low water content are either pumped or moved with a screw conveyer into the incinerator depending on their fluid characteristics.

(6) Wastes are also collected in drums, and the entire container is fed into the incinerator. This is sometimes desirable for toxic or "sensitive" material to avoid excessive handling or exposure to the material. Although this is a simple handling method, it complicates combustion. High temperatures are usually required to destroy the container, the combustion chamber is subjected to extreme temperature variations, and great effort must be taken to avoid overloading the chamber over short periods. The process is eased by using combustible drums instead of steel drum.

(7) Where applicable, drums are sometimes transported with drum handling equipment and bucket type conveyors, emptied into feed hoppers, discharging either directly to the incinerator or to a continuous incinerator feed conveyor.

Liquid Wastes

A liquid can be effectively atomized if it is pumpable, having a viscosity of less than 750 SSU or can be steam atomized to reduce the viscosity. In such cases, liquid wastes are pumped to combustion chamber nozzles, atomized to a fine mist, subjected to the hot gases, and evaporated. However, besides viscosity, nozzle design requires considering the waste solid content and solid properties such as size, melt point, and abrasion. As previously discussed, nozzles can be classified as internal mechanical, external mechanical, or sonic. The selection is dependent on the solid content.

Incinerator performance, and especially nozzle operation, are greatly affected if the solids form molten droplets at the temperatures near the nozzle and are a greater problem if they remain in a molten or semimolten form in the combustion chamber. In such cases, excessive slags can plug or damage the nozzle, destroy the spray pattern, and form slag in critical incinerator areas. Abrasive solids can also be destructive to the nozzle by eroding the nozzle orifices, the nozzle itself, or both. Corrosive materials will have the same effect. Some related nozzle design considerations include

(1) The combustion chamber must be designed with multinozzles, which can be readily removed and checked on-line.

(2) Provisions should be made for nozzle purging with water, steam, or high-pressure air.

(3) The atomizing pattern should be investigated to assure that refractory impingement is avoided, especially when the nozzles are partially worn.

(4) Where erosion, corrosion, or plugging is possible provisions should be made for on-line maintenance, flushing, and nozzle replacement.

(5) Provision should be made to be able to visually monitor nozzle performance and to measure and record some critical nozzle flow parameter such as flow or pressure.

The number of nozzles employed and their location are important design considerations. Multiple nozzles should always be employed to assure operating redundancy. The actual number required is based on flow rate, with a unit commonly effectively operating at 2 to 19 liters per minute (0.5 to 5 gpm). Excessive rates result in larger orifice openings, and corresponding larger droplet sizes. Large droplets can be driven out of the influence of the combustion environment by gravitational forces and eventually carried over as a finer mist with the exhaust. What is important is the spray pattern, which should form fine droplets resulting in rapid volatilization at the combustion temperature and rapid oxidation in 2- to 3-second residence time.

Another important feed system design consideration is the allowable incinerator nozzle turndown ratio. If the flow volume or the heating value of the feed can vary significantly, usually more than 4 to 6:1, the turn-down range will be exceeded, and the combustion process will be unstable. It is important to allow enough upstream storage and equalization to avoid these conditions.

Blowers and Fans

Broadly defined, blowers imply air injection, while fans imply gas exhaust systems, both commonly defined in terms of output volume and pressure. Based on output pressure, compressors can be employed at discharge pressures greater than 275 kPa (40 psig), blowers up to 275 kPa, and fans exhausting high volumes at less than 3.5 kPa (0.5 psig). Blowers are commonly used to supply primary air to the burners. Fans are used to collect the combustion gases, impose a negative pressure in the combustion chamber, and direct the gases through specific downstream equipment.

Where high-pressure outputs are required compressors can be employed. However, compressors are pulsating machines, whose output must be dampened to assure a constant air supply. For that reason compressors are not commonly employed. Typically, blowers and fans are used as follows:

(1) Combustion air is directly applied (forced) to gas burners for all incinerators.

(2) Secondary air is generally directly applied (forced) to liquid incinerators, primary kilns, and secondary chambers.

(3) Induced draft fans are used to exhaust combustion products when a slightly negative draft is required in the incinerator.

(4) Induced draft fans are used to provide draft to pull the quenched combustion gases through air pollution control equipment.

Some special incinerator conditions include

(1) Pressurized air is supplied to a fluidized bed to maintain the bed fluid, the total air input must balance the required combustion air requirements.

(2) Multiple hearth use induced air fans to exhaust the combustion gases and maintain a negative pressure in the hearths. Draft is controlled in the individual hearths to balance the air requirements.

Fan or blower characteristics must be compatible with the required combustion performance, such as

(1) The fan or blower must have a capacity to operate at the required system pressure and volume range.

(2) When a constant pressure output is required regardless of volume, such as with combustion air, the blower should have a flat performance curve.

(3) In other cases, a more sensitive pressure-volume response curve may be required.

(4) In selecting a fan or blower, the gas temperature, and gas cleanliness and corrosiveness are critical factors in choosing the blade configuration.

In any of these cases the entire performance range and reaction sensitivity must be evaluated before a fan or blower is selected.

Forced Air Systems

A forced air blower is employed at the incinerator inlet, pushing air into the system. Construction details cited for municipal blowers include [24]

(1) Black steel is a common material of construction for the plate casing and impeller.

(2) Allowance should be made for easy access to allow impeller inspection.

(3) The bearings should be externally mounted for easy inspection and lubrication.

(4) A horizontally split casing should be provided for impeller sizes greater than 30 inches for easy wheel maintenance.

(5) The impellers should be statically and dynamically balanced so that vibration velocity does not exceed 0.2 meter per minute (0.6 fpm).

(6) The motor should be sized for ambient air volume adjustments for winter and summer intake conditions, and the complete range of incinerator operating conditions.

(7) A V-belt drive should be used to allow field capacity adjustment.

(8) Noise should be limited to a maximum 85 dBA, or to meet local requirements. This may require lower speeds, noise isolation design, or enclosures.

Induced Draft Fans

Induced air fans are included to maintain a draft in the combustion chambers, and supply the draft requirements of downstream pollution control systems. The draft required to support air pollution equipment is dependent on the specific equipment. Where low-pressure drops are required the combustion chambers are sometimes operated at increased pressure to supply the required head. Induced air fan impose a serious design problem because the machine must be capable of handling high temperature, corrosive, wet, dust laden exhaust gases. In many cases stainless steel, or special alloys such as inconel and hastelloy must be specified for process conditions. Stainless steel may not be suitable if chlorides are present in the gases.

Hazardous or Toxic Waste Piping

Special design considerations are required for hazardous or toxic waste piping systems [3]. These include

(1) Indoor hazardous waste installations should be built with automatic sprinkler protection as well as suitable drainage and ventilation capacity. The facility sewer and exhaust systems should be designed to allow rapid removal of any spill or exhaust to proper treatment facilities. The facility should be designed for prevention of leaks, early detection of leaks, and immediate response to any incidence.

(2) All hazardous waste piping should be designed with readily accessible draining and cleanout provisions for maintenance or during long shutdown. Any drainage must be directed to suitable containment and disposal facilities.

(3) Selected piping materials must be suitable for the intended maximum service conditions based on temperature, pressure, and corrosiveness. The use of stainless steel, nickel alloys, steel lined, or similar rigid piping may be necessary for corrosive wastes.

(4) All piping must be clearly marked and identified as to contents and process conditions. Coded piping is recommended for this purpose.

(5) All tank entering piping should be above liquid levels. A one-way flow check valve should be located as close to the tank as possible for liquids entering a tank below the operating level.

(6) Vessels should be valved so that exit piping can be maintained regardless of the tank liquid level. To protect against frozen valves, large vessels outlets should be double valved, and have more than one exit nozzle installed to allow isolating one and emptying from another.

(7) Above-ground piping carrying hazardous materials may have to be double piped, flange shields used, and a leak detection system employed.

(8) All elevated piping should be designed for easy access and inspection.

(9) Buried piping should be avoided. Trenched piping should be installed with removable steel plate covers, draining to a safe and suitable discharge, and properly ventilated.

Refractory Selection

Incinerators are constructed of steel shells, the thickness and reinforcements designed to assure mechanical and structural integrity, but seldom designed to withstand the chamber temperatures. The refractory serves as a protective membrane between the combustion reaction products and the shell, shielding the shell from high thermal, corrosive, and oxidizing conditions. The refractory also serves to minimize heat losses. System integrity depends on the refractory not being penetrated. Refractory life is a vital economical consideration, dependent on proper selection, application, and maintenance.

Refractories are nonmetallic materials employed primarily as high temperature protective linings, but they must also be selected to protect against chemical and physical stress resulting from process conditions. Physically, the refractory must protect against thermal and mechanical stresses, as well as erosion and abrasion. Chemically, the refractory must provide corrosion resistance to the reaction environment, encompassing both chemical and thermal conditions. Thermally, the refractory must provide heat resistance to protect the shell, minimize heat losses, be resistant to thermal shock, and provide heat radiation to enhance the combustion processes. Refractory construction can include baffling to enhance turbulence. The refractory, along with any external insulation, provides personnel protection from high temperature contact.

Primary refractory materials include alumina (Al_2O_3) and silica (SiO_2), with additions of components such as lime (CaO), magnesite (MgO), titania (TiO_2), or other inorganic oxides. The system strength is a result of the bonding material, method, and temperature resistance. Refractory materials are produced and available as a brick or cement castable materials. Bricks are preformed materials in standard brick size and a variety of shapes. The bricks are attached to each other with bonding cements that are of the same composition as the primary brick. Cement castables, commonly referred to as monolithic refractories, include a variety of dry materials which are mixed with water before using, or preprepared mixes ready to be used. Preprepared mixes are manufactured as "workable" cements that can be poured or sprayed, and molded to shell.

Refractory effectiveness depends on the system temperature, and the temperature fluctuations to which they are exposed. In addition, both the refractory and bonding material must be able to withstand exposed conditions. The impact of exposure to combustion products include

(1) Combustion products such as carbon dioxide, water, oxygen, and nitrogen are usually not a major concern, although water vapor can increase acid fume corrosiveness.

(2) Carbon monoxide and acid gases such as sulfur dioxide, hydrochloric acid, and hydrofluoric acid, could inflict corrosive problems if condensation occurs and the condensate and gases reach the steel enclosure by penetrating high porosity or exposed refractory. Chlorine and fluoride can react with silica and alumina, reducing the refractory strength or that of the bonding matrix.

(3) Ash containing alkali materials such as sodium, potassium, or calcium or waste feed materials such as glass can react with the aluminum and silicon oxide refractory constituents. As a result a glassy slag is formed, reducing and eventually destroying refractory integrity.

Refractory effectiveness is also dependent on its physical characteristics, as it relates to its ability to withstand mechanical shock, rotation, and vibration.

Refractory Classification

Refractory can be classified into six major categories such as superduty fireclay, high alumina, silicon carbide, silica, chrome, and insulating refractory. Typical properties are listed in Table 1.18 [4,8,15,29].

(1) Superduty fireclay is basic clay material consisting of blends of alumina and silica and impurities such as iron oxide, lime, magnesia, titania, and alkalies. They are available at various grades, with the properties cited in Table 1.18. These materials have characteristics that result in high refractiveness, good spall resistance, good load bearing properties, and stable volumes, with some resistance to acid and basic slagging. An increase in the silica content to 70 to 80% results in high resistance to structural spalling and volatile alkali attack.

(2) High alumina refractory is alumina-silica material containing from 50 to 99+% of aluminum oxide, improving its refractory characteristics over that of fireclay. The refractory qualities increase with increasing alumina

TABLE 1.18. Refractory Characteristics (adapted from References [4, 8, 15, 29]).

	Composition			Limiting	Hot	Thermal	Chemical Resistance to	
	SiO$_2$	Al$_2$O$_3$	TiO$_2$	Temp, °F	Strength	Shock	Acid	Alkali
Fireclay								
Super duty	40–56	40–44	1–3	3185	Fair	Good	Good	(1)
High duty	51–61	40–44	1–3	3175	Fair	Fair	Good	(1)
Medium duty	57–60	25–38	1–2	3040				
Low duty	60–70	22–33	1–2	2815				
Semi-silica	72–80	18–26	1–2	3040				
High Alumina								
45–48%	44–51	45–48	2–3	3245				
60%	31–37	58–62	2–3	3295	Good	Good	Good (3)	(2)
80%	11–15	78–82	3–4	3390				
90%	8–9	89–91	0.4–1	3500	Excellent	Good	Good (3)	(2)
Mullite	18–34	60–78	0.5–3	3360	Good	Good	(5)	(4)
Correndum	0.2–1	98–99	trace	3660				
Silica								
Superduty	95–97	0.1–0.3	Some	3100	Excellent	Poor,	Good	(1)
Conventional	94–97	0.4–1.4	lime	Fusion		Good		
Light weighty	94–97	0.4–1.4		point		above 1200		
Silicon Carbide Bonded	60 to 90% Silicon Carbide			3360	Excellent	Excellent	(3)	(6)
Chrome Fired	28–38 Cr$_2$O$_3$, 14–49 MgO				Good	Poor	Fair/Good	Poor
	15–34 Al$_2$O$_3$, 11–17 Fe$_2$O$_3$							
Insulating (1600–3000°F)	Varies				Poor	Excellent	Poor	Poor

Notes:
(1) Good at low temperatures.
(2) Very slight attack with hot solutions.
(3) Some attack with HF.
(4) Slight reaction.
(5) Insoluble with most acids.
(6) Attacked at high temperatures.

content. In addition, they exhibit high mechanical strength, fair to excellent spall resistance, fair resistance to acid slags, and high resistance to basic slags.

(3) Silicon carbide refractory is produced by fusing coke and sand. Although they oxidize at critical temperatures, they exhibit hardness, low thermal expansion, high thermal conductivity as well as high resistance to abrasion, thermal shock, spalling, and corrosion.

(4) Silica refractory contains over 95% silica, with the remainder primarily lime. They exhibit low thermal spalling above 650°C, high resistance to acid slag, and low resistance to basic slag. They possess high refractiveness, abrasion resistance, and mechanical strength.

(5) Chrome refractory contains 28 to 38% Cr$_2$O$_3$, 14 to 19% MgO, 15 to 34% Al$_2$O$_3$, and 11 to 17% Fe$_2$O$_3$. They exhibit high corrosion resistance to basic and moderately acid slags. Spalling can be caused by iron oxide.

(6) Insulating refractories exhibit low conductivity, used to reduce heat loss, and employed in combination with other refractories. The insulation choice is based on the interface temperature. They are generally light weight materials, categorized as brick, monolithic, or fibrous.

Alone they have poor abrasion, corrosion, or erosion resistance.

Guidelines to Refractory Selection

The major considerations in refractory selection are system *temperature* and *waste composition*. This usually results in refractory systems being evaluated for temperature, slag reactions, and chemical environment. In addition, important refractory design considerations include mortar selection, installation, thermal gradient, expansion allowance, and brick configuration. *Slag characteristics,* especially its viscosity and chemical reactivity, are important factors affecting refractory life. Low viscosity slag flows quickly, maximizing refractory wear by washing away reacted slag and mortar. A slow moving, viscous slag results in a protective coating, but increasing orifice clogging tendencies of internal equipment. A preliminary refractory selection can be made on the basis of the potential slag reactions, relating the slag chemistry with the mortar constituents. A general criterion "not an absolute criterion," is to match the brick with the slag to reduce the potential for chemical reaction [4,15].

TABLE 1.19. Brick Characteristics (adapted from References [4,15]).

Acidic Brick	Basic Brick
Fireclay	Chrome
High-alumina	Magnesite
Silica	Forsterite
Acidic Slag	Basic Slag
SiO_2	FeO
Al_2O_3	CaO
TiO_2	MgO
	K_2O
	Na_2
	Cr_2O_3

Refractory can be classified as acidic or basic, based on the principal constituent, as indicated in Table 1.19 [4,15].

Slag can be classified according to the predominant constituent, as indicated in Table 1.19.

Guidelines published by Caprio [28] listed some *temperature*-related criteria for waste incinerator refractory selection. As a rule of thumb, it is suggested that the temperature dictate the aluminum oxide (Al_2O_3) content of the lining, as indicated in Table 1.20.

This temperature-aluminum oxide guideline must be tempered with observed laboratory study results and some industrial experience indicating the following exceptions [6]:

(1) Some laboratory testing indicate the use of high alumina (80 to 90% Al_2O_3) phosphate bonded brick exhibit higher brick-slag reactions than those mulite bonded.

(2) Calcium aluminate cement bonded castables and gun mixes exhibit poor resistance to most hazardous wastes slags and halogenated compounds.

(3) The use of basic brick (magnesite and magnesite-chrome) in high temperature may be a suitable alternative. However, the design must compensate for the bricks tendency to hydrate when exposed to moisture and a spalling potential resulting from brick penetrated with reacting slag expanding differently than unreacted portions.

It should be noted that service temperatures are different in the various sections of an incinerator system; decreasing downstream of the burner, through the various combustion unit sections and stages, to the stack. Refractory selection can be downgraded accordingly, but interaction and compati-

bility in adjoining lining sections must be evaluated. In addition, some times when a refractory is upgraded the heat transfer increases, thermal resistance decreases, so that insulating refractories may have to be upgraded to protect the steel.

Specific combustion chamber *Chemical environments* will influence the refractor selection, especially when inorganic alkali or halide compounds are present [28].

(1) The reaction of alkalies with fireclay and high alumina refractories can be damaging, with results worsening with increase temperature and alumina concentration.

(2) The reaction of chlorine with silica and alumina can be severe, forming volatile chlorides, reducing refractory strength.

(3) Halides can be corrosive to most refractories.

Finally, it should be noted that factors such as rotating equipment, vibration, abrasion, and mechanical shock may require special upgrading considerations.

The refractory information discussed is intended as background information, not as definitive design or selection criteria. The Process Engineer should contact a refractory expert to establish the requirements based on *maximum operating temperature* as well as a *waste characterization*. The Process Engineer should be aware that refractory selection is an *economic* consideration based on refractory cost, expected life, and value added for upgraded refractory selection.

Process Controls

The degree to which an incinerator is automated requires a balance between manpower, and added instrument costs. Because of environmental and public pressures driving incineration projects, the perception is that automation reduces the risk of an "incident." This is true up to the "pinch-point" of the process. Good automation will not correct a bad design. Many plant management personnel are shocked by the resources required to maintain an incinerator system "legally" operable. The initial feeling is that what is being installed is a boiler, monitored by part-time personnel. The differences between a waste incinerator and other industrial combustion equipment are:

(1) Most waste incinerators are stringently regulated.

TABLE 1.20. Refractory Guidelines (adapted from Reference [28]).

	Incinerator Section Temperature		
Temperature, °C	982–1093	1149–1371	1371–1538
°F	1800–2000	2100–2500	2500–2800
Quality of Al_2O_3 Brick	High fired superduty	High purity	High purity, mullite bonded
Al_2O_3, %	40–50%	60–70%	90%

(2) Because of costs, most incinerators have little spare capacity, and few installations have spare incinerators in full back-up operation. Long incinerator downtime requires either curtailing the production facilities generating the wastes, or costly off-site disposal.

(3) Waste incinerators are considered by regulators more as an *air pollution device* than a waste or sludge treatment system, with emissions an immediate environmental problem, requiring an immediate response.

Manual Control or Automation

Incinerator control design requires an understanding that the combustion reaction is rapid, uncorrected upsets will immediately move the process toward poor combustion conditions, increased stack emissions, regulatory noncompliance, process interruption, and mandatory shut-down until the condition is corrected. Any corrective action requires identifying process causes, establish corrective action, implement the remedy, and resume the incineration process. Automation is designed to correct upset situations to a "selected" set of conditions, but not necessary address causes. Therefore, it is an illusion that complete automation will reduce the operator skill required—or the operator. Computerized distribution systems (CDS), currently a trend in incineration systems, can further complicate a problem if not properly applied. In such cases a field technician is replaced with a desk engineer manipulating software changes to fine tune process loops to correct basic process problems.

Regulatory requirements may dictate specific incinerator controls and monitoring. Unlike any other waste treatment system, incineration is regulated on a continuous performance basis as opposed to monthly reporting. Short-term deviations from permit requirements may have to be reported. This requirement can be a persuasive incentive for excessive incinerator automation. The belief that high incinerator automation reduces the potential for operator error results in a false sense of security. In fact, automation immediately records operator error and any possible regulatory violation.

For the reasons mentioned, critical operating parameters must be identified, and a judgment must be made as to whether they can be best operator managed, automated, or both. Such decisions will require establishing the following:

(1) Will individual operating sequences be designed for operator verification, or will operating lights simply indicate the current operational stage?

(2) What variables should be automatically monitored and recorded, and what should be locally measured and operator recorded?

(3) Should monitored variables automatically control the process, or is operator judgment required?

(4) Should exceeded process limits be alarmed and operator correction be required, should other variables be altered to attempt to correct the problem, or should the system shut down?

It should be pointed out that safety decisions associated with the burner operation, or pilot flameout, are always automated, as mandated by insurance underwriters, local codes, and safety institution code requirements.

Incineration Control Sequence

Process control evaluation requires considering a natural incinerator operating sequence, which must be maintained to assure stable operation. Ancillary storage, transfer, pollution control, and heat recovery operations have their own control requirements, but their performance is still highly dependent on continuous incinerator operation. The incinerator process must include capabilities to perform a sequential series of verifiable operational measures, the successful completion of each ensuring a safe and stable process. Some of these measures will be discussed in general terms assuming a simple stage unit, specific details depend on the incinerator selected and process design. A common incineration sequence includes the following:

(1) A system check prior to the burner firing, verifying that all feed systems, air pollution control equipment, and ancillary equipment are in an operating mode, and all system safety checks are completed.

(2) The system is purged of any residuals prior to startup, or as needed during the start-up.

(3) Before the burner firing sequence is initiated a positive response must be received that critical incinerator elements are operational or the waste and fuel feed lines will be shut down and the system will revert back to a fail safe mode.

(4) The combustion burner start-up sequence is initiated and the combustion chamber temperature is slowly elevated. During this process the gas and air pressure must be maintained, and the flame must be active to continue the process.

(5) The system combustion temperature is reached.

(6) If applicable, the induced fan is activated and a slight incinerator vacuum maintained by controlling the inlet fan damper.

(7) Before the waste feed sequence is initiated, or can be sustained, a positive response must be received that critical items are operating, or the waste and fuel feeds must be terminated, and the system revert back to a

fail safe inactive mode. This includes *all* the incinerator operating elements, as well as downscreen recovery and pollution control equipment.

(8) Waste is fed to the incinerator and the system maintained operable by monitoring and controlling variables such as waste feed(s) and fuel flow, feed and fuel pressure, air flow, air pressure, combustion temperature, combustion pressure, nozzle pressure, nozzle flow, refractory temperature, and exhaust temperature.

(9) Maintaining vital system variables is important to assuring a continuing stable oxidation process. Some significant variables include feed flow, combustion chamber(s) temperature and oxygen levels. In addition, exhaust carbon monoxide level is a measure of combustion efficiency. Commonly the combustion temperature is automatically monitored by adjusting the supplementary fuel feed, or if the system is autogenous the waste feed flow. Oxygen concentration is monitored in the chamber exhaust and the combustion chamber blower adjusted accordingly.

(10) In addition, critical air pollution control variables must be monitored to assure optimum performance. Some variables considered critical by regulators include

  a. Pressure drop across particulate venturi scrubbers
  b. Minimum liquid to gas ratio in scrubbers, as well as pH
  c. Caustic feed to a dry scrubber
  d. Minimum scrubber blowdown, recirculation stream concentration, and makeup rate
  e. KVA settings to a wet or dry ESP
  f. Pressure drop across a fabric filter, influent gas temperature, and gas velocity
  g. Scrubber nozzle pressure
  h. Quench tank water makeup and outlet gas temperature

(11) Added to this sequence may be regulatory requirements that the incinerator process be terminated when *critical* permit limits are exceeded. This is always understood to mean termination of the feed until the problem is remedied. Actual incinerator shutdown is a lengthy process requiring gradual temperature reductions.

All incinerators will not necessarily operate in this same manner or sequence. This process sequence is general in nature and not intended as a definitive control philosophy, but discussed to illustrate the need to carefully evaluate the processing steps, including operations not discussed such as feed preparation, ash removal, heat recovery, and air pollution systems. *Each* system must be tailored to the total facility needs, coordinated with the operator's responsibility, and to the regulatory mandates.

AIR POLLUTION CONTROLS

Incinerator pollution emissions can be classified as point source or fugitive. Point source emissions are direct emissions which are emitted from the combustion operation, "collected," and transferred to an air pollution control device. Fugitive emissions are "leaked" from the system as unintentional releases, difficult to identify, and not as easily controlled.

Applicable air pollution controls are primarily based on the pollutant's physical state—particulate or gas. In addition, the pollutant's stability will impact the removal efficiency. Will the pollutant condense, the solid melt? Finally, the pollutant's properties influence the specific control device selection. This includes the particulate size distribution, concentration, and the pollutant physical or chemical property driving the removal; such as solubility, settling velocity, vapor pressure, etc. The effects of waste properties on pollution control requirements are illustrated in Table 1.21. As illustrated in Figure 1.28, air pollution devices can be divided into those for particulate removal and those for gaseous pollutants. The air pollution device is commonly purchased as an integral part of the incineration system. This is neces-

TABLE 1.21. Impact of Waste Composition on Pollution Control Requirements.

Dominant Component	Dissolved Solids	Ash Content	Acid Components	Combined Nitrogen	Sulfur Compounds	Volatiles
Pretreatment	Primary consideration for reduction of feed producing pollutants.					
Temperature control, and/or				X		
Gas recycle, and/or				X		
Stage combustion				X		
Secondary combustion chamber						X
Exhaust Treatment						
Venturi scrubber, and/or	X	X				
Bag filter, and/or	X	X				
ESP	X	X				
Fume scrubber			X	Option	X	

Primary incinerator configuration adjusted, or exhaust cleaning equipment added, based on the primary waste component.

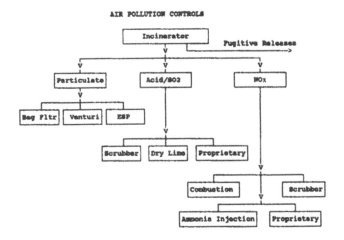

Figure 1.28 Incinerator exhaust treatment.

sary because the incinerator manufacturer can (or should be better able to) characterize the incinerator exhaust.

Noteworthy, the control devices commonly used for 90 hazardous incinerators surveyed are listed in Table 1.22 [17]. The effectiveness of controls for individual pollutants from a refuse incinerator are cited in Table 1.23 [5]. Typically applied pollution controls for hazardous waste incinerators are (1) a high energy venturi followed by a packed tower or (2) flue gas condensation with a multiple throat venturi. Liquid waste incinerators typically employ a packed tower, or a high energy venturi followed by a packed tower if the waste contains a high dissolved solids concentration [5].

COMMON INCINERATOR DESIGN DEFICIENCIES

Some design deficiencies common for municipal sludge incinerators have been investigated by the EPA and are identified for the reader as a checklist of process and mechanical design considerations [22]. Reference is made to the

TABLE 1.22. Typically Applied Controls for Hazardous Incinerators (adapted from Reference [17]).

Control Device	%	Contaminant
Quench chambers	23.3	Particulates
Venturi scrubbers	35.6	Particulates
Wet scrubbers	7.8	Particulates
Wet electrostatic precipitators	5.5	Particulates
Ionizing wet scrubbers	5.5	Particulates
Other wet scrubbers	13.3	Particulates
Packed tower	20.0	Absorber
Spray towers	2.1	Absorber
Tray towers	1.1	Absorber
Other towers	2.2	Absorber
None/unknown	34.4	

Absorbers used for dissolved gases. Percents represent reported use.
Total exceeds 100% because multiple devices are used.

TABLE 1.23. Typical Control Effectiveness for Refuse Incinerators (adapted from Reference [5]).

	Uncontrolled	Controlled
Particulate, gr/dscf	0.5–4.0	0.002–0.015
Heavy metals, mg/nm^3		
Arsenic	<0.1–1	<0.01–0.1
Cadmium	1–5	<0.01–0.5
Lead	20–100	<0.1–1
Mercury	<0.1–1	<0.1–0.7
Gases, ppmdv		
Nitrogen oxides by nonselective catalytic reduction	150–300	60–180
Sulfur dioxide	150–600	5–50
Hydrogen chloride	100–400	10–50
Hydrogen fluoride	0–10	1–2
PCDD/PCDF, *ng*/nm^3	20–500	<1–10

cited reference for a more detailed discussion of these potential problems. The most common deficiencies cited are summarized below.

General

(1) Inadequate consideration of the feed thermal properties and related energy requirements. No provision made for dewatering and predrying to reduce fuel requirements.

(2) No provision made for sludge storage prior to incineration, consistent with the expected downtime, operating effectiveness, and required maintenance time.

(3) City or well water is used instead of plant effluent for pollution control scrubbing systems.

(4) Facilities housing incinerators are inadequately ventilated for operating personnel comfort or to meet OSHA requirements.

(5) Sludge transfer equipment is inadequate to assure steady incinerator feed flow.

(6) The refractory selected is inadequate for sulfide and chloride exposure in highly oxidizing conditions.

(7) The incinerator system reliability was not considered in the design. No provision was made for sludge storage during shutdown, and no redundance of major components to assure minimum shutdown.

(8) The exhaust flow, temperature, and pressure monitoring equipment are incompatible with the exhaust properties and temperature.

(9) Stack sampling ports were not provided.

(10) Hot piping was not insulated for personnel protection.

(11) Ash and flyash disposal provisions are inadequate. Residue characteristics, disposal effects, or the quantity to be disposed were not properly considered.

Multiple-Hearth Fluidized Bed

(1) Provided sludge storage capacity is inadequate for anticipated shutdowns.
(2) The solids loading equipment range is not consistent with incinerator capacity.
(3) No provision was made for heat recovery.
(4) No provision was made for a reliable auxiliary fuel source.
(5) Flyash control equipment was not adequately sized.
(6) The equipment drive is subjected to excessive temperatures.
(7) No provision made for monitoring oxygen concentrations in the system and exhaust, resulting in no control of the oxygen input.
(8) Operator access to the incinerator ash container is difficult.
(9) No provision was made for adequate odor control of incinerator exhaust gases.
(10) Air pollution control equipment installed is inadequate.

CASE STUDY NUMBER 24

Design a *single-stage liquid waste incinerator* to destroy spent solvents, aqueous wastes, and process rinses generated from a manufacturing process. Oil will be used as the supplementary fuel. The feed streams have the following characteristics:

	Spent Solvents	Aqueous Wastes	Rinses
Tons/day	6.0	10.0	44.0
Pounds/hr	500	833.3	3666.7
Carbon, %	44.3	15.4	2.4
Hydrogen, %	11.7	3.6	0.6
Oxygen, %	40.0	11.0	1.5
Water, %		68.0	95.0
Ash, %	4.0	2.0	0.5
Heat combustion, BTU/lb	10,600	6600	−420
Specific heat, BTU/lb/°F	1	1	1
Specific gravity, mg/L	0.8	0.9	1

The incinerator is operated at the following conditions:
Combustion temperature, °F: 1800
Combustion pressure, psiA: 14.7
Excess oxygen, %: 25
Residence time, seconds: 2
The ash will have a specific heat of 0.3 to 0.4 BTU/lb/°F.

PROCESS CALCULATIONS

Preliminary Material Balance

(1) Establish influent mass flows

Tons/day · 2000 · % constituent/100 = lb/day

6 · 2000 · 44.3/100 = 5316 lb/day carbon in solvent

10 · 2000 · 15.4/100 = 3080 lb/day carbon in aqueous

44 · 2000 · 2.4/100 = 2112 lb/day carbon in rinses

xf (1) Total = 10,508 lb/day or 437.8 lbs/hr

Totals for each influent constituent, following the same steps used for the carbon estimate.

Carbon, lb/hr:	437.8
Hydrogen, lb/hr:	110.5
Oxygen, lb/hr:	346.7
Water, lb/hr:	4050.0
Ash, lb/hr:	55.0
Totals, lb/hr:	5000

(2) Establish waste combustion products
(3) Carbon balance

Lb carbon/hr/MW carbon = moles carbon/hr

[437.8/12] = 36.5 atoms carbon injected

Moles carbon/hr · MW CO_2 = lbs CO_2 produced/hr

36.5 · 44 = 1606 lb/hr CO_2 produced

Moles carbon/hr · MW O_2 = Lbs O_2 required/hr

36.5 · 32 = 1168 lb/hr O_2 required

(4) Hydrogen balance

Lb hydrogen/hr/[2 · MW hydrogen] = moles hydrogen/hr

[110.5/2] = 55.3 moles hydrogen (H_2) injected

Moles hydrogen/hr · MW H_2O = lb H_2O produced/hr

55.3 · 18 = 995 lb/hr H_2O produced

995 produced + 4050 feed water = 5045 total water

5045 lb/hr/MW 18 = 280.3 moles/hr

Moles hydrogen/hr · MW O_2/2 = lb O_2 required/hr

55.3 · 32/2 = 885 lb/hr O_2 required

(5) Establish oxygen requirements

Oxygen required for CO_2:	1168
Oxygen required for H_2O:	885
Theoretical oxygen required:	2053 lb/hr
Excess oxygen @ 25%:	513 lb/hr
Total required oxygen:	2566 lbs/hr
Oxygen in feed:	347 lb/hr
Deficit Oxygen:	2219 lb/hr

(6) Oxygen and nitrogen in exhaust

Excess oxygen in exhaust (above): 513 lb/hr (16 moles/hr).
Nitrogen from supplementary (deficit) oxygen:

$$Lb\ O_2/MW\ O_2 \cdot 79\%N2/21\%\ O_2 \cdot MW\ N2 = Lb\ N2/hr$$

$$[2219/32] \cdot [79/21] \cdot 28 = 7304\ lb/hr$$

Feed nitrogen $= 0$

Total nitrogen $= 7304$ lbs/hr (260.9 moles/hr)

	Exhaust	
	lb/day	moles/day
CO_2	1606	36.5
H_2O	5045	280.3
O_2	513	16
N_2	7304	260.9
ASH	55	solid
Totals	14,523	594

ENERGY BALANCE

An energy balance requires that the heat released equal the heat absorbed.

Heat released ——> must equal ——> heat absorbed

1. Heat content of feed	1. Sensible heat absorbed by exhaust gases.
2. Heat released from waste constituents	
3. Heat released from fuel Adjusted	2. Quench from excess air, water, or steam.

a. If heat released results in an excess heat content in the exhaust products, additional quench is required.
b. If heat released results in a deficit heat content in the exhaust products, additional fuel is required.
c. Since any addition of quench or fuel alters the material balance, the material and energy balances are completed by trial-and-error.

(7) Calculate preheat energy input

Reference temperature will be taken at 60°F, equal to the feed temperature, so that the input energy is *zero*.

(8) Calculate combustion energy release

		Combustion	
Wastes	lb/hr	BTU/lb	BTU/hr
Solvent	500.0	10,600	5,300,000
Aqueous	833.3	6,600	5,500,000
Rinses	3666.7	−420	−1,540,000
		Total (net) input	9,260,000

Total heat release: $5,300,000 + 5,500,000$

$$= 11,000,000\ BTU/hr$$

(9) Calculate total input energy

Preheat:	0
Combustion release:	9,260,000 BTU/hr
Net input energy:	9,260,000 BTU/hr

(10) Calculate heat absorbed

Enthalpy reference temperature (RT), 60°F. Enthalpy at combustion temperature (CT), 1800 °F. (based on the Gas phase)

		BTU/lb		
	lb/hr	CT	RT	BTU/hr
CO_2	1606	*469	0	753,000
H_2O	5045	*888	0	4,480,000
O_2	513	*430	0	221,000
N_2	7304	*465	0	3,396,000
Ash	55	600.	0	33,000
	* Heat losses @ less than 5%			377,000
	Total heat absorbed, BTU/hr =			9,260,000

* Notes:

(1) The heat loss is an estimate, usually less than 5% for a well insulated, closed system. *In this case*, heat losses less than the quantity indicated would require excess air or water quench to absorb all the heat released from the feed. In an operating system the air (water, or steam) input is usually automatically adjusted to maintain the required temperature, in this case 1800°F.

(2) The CO_2, H_2O, O_2, and N_2 BTU/lb are from Reference [4].

Exhaust Treatment

(11) Venturi scrubber performance

Assume that a Venturi scrubber will remove 99% of the particulates, reducing the incinerator exhaust temperature from 1800°F to 125°F, utilizing 7 gallons of water per 1000 cubic feet of exhaust gases. The actual water rate must be adjusted to assure that the required gas exhaust temperature is achieved.

Exhaust dry gas, lb/hr: $(14,523 − 5045 − 55) = 9,423$

Humidity @125°F: 0.0954 lbs water/lb d.a.

Water in exhaust: $0.0954 \cdot 9,423 = 899$ lb/hr

Water removed: $5045 − 899 = 4,146$ lb/hr

	Incinerator Exhaust		Venturi Exhaust		Blowdown
	Lb/hr	Moles/hr	Lb/hr	Moles/hr	Lb/hr
Temperature, °F	1,800		125		
Carbon dioxide	1606	37	1606	37	
Water	5045	280	899	50	4146
Oxygen	513	16	513	16	
Nitrogen	7304	261	7304	261	
Ash	55		0.55		54.45
	14,523	594	10,322.55	364	4200.45

(12) Incinerator gas rate

$$594\ moles/hr \cdot 359\ std\ ft^3/mole \cdot \frac{1800 + 460}{460 + 32} \cdot \frac{1}{3,600}$$

$$= 272\ ft^3/s\ gas$$

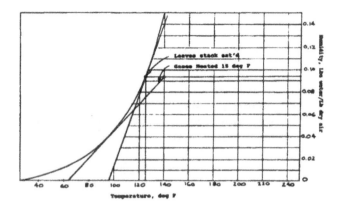

Figure 1.29 Plume evaluation (for calculation).

(13) Venturi water system

(14) Scrubber water: Assume 7 gallons of water is required per 1000 cubic feet of exhaust gases.

$$272 \text{ cfs} \cdot 60 \text{ secs} \cdot 7 \text{ gal}/1000 \text{ ft}^3 = 115 \text{ gpm water}$$

(15) Makeup water: Use 5 gpm makeup = 2500 Lb/hr

(16) Blowdown characteristics

Water removed,	4146	(from exhaust)
Bleed water,	2500	(equal to makeup)
Ash	54.45	
	6700 lbs/hr	

Blowdown 13 to 14 gpm with 8127 mg/L solids.

COMBUSTION EQUIPMENT SIZING

(17) Select a residence time: Assume two seconds.
(18) Combustion chamber volume

Volume = 272 cfs · 2 seconds = 544 cubic feet volume

Burn rate, BTU/hr/cu ft
$$= 11,000,000 \text{ BTU/hr heat release}/544 \text{ ft}^3$$

$$= 20,000 \text{ BTU/hr/cubic foot}$$

(19) Combustion chamber dimensions
Assume a length to diameter ratio of 4 to 5, and adjust to a convenient configuration, to approximately 544 ft³. Diameter is equal to 6 feet and length is equal to 20 feet.

EXHAUST CHARACTERISTICS

$$364 \text{ moles/hr} \cdot 359 \text{ std ft}^3/\text{mole} \cdot \frac{125 + 460}{460 + 32} \cdot \frac{1}{60}$$
$$= 2590 \text{ ft}^3/\text{min gas}$$

Venturi exhaust 364 moles/hr @ 125 °F

0.55 lb/hr · 7,000 grains/lb/60 = 64.2 grains/minute

64.2 grains/2590 = 0.025 grains/cubic foot

PLUME EVALUATION

Referring to Figure 1.29, an exhaust condition at 125°F and a moisture content of 0.0954 pounds of moisture per lb of dry exhaust will result in a visible plume when ambient conditions to the left of the operating line occur. The potential for not preventing a plume is improved by heating the exhaust 15°F, as indicated by the new operating line.

CASE STUDY NUMBER 25

Design a *two-stage rotary kiln waste incinerator* to destroy plant wastes, residues and sludges generated from a manufacturing process. Oil is be used as the supplementary fuel. The feed constituents have the following characteristics:

	Plant Wastes	Residue Wastes	Residue Rinses	Fuel Oil
Tons/day	1.0	2.0	20.0	
Pounds/hour	83.3	166.7	1,666.7	
Carbon, %	62.0	60.0	15.30	87.3
Hydrogen, %	14.0	6.0	3.50	12.5
Sulfur, %	0.1		0.02	0.2
Oxygen, %	15.0	12.0	3.68	
Chlorine, %	1.9		0.50	
Water, %	1.0	5.0	75.50	
Ash, %	6.0	5.0	1.50	
Heat combustion, BTU/lb	15,000	10,000	1,800	17,855
Specific heat, BTU/lb/°F	1	1	1	0.55
Specific gravity, mg/L	1	1	1	0.865

The incinerator is operated at the following conditions:

Primary combustion temperature, °F:	1800
Combustion pressure, psiA:	14.7
Minimum excess oxygen, %:	25, adjust to the specific configuration
Primary chamber heat release, BTU/hr/ft³:	15,000
Secondary chamber temperature, °F:	2200 °F
SCC residence time, seconds:	2

The ash will have a specific heat of 0.3 to 0.4 BTU/lb/°F.

PROCESS CALCULATIONS

Rotary Kiln Preliminary Material Balance

(1) Establish influent mass flows
tons/day · 2000 · % constituent/100 = lb/day

1 · 2000 · 62.0/100 = 1240 lb/day carbon in solvent

2 · 2000 · 60.0/100 = 2400 lb/day carbon in aqueous

20 · 2000 · 15.3/100 = 6120 lb/day carbon in rinses
Total = 9760 lb/day or 406.7 lb/hr

Totals for each influent constituent, using same format as above,

Carbon, lb/day:	406.7
Hydrogen, lb/day:	80
Oxygen, lb/day:	93.8
Sulfur, lb/day:	0.5
Chlorine, lb/day:	29.9
Water, lb/day:	1267.5
Ash, lb/day:	38.3
Totals, lb/day:	1916.7

(2) Establish waste combustion products

(3) Carbon Balance

Lb carbon/hr/MW carbon = moles carbon/hr

[406.7/12] = 33.9 atoms carbon injected

Moles carbon/hr · MW CO_2 = Lbs CO_2 produced/hr

33.9 · 44 = 1492 lb/day CO_2 produced

Moles carbon/hr · MW O_2 = lbs O_2 required/hr

33.9 · 32 = 1085 lb/day O_2 required

(4) Hydrogen balance

Lb hydrogen/hr/[2 · MW hydrogen] = moles hydrogen/hr

[80/2] = 40 mol hydrogen (H_2) injected

Lb chlorine/hr/[AW chlorine] = atoms chlorine/hr

[29.9/35.5] = 0.84 atoms chlorine injected

Atoms chlorine · MW HCl = lb HCl produced/hr

0.84 · 36.5 = 31 pounds of hydrogen chloride produced

Atoms chlorine · AW H = lb H used/hr

0.84 · 1 = 0.8 pounds hydrogen used = 0.4 moles H_2

[40 − 0.4] · 18 = 712.8 lb/day H_2O produced

712.8 produced + 1267.5 feed water = 1980 total water/hr

Moles hydrogen/hr · MW $O_2/_2$ = lb O_2 required/hr

[40 − 0.4] · 32/2 = 633 lb/day O_2 required

(5) Sulfur balance

Lb sulfur/hr/MW sulfur = moles sulfur/hr

[0.5/32] = 0.016 moles sulfur injected

Moles sulfur/hr · MW SO_2 = lb SO_2 produced/hr

0.016 · 64 = 1 lb/day SO_2 produced

Moles sulfur/hr · MW O_2 = lb O_2 required/hr

0.016 · 32 = 0.5 lb/day O_2 required

(6) Establish oxygen requirements

Oxygen required for CO_2:	1085
Oxygen required for H_2O:	633
Oxygen required for SO_2:	0.5
Theoretical oxygen required:	1719 lb/hr
Excess oxygen @ 25%:	430 lb/hr
Total *required* oxygen:	2149 lb/hr
Oxygen in feed:	94 lb/hr
Deficit Oxygen:	2055 lb/hr

(7) Oxygen and nitrogen in exhaust
Excess oxygen in exhaust (above): 430 lb/hr. Nitrogen from supplementary (deficit) oxygen:

Lb O_2/MW O_2 · 79% N_2/21% O_2 · MW N_2 = lb N_2/hr

[2055/32] · [79/21] · 28 = 6,764 lbs/hr

feed nitrogen = 0

total nitrogen = 6,764 lb/hr

(8) Ash partitioning
Assume 50% of the ash is discharged with the ash, so that 19 lb/hr is in the exhaust and 19 lb/hr as kiln ash, although 38/hr will be used for the kiln energy balance.

	Kiln Exhaust	
	lb/day	moles/day
CO_2	1492	33.9
H_2O	1980	110
SO_2	1	0.02
HCl	31	0.8
O_2	430	13.4
N_2	6764	241.6
Ash	19	solid
Totals	10,717	400

ENERGY BALANCE

See Case Study 24, beginning paragraph under "Energy Balance," for a detail explanation of energy balance criteria.

(9) Calculate preheat energy input
Reference temperature will be taken at 60°F, equal to the feed temperature, so that the input energy is *zero*; so that the reference temperature for all stream heat values is 60°F.

(10) Calculate combustion energy release

Wastes		Combustion	
	lb/hr	BTU/lb	BTU/hr
Plt waste	83.3	15,000	1,250,000
Residue	166.7	10,000	1,670,000
Rinses	1666.7	1800	3,000,000
		Total release	5,917,000

(11) Calculate total input energy

Preheat: 0

Combustion release: 5,917,000 BTU/hr

Total input energy: 5,917,000 BTU/hr

(12) Calculate heat absorbed

Enthalpy reference temperature (RT), 60°F. Enthalpy at combustion temperature (CT), 1800 °F. (based on the gas phase)

	lb/hr	BTU/lb		BTU/hr
		CT	RT	
CO_2	1492	469*	0	700,000
H_2O	1980	888*	0	1,758,000
O_2	430	430*	0	185,000
N_2	6764	465*	0	3,145,000
SO_2	1	313*	0	
HCl	31	365*	0	11,000
Ash	38	600	0	23,000
Heat losses				95,000
		Total Heat Absorbed, BTU/hr =		5,917,000

*Notes

(1) The heat loss is an estimate, usually less than 10%.

(2) The CO_2, H_2O, O_2, and N_2 BTU/lb are from Reference [4].

(3) The HCl heat capacity is based on an *average* heat capacity of 0.21 BTU/hr/°F over the 60 to 1800°F range; and an SO_2 heat capacity of 0.18 for the same operating range. The values used are within the accuracy of the calculation, considering the relative small quantities of these constituents. More accurate data may be required for some incinerator calculations.

Secondary Combustion Chamber

(13) Supplementary fuel combustion products

Supplementary SCC fuel will be added to increase the kiln exhaust temperatures from 1800°F to 2200°F. The combustion products from the supplementary fuel will increase the SCC mass and gas flow. A fuel value of 249 lb/hr was selected by *trial-and-error* to achieve an energy balance at 2200°F.

(14) Fuel mass flows

Lb/hr · % constituent/100 = lb/day

249 · 87.3/100 = 217 lb/day carbon

249 · 12.5/100 = 31 lb/day hydrogen

249 · 0.20/100 = 0.5 lb/day sulfur

(15) Fuel combustion exhaust products

	lb/hr	Mole/hr	Theoretical Oxygen lb/hr
CO_2	796	18.1	579
H_2O	279	15.5	248
SO_2	1	0.02	0.5
	1076	33.6	827.5
	1076	33.6	Combustion products
Excess			
O_2	207	6.5	@ 25% of 827.5
N_2	3405	121.6	[(827.5+207) · (28/32) · (79/21)]
Totals	4688	161.7	

(16) Calculate combustion energy release

Wastes	Combustion		
	lb/hr	BTU/lb	BTU/hr
Fuel	249	17,855	4,446,000

(17) Calculate heat absorbed

(18) Heat absorbed by SSC fuel gas exhaust

Fuel feed reference temperature (ft), 60°F. Fuel gases at SCC temperature (SCC), 2200 °F.

	lb/hr	SCC	Ft	BTU/hr
CO_2	796	594	0	473,000
H_2O	279	1130	0	315,000
O_2	207	539	0	112,000
N_2	3405	582	0	1,982,000
SO_2	1	410	0	Neglect
		Total heat absorbed, BTU/hr =		2,882,000

(19) Heat absorbed by kiln gases in SCC: Total heat absorbed in the SCC will equal that required to heat the supplementary fuel gas combustion products from 60 to 2200°F, *plus* the heat required to heat the kiln gas exhaust from 1800 to 2200°F.

Enthalpy *kiln* temperature (KT), 1800°F. Enthalpy at SCC temperature (SCC), 2200°F.

	Kiln Gases Heat Absorbed			
	lb/hr	SCC	KT	BTU/hr
CO_2	1492	594	469	187,000
H_2O	1980	1130	888	479,000
O_2	430	539	430	47,000
N_2	6764	582	465	791,000
SO_2	1	410	313	Neglect
HCl	19	450	365	3,000
Ash	19	730	600	2,000
				1,509,000

(20) Total heat absorbed in SCC

SCC fuel gases	2,882,000
Kiln gases	1,509,000
Losses	55,000
Total	4,446,000

(21) Total SCC mass flow

	SCC Exhaust	
	lb/hr	moles/hr
CO_2	2288	52
H_2O	2259	126
O_2	637	20
N_2	10,169	363
SO_2	2	
HCl	31	1
Ash	19	solid
Totals	15,405	562

(22) Exhaust gas rate

$$562 \text{ moles/hr} \cdot 359 \text{ std ft}^3/\text{mole} \cdot \frac{460 + 2,200}{460 + 32} \cdot \frac{1}{3,600}$$
$$= 303 \text{ ft}^3/\text{s gas, based on 25\% excess air}$$

HEAT RECOVERY

Assume a waste boiler for flue gas heat recovery, operating to reduce the flue gas temperature from 2200°F to 500°F.

	lb/hr	H_{2200}	H_{500}	BTU/hr
CO_2	2288	594	99	1,133,000
H_2O	2259	1130	200	2,101,000
O_2	637	539	100	280,000
N_2	10,169	582	110	4,800,000
SO_2	2			neglect
HCl	31			neglect
Ash	19			neglect
	15,405			8,314,000

Assume 50% recovery efficiency, so that about 4 million BTU can be recovered per hour.

EXHAUST TREATMENT

(23) ESP performance
Assume that 99% of the flyash is removed using an electrostatic precipitator, equivalent to approximately 1/4 ton of dust per day; so that the exhaust contains 0.19 lb/hr. As an alternate consideration, dust and acid removal can be combined using a high energy Venturi scrubber with an alkaline water recycle.

Removal	Inlet	Outlet	Ash
99%	19	0.19	19

(24) Acid scrubber performance
Assume that a low energy acid scrubber will be used to remove some of the remaining particulates and most of the acid components, reducing the incinerator exhaust temperature from 500 to 125°F. Under the right conditions some of the carbon dioxide may react with *excess* sodium hydroxide scrubbing solution to form sodium carbonate. This will be neglected in this calculation.

Exhaust dry gas, lb/hr:	13,094 (exclude H_2O, SO_2, HCl, and ash)
Humidity @ 125°F:	0.0954 lb water/lb d.a.
Water in exhaust:	0.0954 · 13,094 = 1249 lb/hr
Water removed:	2259 − 1249 = 1010 lbs/hr

Removal %		Inlet	Scrubber Exhaust		Blowdown	NaOH (3)
		lb/hr	lb/hr	moles/hr	lb/hr (1)(2)	lb/hr
°F		500	125			
CO_2		2288	2288	52		
H_2O		2259	1249	69	1010	
O_2		637	637	20		
N_2		10,169	10,169	363		
SO_2	80	2	0.4		4 as Na_2SO_3	3
HCl	99	31	0.3		50 as NaCl	34
Ash	20	0.19	0.15			
		15,386	14,344	504	1064	37

Notes
(1) Blowdown will include additional water blowdown, the quantity dependent on the scrubber design, and that required to achieve the exhaust temperature.
(2) Salts formed assume SO_2 to Na_2SO_3, and HCl to NaCl. Requiring 1.25 grams NaOH per gram of SO_2, and 1.1 per gram of HCl; producing 1.97 grams of Na_2SO_3 per gram of SO_2, and 1.6 grams of NaCl per gram of HCl.
(3) As 100% NaOH.

(24) Exhaust gas rate

$$504 \text{ moles/hr} \cdot 359 \text{ std ft}^3/\text{mole} \cdot \frac{460 + 125}{460 + 32} \cdot \frac{1}{3600}$$
$$= 60 \text{ cfs gas @ 125°F}$$

COMBUSTION EQUIPMENT SIZING

(26) Rotary kiln heat release rate
Design burn rate @ 15,000 BTU/hr/ft³
Heat release @ 5,917,000 BTU/hr
5,917,000 BTU/hr/15,000 = 394 ft³ kiln volume
Length to diameter ratio of 4:1
Diameter = 5 feet
Length = 20 feet

(27) SCC characteristics
Gas rate @ 303 ft³/second
Gas residence time = 2 seconds
303 ft³/s · 2 seconds = 606 ft³ secondary chamber
Length to diameter ratio of 4:1
Diameter = 5 feet, 6 inches
Length = 26 feet

EXHAUST CHARACTERISTICS

Acid scrubber exhaust 504 moles/hr @ 125°F.
Gas rate at 60 ft³/s, based on 25% excess air.
SO_2 @ @ 0.4 lb/hr = 0.0063 moles/hr
HCl @ 0.3 lb/hr = 0.0082 moles/hr
0.15 lb/hr particulate

$$(0.0063 / 504) \cdot 1,000,000 = 13 \text{ ppmv of } SO_2$$

$$(0.0082 / 504) \cdot 1,000,000 = 16 \text{ ppmv of HCl}$$

0.15 lb/hr · 7000 grains/lb/60 = 17.5 grains/minute
17.5 grains/min/(60 cfs · 60 secs) = 0.0049 grains/ft³

CASE STUDY NUMBER 26

Design a 10-ton per day *single-stage multiple hearth incinerator* to destroy excess *sludge* generated from an industrial activated sludge waste treatment facility, using available plant rinses with minute contaminant levels as a quench medium. The exhaust is to be designed with a venturi scrubber for particulate removal. The waste streams have the following characteristics:

	Waste Sludge	Aqueous Wastes	
Tons/day	10.0	0.3	
Lb/hr	833.3	25	
Carbon, %	5.40	Tr	(Varying trace amounts
Hydrogen, %	0.60	Tr	neglected in the
Oxygen, %	2.80	Tr	calculations)
Nitrogen, %	1.20		
Water, %	89.00	100.00	
Ash, %	1.0	Tr	
Heat commbustion, BTU/lb	1800	−1000	
Specific heat, BTU/lb/°F	1	1	
Specific gravity, mg/L	0.8	0.9	

The sludge is self sustaining, fuel is used only as a standby. The incinerator is to operated at the following conditions:

Combustion temperature, °F: 1700 Maximum
Combustion pressure, psiA :   14.7
Excess oxygen, %:   125

Overall hearth loading, lbs cake/HR/square foot: 8. The ash generated will have a specific heat of 0.3 BTU/lb/°F.

PROCESS CALCULATIONS

Preliminary Material Balance

(1) Establish influent mass flows

Tons/day · 2000/24 · % constituent/100 = lb/hour

$10 \cdot 2000/24 \cdot 5.4/100 = 45$ lb/hr carbon in sludge

$10 \cdot 2000/24 \cdot 0.6/100 = 5$ lbs/hr hydrogen in sludge

$10 \cdot 2000/24 \cdot 2.8/100 = 23.3$ lb/hr oxygen in sludge

$10 \cdot 2000/24 \cdot 1.2/100 = 10$ lb/hr nitrogen in sludge

$10 \cdot 2000/24 \cdot 1/100 = 8.3$ lb/hr ash in sludge

$10 \cdot 2000/24 \cdot 89/100 = 741.7$ lb/hr water in sludge

$0.3 \cdot 2000/24 \cdot 100/100 = 25$ lb/hr water in rinses

(2) Establish waste combustion products

(3) Carbon balance

Lb carbon/hr/MW carbon = moles carbon/hr

$[45/12] = 3.75$ atoms carbon injected

Moles carbon/hr · MW CO_2 = lb CO_2 produced/hr

$3.75 \cdot 44 = 165$ lb/day CO_2 produced

Moles carbon/hr · MW O_2 = lb O_2 required/hr

$3.75 \cdot 32 = 120$ lb/day O_2 required

(4) Hydrogen balance

Lb hydrogen/hr/[2 · MW hydrogen] = moles hydrogen/hr

$[5/2] = 2.5$ moles hydrogen injected

Moles hydrogen/hr · MW H_2O = lb H_2O produced/hr

$2.5 \cdot 18 = 45$ lb/day H_2O produced

45 produced + 766.7 feed water = 811.7 total water

Moles hydrogen/hr · MW $O_2/2$ = lb O_2 required/hr

$2.5 \cdot 32/2 = 40$ lb/day O_2 required

(5) Establish oxygen requirements

Oxygen required for CO_2:	120
Oxygen required for H_2O:	40
Theoretical oxygen required:	160 lb/hr
Excess oxygen @ 125%:	200 lb/hr
Total *required* oxygen:	360 lb/hr
Oxygen in feed:	23.3 lb/hr
Deficit oxygen:	336.7 lbs/hr

(6) Oxygen and nitrogen in exhaust

Excess oxygen in exhaust (above):	200 lbs/hr
Nitrogen in feed (above):	10 lbs/hr

Nitrogen from supplementary (deficit) oxygen:
Lb O_2/MW O_2 · 79% N_2/21% O_2 · MW N_2 = lb N_2/hr
$[336.7/32] \cdot [79/21] \cdot 28 = 1108.3$ lb/hr
Feed nitrogen = 10.0
Total nitrogen = 1118.3 lb/hr

	EXHAUST	
	lb/day	moles/day
CO_2	165	4
H_2O	812	45
O_2	200	6
N_2	1118	40
Ash	8	solid
Totals	2303	95

ENERGY BALANCE

See Case Study 24, beginning paragraph under "Energy Balance," for a detail explanation of energy balance criteria.

(7) Calculate preheat energy input

Reference temperature will be taken at 60°F, equal to the feed temperature, so that the input energy is *zero*.

(8) Calculate combustion energy release

Wastes	lb/hr	BTU/lb	Combustion BTU/hr
Sludge	833.3	1800	1,500,000
Rinses	25.0	−1000	− 25,000
		Total	1,475,000
		Total heat release:	1,500,000 BTU/hr

(9) Calculate total input energy

$$\text{Preheat: } 0$$
$$\text{Combustion release: } 1,475,000 \text{ BTU/hr}$$
$$\text{Total input energy: } 1,475,000 \text{ BTU/hr}$$

(10) Calculate heat absorbed

Enthalpy reference temperature (RT), 60°F. Enthalpy at combustion temperature (CT), 1700 °F. (based on the gas phase)

	lb/hr	BTU/lb CT	RT	BTU/hr
CO_2	165	438	0	72,000
H_2O	830	830	0	674,000
O_2	200	404	0	81,000
N_2	1118	436	0	487,000
Ash	8	600	0	5,000
Heat losses @ 10%				156,000
		Total heat absorbed, BTU/hr =		1,475,000

Exhaust Treatment

(11) Venturi scrubber performance

Assume that a Venturi scrubber will remove 99% of the particulates, reducing the incinerator exhaust temperature from 1800°F to 125°F, utilizing 7 gallons of recirculation water per 1000 cubic feet of exhaust gases.

Exhaust dry gas, lb/hr:	1483
Humidity @ 125°F:	0.0954 lb water/lb d.a.
Water in exhaust:	0.0954 · 1483 = 141 lb/hr
Water removed:	812 − 141 = 671 lb/hr

	Incinerator Exhaust		Venturi Exhaust		Blow-down
	lb/hr	Moles/hr	lb/hr	Moles/hr	lbs/hr
Temperature, °F	1700		125		
Carbon dioxide	165	4	165	4	
Water	812	45	141	8	671
Oxygen	200	6	200	6	
Nitrogen	1118	40	1118	40	
Ash	8		0.08		8
	2303	95	1624	58	679

(12) Gas feed rate

$$95 \text{ moles/hr} \cdot 359 \text{ std ft}^3\text{/mole} \cdot \frac{1800 + 460}{460 + 32} \cdot \frac{1}{3600}$$
$$= 44 \text{ ft}^3\text{/s gas}$$

(13) Venturi water system

(14) Scrubber water: Assume 7 gallons of water is required per 1000 cubic feet of exhaust gases.

$$44 \text{ cfs} \cdot 60 \text{ secs} \cdot 7 \text{ gal/1000 ft}^3 = 18 \text{ gpm circulating water}$$

(15) Makeup water: Use 2 gpm makeup ≈ 1,000 lb/hr

(16) Blowdown characteristics

Water removed	671	(from exhaust)
Bleed water	1000	(equal to makeup)
Ash	8	
	1679 lb/hr	

Blowdown ≈ 3.5 gpm with 4765 mg/L solids

COMBUSTION EQUIPMENT SIZING

(17) Select a loading rate: Assume 8 pounds cake/hr/square foot.

(18) Hearth cake loading

10 tons/day

10 · 2000 /24 = 833 pounds per hour

833/8 = 104 total ft² of hearth area

(19) Hearth dimensions

7 ft 6 inch hearth with 6 hearths has a total area of 100 sq ft, 7 ft 6 inch hearth with 7 hearths has a total area of 114 sq ft.

EXHAUST CHARACTERISTICS

Venturi exhaust 58 moles/hr @ 125 °F.

$$58 \text{ moles/hr} \cdot 359 \text{ std ft}^3\text{/mole} \cdot \frac{125 + 460}{460 + 32} \cdot \frac{1}{60} = 413 \text{ ft}^3\text{/m gas}$$

0.08 lb/hr · 7000 grains/lb/60 = 9.3 grains/minute

9.3 grains/413 = 0.02 grains/ft³

APPENDIX: MATERIAL BALANCE CALCULATIONS AND HEAT LOSS CALCULATIONS

MATERIAL BALANCE CALCULATIONS

The basis of a material balance is the weight of an element (atomic weight) or the weight of a compound (molecular weight) and its stoichiometric relationship in the combustion equation. A step-by-step material balance is accomplished as follows:

(1) The hourly mass rate of carbon fed to the incinerator is estab-

lished from the feed composition, and the related reactants calculated as follows:

$$1 \; C + 1 \; O_2 = 1 \; CO_2$$

$$12 \qquad 32 \qquad 44$$

It can be seen that one atom of carbon of atomic weight 12 will combine with one mole of oxygen with a molecular weight of 32, to form one mole of carbon dioxide with a molecular weight of 44.

(2) The hourly mass rate of hydrogen fed to the incinerator is established from the feed composition, *the equivalent quantity combined to form acids deducted* [see (4)] and the remaining *oxidized* as follows:

$$2 \; H + 1/2 \; O_2 = 1 \; H_2O$$

$$H + 1/4 \; O_2 = 1/2 \; H_2O$$

$$1 \qquad 8 \qquad 9$$

It can be seen that one atom of hydrogen of atomic weight 1 will combine with one quarter mole of oxygen with a weight of 8 to form one half mole of water with a weight of 9. (The fraction of a mole is for convenience so that the requirements are based on one unit weight of the feed constituent.)

(3) The hourly mass rate of sulfur fed to the incinerator is established from the feed composition, and the related reactants calculated as follows:

$$1 \; S + 1 \; O_2 = 1 \; SO_2$$

$$32 \qquad 32 \qquad 64$$

(4) The hourly mass rate of halide (chlorine or fluorine) fed to the incinerator is established from the feed composition, and the related reactants are calculated as follows:

$$1 \; Cl + \; H \; = 1 \; HCl$$

$$35.5 \qquad 1 \qquad 36.5$$

$$1 \; F + \; H \; = 1 \; HF$$

$$19 \qquad 1 \qquad 20$$

The quantity of hydrogen used to form the halide acids is *deducted* from the total hydrogen feed quantity, the rest is available for the oxidation reaction in step 2.

(5) The hourly mass rate of other reactants fed to the incinerator are calculated, and the related reactants calculated as follows:

$$1 \; M^{++} \qquad + \qquad O_2 \qquad = \qquad MO_2$$

Atomic weight $\qquad$ 32 $\qquad$ Molecular weight

$$1 \; M^{+++} \qquad + \qquad 3/2 \; O_2 \qquad = \qquad 1/2 \; M_2O_3$$

Atomic weight $\qquad$ 48 $\qquad$ 1/2 molecular weight

(6) The hourly rate of water in the feed is calculated, and the quantity added to the amount formed from the hydrogen oxidation in step 2.

(7) The quantity of oxygen in the exhaust is equivalent to the excess oxygen fed to the incinerator, which can be from 20 to 150%, depending on the incinerator.

The sum of oxygen calculated in steps 1 through 5 is the *theoretical* quantity of oxygen required for the combustion process. The *theoretical* plus the *excess* oxygen equals the *actual* oxygen quantity required for the system. The quantity that is *supplied* to the incinerator system is equal to the *actual* oxygen required *minus* the oxygen *content* of the *waste*. The excess oxygen fed to the incinerator will be directly discharged as an exhaust product.

(8) The nitrogen in the exhaust products is directly related to the quantity of *atmospheric* oxygen fed to the incinerator. Assuming that air is used as the oxygen source (as opposed to pure oxygen), the quantity of nitrogen is based on air containing approximately 21 volume (or mole) percent oxygen and 79 volume (or mole) percent nitrogen, which based on a molecular weight of 32 for oxygen and 28 for nitrogen, calculates to a weight ratio of nitrogen to oxygen of 3.29.

(9) The weight rate of ash generated from the feed is calculated directly from the ash content of the waste.

The stoichiometric ratios discussed in steps 1 through 9 are summarized in Table 1.24.

TABLE 1.24. Combustion Stoichiometric Ratios.

		+	O_2	=	CO_2
Moles	1		1		1
Weight	12.02		32		44.02
	H_2	+	$\frac{1}{2}O_2$	=	H_2O
Moles	1		$\frac{1}{2}$		1
Weight	2.016		16		18.016
	S	+	O_2	=	SO_2
Moles	1		1		1
Weight	32.066		32		64.066
	H_2S	+	$1\frac{1}{2} O_2$	=	H_2O + SO_2
Moles	1		$1\frac{1}{2}$		1 $\quad$ 1
Weight	34.082	48			18.016 $\quad$ 64.066
	Cl_2	+	H_2	=	2HCl
Moles	1		1		2
Weight	70.914		2.016		72.930
	Air = 0.21 O_2 + 0.79 N_2				
Moles	1		0.21		0.79
Weight	28.85		6.72		22.13
	H_2O (water)		=		H_2O (steam)
Moles	1		1		
Weight	18.016		18.018		
	2 · N		=		N_2
Moles	2				1
Weight	28.016				28.016

EXHAUST VOLUMES

Exhaust volumes are directly proportional to the total exhaust moles, which in turn is related to the mass rate exhaust discharged as follows:

Moles/unit time = wt A/MW A + wt B/MW B + wt X/MW X

where wt A, B, X is the mass rate per unit time in the exhaust ofcomponent A, B, X. MW A, B, X is the molecular weight of A, B, X.

The volume of gas in the exhaust is then calculated using the ideal gas laws, based on the following relationship:

(1) One mole of a gas at 32°F and one atmosphere pressure occupies 359 cubic feet.
(2) One mole of a gas at 0°C and one atmosphere pressure occupies 22.4 liters.

Therefore, the volume of exhaust cases at standard conditions can be calculated as follows:

Cubic feet/time = total moles/time × 359

Liters/time = total moles/time × 22.4

The volume can be corrected to any other condition as follows:

Actual ft^2 = 359 × (460 + T)/492 × 1/(1 + P)

Actual Liters = 22.4 × (270 + t)/273 × 1/(1 + P)

where T is the new temperature in °F, t is the new temperature in °C, P is the new pressure in atmospheres.

SYSTEM ENERGY BALANCE

System Energy Balance

The basis for an incinerator combustion energy balance is illustrated in Figure 1.30 and Table 1.25. Basically, the heat input and release by the feed must equal the heat loss from the system and to the ash, as well as the heat absorbed (sensible heat content) of the exhaust gases. The balance is conducted from the material balance quantities, based on the heat content of each of the components. Definitions of heat capacity and heat release are detailed in Basic Concepts: Energy Transfer.

Supplementary Fuel (or Quench) Requirements

When a preliminary energy audit is conducted, as illustrated in Table 1.25, an energy balance will most likely not be achieved! The energy input may be in excess, the combustion temperature too high, indicating that the system requires more quench capacity, or the heat absorbed may result in too low a combustion temperature, indicating that supplementary fuel is required. In either case, adjustments are made to the feed, a new material balance determined, and a new energy audit calculated. This trial-and-error procedure is repeated until a material and energy balance is achieved.

The fuel or quench requirements to balance a system is estab-

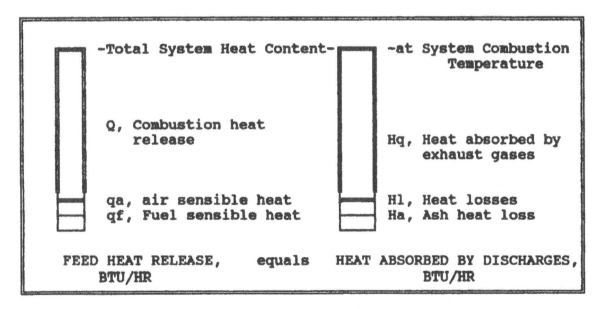

Figure 1.30 System energy balance.

TABLE 1.25. System Energy Audit.

Energy Input	Energy Absorbed
Heat release from waste	Sensible heat absorbed by
Heat release from fuel	Carbon dioxide
Sensible heat content fuel	Water vapor
Sensible heat content of air	Sulfur dioxide
	Halides or acid fumes
	Excess oxygen
	Nitrogen
	Sensible heat absorbed by ash
	Heat losses from system walls

lished by first determining the system *surplus* or *deficit heat load*, as follows:

$$\text{Net Heat} = \text{energy absorbed} - \text{energy input}$$

When the calculated net heat is a positive value the required supplementary fuel can be approximated as follows:

$$\text{Fuel, lbs/hr} = \text{net heat/heat combustion fuel}$$

When the calculated net heat is a negative value air, water, or steam quench will have to be added. The required amount can be approximated as follows:

$$\text{Quench, lbs/hr} = \text{net heat/[enthalpy vapor} - \text{enthalphy feed]}$$

Material and energy balances are illustrated in Case Studies 23, 24, and 25.

REFERENCES

1. Acharya, P, DeCicco, S.G., Novak, R.G.: "Factors That Can Influence and Control the Emissions of Dioxins and Furans from Hazardous Waste Incinerators," *J. Air Waste Management Assoc*, V 41, N 12, Pg 1605, December, 1991.

2. Barton, R.G., Clark, W.D., Lanier, W.S., Seeker, W.R.: "Dioxin Emissions During Waste Incineration," Presentation at 1988 Western States Section (Combustion Institute), March, 1988.

3. Bonner, T. et al: *Hazardous Waste Incineration Engineering, Pollution Technology No 88*, Noyes Data Corporation, 1981.

4. Brunner, C.R.: *Incineration Systems, Selection and Design, Incinerator Consultants Inc., Virginia,* 1984.

5. Buonicore, A.J., Davis, W.T.: *Air Pollution Engineering Manual,* Air and Waste Management Association, 1992.

6. Caprio, J.A., Wolfe, H.E.: "Refractories for Hazardous Waste Incineration, an Overview," *1982 National Waste Processing Conference,* American Society of Mechanical Engineers, 1982.

7. Dean, J.A. (Editor): *Lange's Handbook of Chemistry,* 13th Edition, McGraw-Hill, 1985.

8. Harbison-Walker: *Handbook of Refractories for Incineration Systems,* Dresser Industries, Inc., 1991.

9. LaGrega, M.D., Buckingham, P.L., Evans, J.C.: *Hazardous Waste Management,* McGraw Hill, 1994.

10. Lee, K.C.: "Research Areas for Improved Incineration Performance," *J. Air Pollution Control Association,* V 38, Pg 1542, 1988.

11. Lewis, W.K., Radasch, A.H., Lewis, H.C.: *Industrial Stoichiometry,* McGraw Hill, 1954.

12. Lide, D.R. (Editor): *Handbook of Chemistry and Physics,* CRC Press, Inc, Florida, 1993.

13. 13. Lis, R.E., Korn, S.R.: "Prevent the Formation of Steam Plumes," *Chemical Engineering Progress,* Pg 55, August, 1991.

14. McGrath, T.P., Seeker, W.R., Chen, S.L.: "Pilot Scale Studies on PCDD/PCDF Formation and Control in Municipal Waste Combustion Systems," *Combustion Science and Technology,* V 85, N 1/6, Pg 187, 1992.

15. Niessen, W.R.: *Combustion and Incineration Processes: Applications in Environmental Engineering,* Marcel Dekker, Inc, New York, 1978.

16. Oppelt, E.T.: "Incineration of Hazardous Wastes, a Critical Review," *Journal Air Pollution Control Association,* V 37, No 5, Pg 558, 1987.

17. Rickman, W.S.: *Handbook of Incineration of Hazardous Wastes,* CRC Press, Inc, Florida, 1991

18. Siebert, P.C., Alston, D.R., Jones, K.H.: "Toxic Trace Pollutants from Incineration," *Environmental Progress,* V 10, No 1, Pg 1, February, 1991.

19. Sittig, M.: *Incineration of Industrial Hazardous Wastes and Sludges,* Noyes Data Corporation, 1979.

20. Srivastava, R.K., Mulholland, J.A.: "Low NO_x, High Efficiency Multistaged Burner: Gaseous Fuel Results," *Environmental Progress,* V 7, No 1, Pg 63, February, 1988.

21. U.S. Environmental Protection Agency: *Air Pollution Engineering Manual,* 2nd Edition, May, 1973.

22. U.S. Environmental Protection Agency: *Handbook for Identification and Correction of Typical Design Deficiencies at Municipal Wastewater Treatment Facilities,* EPA-625/6-82-007, 1982.

23. Vogg, H., Stieglitz, L.: "Thermal Behavior of PCDD/PCDF in Fly Ash from Municipal Incinerators," *Atmosphere,* V 15, No 9–12, Pg 1373–1378, 1986.

24. Water Environment Federation MOP-OM11: *Incineration,* WEF, 1988.

25. Water Environment Federation MOP-FD19: *Sludge Incineration: Thermal Destruction of Residues,* WEF, 1992.

26. Water Environment Federation Manual of Practice 8: *Design of Municipal Wastewater Treatment Plants,* WEF, 1992.

27. Barton, R.G., Clark, W.D., Seeker, W.R.: "Fate of Metals in Waste Combustion Systems," *Combustion Science and Technology,* 1990.

28. Caprio, J.A.: "Refractory Practice Survey in Hazardous Waste Incinerators," *76th Annual Meeting of the Air Pollution Control Association,* June, 1983.

29. Perry, R.H., Green, D.: *Perry's Chemical Engineers' Handbook,* Sixth Edition, McGraw-Hill, 1984.

30. Tillman, D.A., et al: "Rotary Incineration Systems for Solid Hazardous Wastes," *Chemical Engineering Progress,* July, Page 19, 1990.

31. U.S. Environmental Protection Agency: Process Design Manual for Sludge Treatment and Disposal, EPA-625/1-79-011, 1979.

32. U.S. Incinerator regulations: EPA: 40 CFR 264-264 defining Treatment, Storage, and Disposal (TSD) Facilities; EPA: 40 CFR RCRA Parts 124, 260-267, 270, 271; EPA: 40 CFR Incinerator Parts 124, 260-265; EPA: 40 CFR Incinerator Subpart O; EPA: Technical Implementation Document for EPA's Boiler and Industrial Furnace Regulation, March 1992, PB92-154 947; EPA BIF RULE: 56FR 7134 (February 21,1991); amendment 56FR 32688 (July 17,1991), 56 42504 (August 27,1991), and 56 FR 43874 (September 5, 1991).

Adsorption

Adsorption is employed to remove select dissolved waste components.

Adsorption is used to remove select dissolved waste contaminants, usually organics, employing adsorbent materials such as activated carbon. Because activated carbon is the most common waste treatment adsorbent, it will be the focus of this chapter. However, the principles discussed apply to any adsorbent and to ion exchange, which can be considered a special case of adsorption.

A fixed bed carbon adsorption system is typical of industrial wastewater treatment systems. Wastewater flows through a vertical column containing activated carbon that adsorbs organics on a highly porous surface. After some time, the carbon bed starts losing its capacity and the effluent concentration approaches that of the influent, indicating that the carbon is spent. Prior to this point, the exhausted unit is put out of service and a second unit placed in operation. The carbon in the first unit is replaced, the spent carbon is either discarded or regenerated.

Adsorption systems are usually designed as continuous processes, configured for cocurrent or countercurrent application, and as a fixed or moving bed. The fixed bed carbon system is common in industrial wastewater treatment, but moving beds are employed for heavy waste loadings, or because of specific waste characteristics. In any adsorption system, performance is measured by *contact time* and *carbon capacity.*

BASIC CONCEPTS

An adsorption system is defined by (1) the achievable effluent quality, (2) its system equilibria characteristics, (3) the required adsorption capacity, and (4) the configuration employed.

ACHIEVABLE EFFLUENT QUALITY

Adsorption performance is established by contact time and carbon dosage. The relationship of effluent quality ver-sus time for a fixed carbon dosage is illustrated in Figure 2.1. Under these conditions effluent quality for any given contact time can only be improved by increased carbon loading. For any carbon dosage (loading) this is the best achievable effluent quality for a given contact time. Theoretically, this effluent quality can be achieved with any process configuration, if the equivalent contact time and ideal batch conditions are duplicated. This performance is seldom achieved in full-scale commercial systems.

SYSTEM EQUILIBRIA DATA

The effect of carbon dosage on effluent quality is illustrated in Figure 2.2, called a carbon isotherm. As illustrated, effluent equilibrium concentration decreases with increasing carbon loading (X/M).

Theoretical isotherm correlations developed for describing adsorption mechanisms include (1) the Freundlich (or van Bemmelen) equation, (2) the Langmuir model developed for single layer adsorption, and (3) the Brunauer-Emmett-Teller isotherm for multilayer adsorption [3]. For dilute wastewater concentrations, involving removal of hydrocarbons with activated carbon, the Freundlich equation is frequently employed.

$$X/M = kC^{1/n} \qquad (2.1a)$$

where X is the weight of material adsorbed; M is the weight of adsorbent material; C is effluent concentration, weight organic/ weight water, etc.; and k and n are constants.

Freundlich isotherms are easily generated from pilot studies, with the data plotted as a log x to log y straight line, represented by Equation (2.1b).

$$\log X/M = 1/n \log C + \log k \qquad (2.1b)$$

where $1/n$ is the slope of the curve.

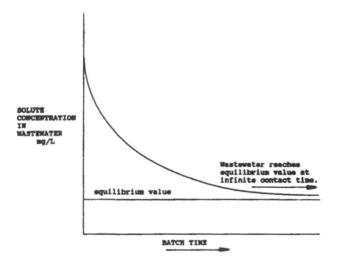

Figure 2.1 Carbon batch performance curve.

That plot establishes the validity (or inappropriate use) of the Freundlich isotherm in representing the system adsorption characteristics. For the sake of convenience the Freundlich isotherm will be assumed in discussing adsorption treatment process principles.

The shape of the curve, defined by the constants (k,n), reveal some significant system characteristics, as illustrated in Figure 2.2. First, any deviation from a straight line indicates inappropriate or restricted use of the Freundlich equation. A large k value is indicative of high adsorption capacity. A flat curve (large n value) indicating a relatively constant adsorption capacity, not significantly dependent upon sub-

strate concentration, and a system effective for batch operation. Conversely, a steep curve indicates increasing adsorptive capacity with concentration, effective for countercurrent continuous operation.

FIXED BED SYSTEM CRITERIA

Contact time in a fixed bed system is governed by mass transfer criteria defined as the mass transfer zone (MTZ), and the adsorbent capacity (or cycle time) governed by bed volume. A fixed bed system is an unsteady state process, in which residence time (bed length) is an important design factor. The condition at any fixed bed position changes with time, and there is a transition zone in which adsorption occurs. Theoretically, this transition zone can be related to the time required to reach batch system equilibrium (Figure 2.1).

Mass Transfer Zone

An understanding of the transition zone is necessary to comprehend a fixed bed operation. The transition section lead edge is defined by the waste concentration and the exit edge value approaching the equilibrium concentration. Figure 2.3 illustrates the action of this transition section, referred to as the mass transfer zone (MTZ), and the "S" shape as the wave front.

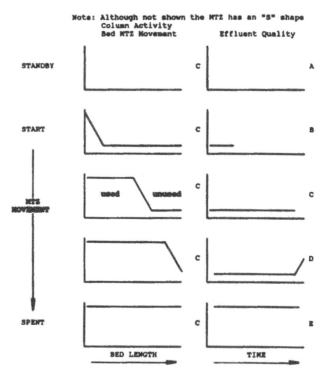

Figure 2.3 Carbon fixed bed mass transfer zone (adapted from Reference [16]).

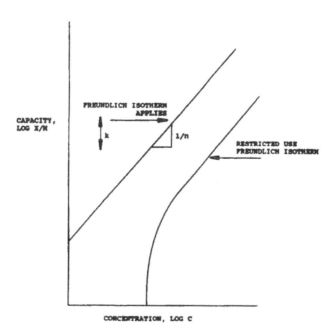

Figure 2.2 Carbon isotherm (adapted from Reference [10]).

Referring to Figure 2.3, the following sequence of conditions occur in the carbon bed:

(1) Initially, the influent entering the bed forms a transition section at the bed entrance, described as the mass transfer zone (MTZ) (B).

(2) Carbon upstream of the MTZ zone becomes saturated, downstream virgin carbon represents the remaining system capacity. The MTZ zone moves into the virgin carbon bed, using the carbon, continually leaving saturated carbon upstream, and producing a constant effluent quality (C). The MTZ zone continues to move intact until the zone reaches the end of the bed (D).

(3) At this point the MTZ zone starts to vanish, and the effluent quality begins to deteriorate (D,E). At some point the MTZ is zero, complete "breakthrough" occurs, and the column is spent (E). Breakthrough is defined as the point in time when influent first "leaks" through and appears in the effluent. From a practical operating standpoint, breakpoint is defined as some arbitrarily acceptable effluent concentration or contact time when the operating cycle is terminated, and regeneration initiated.

Although MTZ was discussed on a theoretical basis, it directly applies to adsorption operations, defining operating characteristics and limitations as follows:

(1) The mass transfer zone is a measure of the waste transfer resistance within the bed, defined by the carbon and contaminant characteristics, and the operating conditions.

(2) The mass transfer zone length represents the minimum bed depth for any configuration, an "active" process volume. Any system with a depth less than the MTZ requirement will produce an inferior effluent quality, the deteriorating effluent quality depending on the available carbon and its saturation limits.

(3) Bed depth above the MTZ provides storage capacity and available carbon inventory.

(4) The breakthrough time is the time required to deplete the carbon inventory prior to MTZ breakthrough, a measure of the system service time.

(5) All adsorption beds have an MTZ zone, or a minimum operating depth requirement. The manner in which inventory carbon is stored and continuously supplied depends on the specific configuration.

Fixed bed adsorber performance depends on column length and volume, and fluid velocity. The column length establishes the available contact time to breakthrough, and must be greater than the MTZ length to assure complete mass transfer. The volume establishes the available bed adsorption capacity. The fluid velocity, besides being directly proportional to residence time, establishes the fluid flow characteristics governing short circuiting potential and mass transfer.

Fixed-Bed Models

Investigators have proposed various models to relate column service time with the waste and operating characteristics [3,6]. Two common design procedures include the empty bed contact time (EBCT) and the Bohart-Adams bed depth service time (BDST), both of which have been proposed for fixed bed process evaluation. Basically, the two are similar in concept, except that the BDST method simplifies evaluating changes in process variables. In either method a suitable depth must be selected, and the effect of the operating variables are analyzed to optimize the system. The reader should review the cited references to obtain detail laboratory procedures to implement the chosen process.

The EBCT uses laboratory data to establish a relation between the bed contact time and the resulting carbon rate. Pilot data are collected using varying fixed bed contact times, correlating effluent COD and gallons processed. The data are processed to determine the following parameters for each data-set for a selected effluent concentration:

$$EBCT = \frac{contractor\ volume}{flow\ rate} = \frac{contractor\ depth}{linear\ velocity} = time$$

In turn,

$$\frac{flow\ rate\ (volume/time)}{linear\ velocity\ (length/time)} = column\ area$$

$$Carbon\ dosage = \frac{weight\ of\ carbon\ in\ column}{waste\ volume\ at\ breakthrough}$$

$$= mass/volume$$

$$Carbon\ usage = \frac{weight\ of\ carbon\ in\ column}{time}$$

$$= pounds/time$$

Analysis of the contact time and carbon usage allows a system cost evaluation to compare capital costs (tower) against operating costs (carbon usage), optimizing the system configuration.

The BDST method assumes that the adsorption rate is proportional to the residual carbon capacity and the waste concentration, as represented by the equation:

$$\ln\left[(C_o/C_b) - 1\right] = \ln\left[\exp\left(K \cdot N_o \cdot D/V\right) - 1\right] \qquad (2.2a)$$

$$- K \cdot C_o \cdot t$$

where C_o is the influent concentration, mass/volume; C_b is the impurity concentration at breakthrough or at a selected design effluent limit, mass/volume; K is the adsorption rate constant, volume of waste water per mass of organic fed to the system per time; N_o is the carbon capacity in mass of organic per volume of carbon; D is the bed depth, length; V is the wastewater linear flow rate, volume/time/area; and t is the service time, time.

This equation can be rewritten in terms of time,

$$t = \frac{N_o}{C_o V} \cdot D - \frac{1}{KC_o} \cdot \ln \left(\frac{C_o}{C_b} - 1 \right) \qquad (2.2b)$$

A modified form of this equation has been proposed to better evaluate the effect of process condition changes [3,6]. Expressing service time equation as a straight line:

$$T = a \cdot D + b \qquad (2.2c)$$

In which case, slope (a) is equal to

$$\frac{(Y \cdot N_o)}{(C_o \cdot V)}$$

The intercept b is equal to

$$\frac{Z}{KC_o} - \ln \left[(C_o/C_b) - 1 \right]$$

In this form the linear flow (V), influent concentration (C_o), or effluent concentration (C_b) effects can be evaluated by changing the slope or intercept constant and developing a new time curve. The value of the constants (Y, Z) depends on the selected units, as indicated in Table 2.1.

In addition, service time and related operating parameters can be estimated for single or multiple fixed beds, as well as moving beds, using a suggested factor to account for configuration effects. Corrective factors (f) indicated in

TABLE 2.1. BDST Equation Units.

	Base Units	English Units	Other Units
t		hours	hours
D		feet	*
N_o	lb/ft³	lb/ft³	**
C_o	lb/ft³	ppm	*
C_b	lb/ft³	ppm	*
V	ft³/hr/sq foot	ppm/ft²	*
K	ft³/lb/hr	ft³/lb/hr	**
Y	1	1990	**
Z	1	16,018	**

*Depth, concentration, and flow units selected to suit.
**Constant units must be consistent with selected units.

TABLE 2.2. Moving or Multiple Adsorption Factors for Correcting BDST Equation (adapted from Reference [6]).

Removal, %	Columns in Series			
	Two	Three	Additional Units, or 5% Pulsing Bed	
50	0.70	0.59	0.50	to	0.55
60	0.63	0.50	0.40	to	0.45
70	0.54	0.40	0.30	to	0.35
80	0.44	0.30	0.20	to	0.25
90	0.31	0.18	0.10	to	0.15
95	0.22	0.11	0.05	to	0.08

The BDST equation defines the wave front movement through a single fixed-bed tower. Performance of multiple towers operating in series should be evaluated utilizing *site specific studies*. Service times of multiple units operating in series, or moving bed performance, can be *approximated* using the following factors:

Table 2.2 are applied to the influent concentration to obtain a new slope constant a', so that

$$a' = (C_o a)/(f C_o) = a/f \qquad (2.2d)$$

A new service time equation can be developed, corrected for multiple or pulsing bed, as represented by Equation (2.2e).

$$T = a' \cdot D + b \qquad (2.2e)$$

For any defined configuration, the manner by which changes in the influent or effluent concentration, depth, or hydraulic loading affects process service time is illustrated in Case Study 27.

The basic Bohart-Adams relationship can be used to estimate the fixed bed minimum height or MTZ, at service time zero. So that

$$D_o = \frac{V}{K \cdot N_o} \cdot \ln (C_o/C_b - 1) \qquad (2.2f)$$

As defined by Equation (2.2f), the minimum height (D_o) is related to the physical characteristics defined by the adsorption rate capacity (K) and the carbon efficiency N_o, and the operating parameters defined by the volumetric loading (V), the feed concentration (C_o), and the acceptable breakthrough concentration (C_b).

CONTINUOUS MOVING BED ADSORBERS

In a moving bed adsorber steady state conditions are achieved, and both wastewater and adsorbent are moving at constant rates. The flow of the two streams, and the resulting overall system balances, are illustrated in Figure 2.4. The theoretical minimum adsorbent to waste ratio (A/W) is represented by the operating line touching the equilibrium line

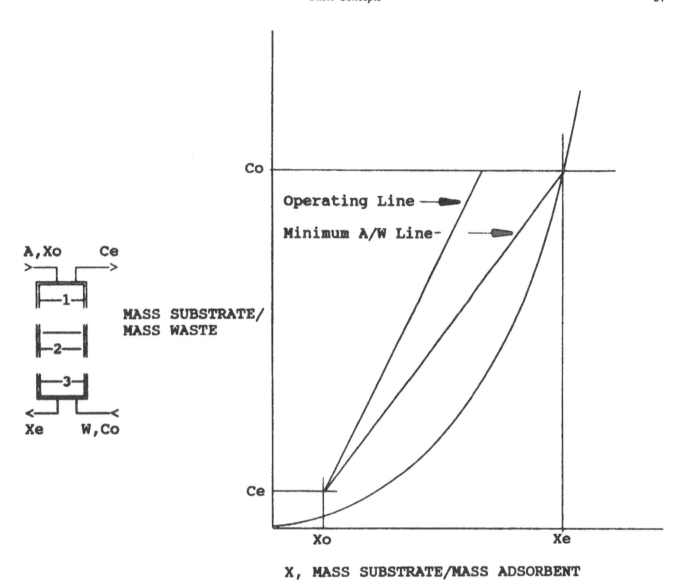

Figure 2.4 Continuous moving bed adsorber.

at the waste influent concentration C_o, the actual ratio represented by the operating line at some point on the ordinate C_o, away from the equilibrium line.

The adsorption mechanisms within a moving bed adsorber can be best understood by assuming a system analogous to single-component continuous stripping discussed in detail in Chapter III-4, with the solid adsorbent replacing the air stream. Theoretically, as with continuous stripping, the number of *ideal* stages of contacts can be estimated on the basis of the inlet and outlet wastewater concentration, and the appropriate adsorbent to wastewater ratio. The adsorbent bed depth (Z) is a product of the height equivalent to a theoretical stage (HTU) and the number of transfer units (NTU).

$$Z = HTU \times NTU \qquad (2.3)$$

However, continuous bed depths are seldom calculated in this manner because of a lack of adequate mass transfer data, and because waste substrates seldom contain single components, further complicating theoretical estimates. Information to establish bed depth, or effluent quality for a bed depth, can only be obtained from specific column testing. However, the Process Engineer should be aware of how the bed depth is established, relating influent, effluent, and related mass transfer rates to the bed depth. In practice, series or pulse beds can be envisioned as an infinite number of fixed bed contactors in series, the equipment size based on the allowable flow velocity, the depth based on that defined by the MTZ, adjusted for an added working inventory. The service time and carbon usage can be approximated using Equation (2.2c) and corrective factors indicated in Table 2.2.

CARBON ADSORPTION SYSTEMS

The appropriate adsorption configuration depends on the waste characteristics, effluent requirements, waste volume, adsorption properties, required carbon quantities, regeneration requirements, and site specific physical limitations. These process considerations must be related to adsorber configuration alternatives, which include

(1) Batch or continuous
(2) Fixed or moving bed
(3) Downflow or upflow
(4) Series or parallel operation

The selected alternative can be installed as a pressure or gravity operation, and modified for any construction limitations.

BATCH OR CONTINUOUS

Batch systems are limited to small daily flow volumes containing low organic loadings. Fixed waste volumes are contacted with a limited carbon quantity, the operation terminated when the required effluent quality is achieved. The limiting operating variable is batch time. Batch system configurations can vary, although frequently a stored waste volume is recirculated through a fixed bed until the desired effluent quality is reached. Batch after batch of waste can be treated in this manner until breakthrough is evident and the carbon is spent. The system usually consists of two batch waste storage tanks, one supplying the carbon bed and the other receiving waste in preparation for the next adsorption cycle.

Another method uses a suspended solids system in which a waste volume is injected with an optimum carbon dosage, in a well mixed vessel, the contact continues until the required effluent quality is obtained. The slurry is then pumped to a separator or a filter where the carbon is removed and effluent discharged. The carbon is added in predetermined quantities, and discarded after the treatment is complete. This is essentially the basis for the PACT® system, where the carbon is injected into an activated sludge aeration basin.

FIXED OR MOVING BED

Successful application of a fixed or moving bed configuration depends on the process characteristics and the carbon requirements. Using the MTZ height as an adsorption characteristic, a broad wave length is best treated in a fixed bed, and a narrow wave length in a moving bed. A narrow wave length produces a relatively rapid break-through from influent to effluent concentration, which is not detrimental in a moving bed performance since fresh carbon is always downstream. A moving bed is also favored when large car-

bon quantities are required as a result of large influent loadings.

A *fixed-bed system* consists of a stationary column in which the waste flows through the bed at a constant rate. A concentration gradient develops in a moving MTZ phase. Fixed-bed systems can be used if adsorption is relatively easy, the available carbon inventory allows an adequate service time, and the MTZ can be contained within the specified contactor depth. Whether single or multiple units are used depends on the required effluent quality and the MTZ characteristics. Theoretically, single fixed-bed contactors can be used if the total bed depth can be sized for 3 to 5 times the MTZ length (or more), employed when the waste exhibits a broad waste length, and the effluent criteria is not stringent. From a practical operating aspect multiple units in series are commonly employed to achieve operating flexibility, when treatment cannot be interrupted, and when high effluent quality must be assured. In providing multiple units the second column acts as a "safety net" preventing poor effluent when the first column is spent, remaining in service while the first unit is being serviced. In some cases, required bed height may not make a single unit practical.

Multiple units in *series* make maximum used of available carbon if configured for countercurrent operation and are generally employed for a waste exhibiting a relatively narrow wave length to guard against rapid effluent deterioration. *Parallel* units are employed for process flexibility, when effluent quality is not stringent, to treat a waste displaying a wide breakthrough curve, or to accommodate high influent mass loadings. Staggering the unit operation cycles assures that all units will not be depleted at the same cycle, simplifying regeneration. Parallel units are also used to accommodate large waste volumes without an excessive pressure drop or employing a large diameter vessel. Combined *series-parallel* trains are used to achieve high effluent quality, accommodate high influent loadings, and provide process flexibility.

A *moving bed* is a steady state system where both the waste and carbon are in motion. At steady state all bed sections remain at constant, but not equal, conditions. The bed is in motion with portions periodically removed from the bottom. This system does not require off-line servicing time since the carbon is constantly replaced, and fresh carbon added. The system size is based on principles similar to an *ab*sorber, with the bed depth dependent on transfer units related to the influent and effluent concentration, as well as carbon-contaminant equilibrium characteristics. The contactor diameter is based on loading limitations, and maintaining a physically stable contactor bed. This system is considered when large carbon quantities are required, allowing for continuous regeneration and minimizing carbon storage requirements. However, where large carbon quantities are required, operating costs may make adsorption prohibitive, favoring competitive alternatives.

UPFLOW OR DOWNFLOW SYSTEM

Selection of an upflow or downflow operation is based on influent solids loading, and maximizing carbon efficiency by approaching counterflow conditions. An *upflow* bed can be fixed or fluidized (expanded), and the system designed for intermittent or continuous carbon removal (and replacement). *Downflow* systems are designed as fixed bed systems, with a contactor removed off-line when spent, carbon removed or regenerated in place, and fresh carbon used when the contactor is placed on-line.

Countercurrent Operation

Countercurrent operation, commonly referring to flow direction of two media in contact systems, is confusing in fixed bed contactors since the carbon is not in motion. In such systems countercurrent refers to the direction of the carbon loading gradient, the column configured to optimize carbon use. As indicated in the general carbon isotherm in Figure 2.2, lower effluent quality is achieved at reduced carbon capacity (weight substrate adsorbed per weight of adsorbent). Conversely, carbon loading can be increased by increasing the acceptable effluent concentration. These two principles can be implemented by allowing the carbon saturated at low concentrations to be used to adsorb high influent concentration as indicated in Figure 2.5.

In *fixed* beds two conditions drive the system towards a countercurrent system, multiple units and/or upflow operation (discussed below). Multiple units can be configured as illustrated in Figure 2.6 to achieve countercurrent flow, with piping arranged to allow interchanging contactor sequence. In *moving* beds, countercurrent is easy to visualize since the carbon *is* moving counter to the waste, the spent carbon removed at the bottom where the influent enters.

Upflow or Downflow Units

Upflow operations approach countercurrent transfer conditions because of the "classifying" nature of the process. Substrate adsorbed by the carbon causes an increased carbon density, settling to the bottom where it encounters the high influent concentration, resulting in more effective use of the total contactor carbon.

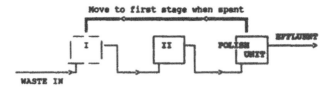

Figure 2.6 Multiple unit countercurrent configuration.

Influent solids concentration affects configuration selection, as follows:

(1) A downflow fixed bed will act as a sand filter, as well as an adsorber, providing for both organic and solids removal. However, as solids loading increases, backwash requirements increase, with air scouring and backwashing required at extreme conditions. The price for this combined treatment is reduced service time, depending on the backwashing frequency.

(2) An upflow fixed bed is usually employed for wastes with turbidity measurements less than 2.5 JTU. At higher solids levels sand filtration pretreatment is employed to reduce influent solids content [12].

(3) Fluidized beds can be employed for relatively high suspended solids influent, utilizing finer and more effective carbon sizes. Expanded beds allow less opportunity for solids rejection, and are therefore less prone to rapid pressure drop deterioration. Solids removal may still be required to meet effluent criteria.

Combining solids and organic removal may have economic advantages at low (water quality) contaminant concentrations; but downflow systems employed for dual service as primary industrial waste systems result in some operational disadvantages:

(1) They may not be as effective in organic removal as a separate carbon treatment unit or as effective in solids removal as a sand filter.

(2) Operator flexibility is greatly reduced.

(3) Capital cost savings may not compensate for increased operating costs resulting from reduced service time and increased pressure drops, both of which could become excessive.

GRAVITY VERSUS PRESSURE CONTACT SYSTEMS

For the most part site physical restrictions, and the resulting plant hydraulic profile, establish whether a gravity or pressure system can be employed. Basically, when a pressurized sewer line is employed, a pressure system will be installed. The advantage of pressure vessel contactors is increased operating flexibility, higher allowable pressure losses, and a wider range of applicable hydraulic loadings; the disadvantage is more expensive construction costs. Gravity contactors are less flexible, having a narrower operating

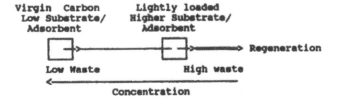

Figure 2.5 Countercurrent configuration.

range, but are usually less expensive to construct, especially for large systems where common wall construction can be used.

SPECIFIC PHYSICAL CHARACTERISTICS

Physical and operating limitations greatly affect the carbon bed shell construction. These include size, configuration, and materials of construction. Some general criteria include:

(1) Pressurized systems are usually steel constructed. The maximum factory fabricated steel tank is 4 meters (12 feet) in diameter and 18 meters (60 feet) long, larger vessels are field fabricated.
(2) The cost of large field fabricated pressure vessels may justify (if hydraulics permit) open tanks, employing a gravity contact system. Open tanks can be conveniently constructed of concrete, with rectangular common wall construction.
(3) The larger the vessel or basin, influent distribution, complete bed utilization, and effluent collection become critical design considerations.

COMMON PROCESS CONFIGURATIONS

The most common carbon adsorption configurations employed include (1) a pressurized upflow packed bed, (2) a pressurized pulse bed, (3) an open upflow packed bed, (4) an open upflow expanded bed, (5) a pressurized upflow expanded bed, (6) a pressurized downflow packed bed, and (7) an open (gravity) downflow packed bed [12]. These systems have the general characteristics discussed above, and detailed in Table 2.3.

INORGANIC REMOVALS

A considerable amount of investigatory effort has been conducted to establish the feasibility of adsorption as a means of removing trace hazardous and inorganic contaminants to meet drinking water requirements, treat contaminated underground water, or meet stringent discharge effluent standards. Studies utilizing carbon as a water treatment operation for removal of contaminants such as arsenic, barium, calcium, chromium, copper, fluoride, mercury, radionuclides, selenium, zinc, chlorine, and hydrogen sulfide have been reported in the literature [3]. In general, except for chlorine and hydrogen sulfide, carbon treatment has proven to be either ineffective, inconsistent, or moderately effective within a definitive pH range. Other investigators [1,8] have reported metals and sulfide removal results using carbon in the presence of complexing agents, pretreating the carbon [2], or using adsorbents other than activated carbon [1,4,5].

Based on these and similar studies, indications are that carbon has some affinity for trace metal removal under controlled conditions, which can be improved by combining adsorption with precipitation or adding complexing agents. Further improvement may be achieved by using specialized adsorbents such as alumina, clay, conditioned pulp, or a preconditioned media coated with active material.

The disadvantages of using activated carbon (or any other adsorbent) to remove trace inorganics include (1) difficulties in regenerating the media to remove the inorganic and recover the carbon, (2) the need to generate a disposable or economically recoverable by-product, or (3) the inability to dispose of the adsorbent saturated with inorganics in a landfill. In all likelihood, the regulatory constraints for solids disposal of adsorbents containing concentrated toxic inorganics may be more stringent than corresponding effluent discharge requirements, and the cost of replacing adsorbents prohibitive. Use of adsorbents must be carefully evaluated relative to competitive technologies such as chemical precipitation or ultrafiltration.

PROCESS ENGINEERING DESIGN

Carbon adsorption is a comparatively simple treatment system with most operating variables related to flow rate, as indicated in Table 2.4. Maintaining a stable system is relatively easy, requiring little operator attention, if the columns are effectively automated. However, achieving high removals at reasonable costs depends on the waste characteristics and service time, over which the operator has little control. System performance is highly contingent on waste adsorbability, with poorly adsorbed influent components reducing removal efficiency, whereas large quantities of adsorbable components reduce service time and increase regeneration requirements. System operating costs are greatly affected by regeneration and virgin carbon costs. In small systems spent carbon is usually disposed of, although this could be expensive if saturated with hazardous components.

REPORTED PERFORMANCE DATA

Carbon removal efficiencies depend on waste and process characteristics. Pilot data reported by the EPA [13] for granular carbon indicate a wide variation in treatment performances, as indicated in Table 2.5. EPA data cited for studies involving powdered carbon additions into activated sludge systems are summarized in Table 2.6.

REQUIRED PROCESS DESIGN DATA

Activated carbon treatment is evaluated in two stages, using carbon isotherms and carbon columns. Carbon isotherms establish adsorption feasibility and capacity. Adsorption performance can be measured as direct organic concentration or in terms of TOC, BOD_5, COD, or color units.

TABLE 2.3. Operating Characteristics of Common Configurations (adapted from Reference [12]).

	Fixed/ Moving Bed	Down/ Upflow	Backwash	Pressure/ Gravity	Inlet	Carbon Removed During Process	Solids Removal (4)	Bed Expansion Process (Backwash)
1. Pressurized upflow, packed bed	F	U	Yes (1)	P	Bottom	Batch (3)	No	Packed (10%)
2. Pressurized pulse bed	MB	U	No	P	Bottom	Continuous	No	Moving bed
3. Open upflow packed bed	F	U	Yes (1)	G	Bottom	Batch (3)	No	10%
4. Open upflow expanded bed	F	U	Yes (1)	G	Bottom	Batch (3)	No	10% (50%)
5. Pressurized upflow expanded bed	F	U	Yes (1)	P	Bottom	Batch (3)	No	10% (50%)
6. Pressurized downflow packed bed	F	D	Yes (1)	P	Top	No	Yes	Note 5
7. Open downflow packed bed	F	D	Yes (2)	G	Top	No	Yes	Note 5

(1) Increased water flow to expand bed and remove solids.
(2) Backwash similar to sand bed, including surface air and water.
(3) Five to 10 removed in batches to simulate counter current operation.
(4) Backwash employed for most fixed beds. Specific solids removal must be employed in packed beds to remove solids.
(5) Operated as packed beds, with minimum 10% expansion in backwash to facilitate dislodging heavy solids loading and to avoid excessive pressure drop and damage to support internals.

Carbon isotherms can also be used to screen the effects of adsorbent type, pH, reaction time, and temperature.

Pilot plant studies are conducted to develop specific process design data, with emphasis on defining the system mass transfer zone and developing service time operating data. Information obtained from these studies includes service time, superficial velocity, carbon loading, pretreatment requirements, column maintenance (backwashing frequency), and achievable effluent quality. The service time curve is a function of superficial velocity and bed depth, both of which can be correlated to effluent quality. This information can be related to the bed depth service time (BDST), utilizing specific test methods developed for these evaluations [3]. The design service time is an optimum value based on minimizing capital and operating costs.

Pilot tests are usually run in columns up to 8 meters (24 feet) in height, with sampling ports at representative heights. The retention time is evaluated by controlling and varying the volume feed to the column, and therefore the superficial velocity. Scale-up from column tests is common. However, it is essential that upset conditions be evaluated to determine whether the data developed are effective through the wide range of influent conditions encountered. Resulting data can be tabulated or integrated into a service time model to obtain process correlations, establishing the required design data listed in Table 2.7.

WASTE EVALUATION

The primary waste parameters affecting process design are the total quantity of active constituents that can be adsorbed, those that must be removed, and the disparity between the two. Adsorption models describe removal of individual organic constituents, which in reality is seldom the case in wastewater treatment. For the most part, wastewater components are never completely identified, highly variant, and always multispecies. For that reason, adsorption performance is commonly based on gross parameters such as BOD, COD, TOC, or color. The design is treated as a single component system, when in reality it will treat multicomponents comprising the measured gross parameter, at varying re-

TABLE 2.4. Operating Characteristics.

Variable	Waste Characteristics Operator Controllable	Critical
Waste generated	No	Yes
Composition	No	Yes
Concentration	No	Yes
Adsorbability	No	Yes
Operating Characteristics		
Flow rate	Minimal	Yes
Hydraulic loading	Limited	Yes
Organic loading	Limited	Yes
Contact time	No	Yes
Service time	No	Yes
System pH	Yes	Yes
System temperature	No	Yes
Distribution	Yes	Yes
Regeneration	Yes	Yes
Backwash	Yes	Yes

TABLE 2.5. Industrial Performance Data (adapted from Reference [12]).

Industry	Influent COD, mg/L	Removal, % Range	Removal, % Mean
Laundry wastes	125–520	0–70	40
Chlorinated wastewaters	995–1570	0–81	35
Chloroethane*	59,000–390,000	0–100	66
1,1-Dichloroethane*	40,000–310,000	42–100	89
1,2-Dichloroethane*	920,000–3,000,000	22–100	88
Refinery wastes	30–130	35—80	65
Textile wastes	30–130	0—55	85

*concentration of chloroethane compounds in μg/L.

moval rates. Therefore, single component theory is at best an approximation, compensated by a conservative design. When adsorption is applied for a specific component removal in a multicomponent waste the design is complicated because adsorbents are not selective. The result is a system designed for material removals greater than required, partial removal of targeted contaminants, or (in rare cases) optimum removal of specific targeted constituents. In most treatment schemes excessive substrate removal is not a problem. However, raw carbon is expensive, and regeneration a significant operating cost. Excessive carbon results in high operating cost, significantly affecting when adsorption can be economically employed. Other critical influent variables include flow rate, concentration, variability, temperature, and pH.

Flow Rate

A selected design flow must represent an average waste generation rate, allowing an operating range compensating for expected variations, assuring conformity with the selected design hydraulic loading and service time conditions. The flow rate selected and the acceptable operating range are critical because they affect process and bed stability. The consequences resulting from excessive flow rate are not only reduced service time and added expenses, but an MTZ that

could be greater than the column height. To the Process Engineer this means selecting a tower configuration consistent with the waste variation, resulting in a reasonable vessel cost. The greatest adsorption system design error is basing the tower size on adsorbent capacity and corresponding tower volume, ignoring service time and MTZ.

Concentration

Unlike flow rate, concentration effects are obvious, since a higher concentration represents a higher system loading, and therefore less service time. At some point the feed concentration and corresponding organic loading will make adsorption prohibitively expensive, requiring evaluation of alternative technologies. Based on a maximum 91 kg/day (200 lb/day) of carbon for a once-through system, or a minimum 680 kg/day (1500 lb/day) for a regeneration system, adsorption feasibility can be related to waste loading as follows:

(1) For substrate with 20% carbon adsorbability, less than 18 kg/day (40 lb/day) of organics would justify a once-through disposable carbon system or require regeneration with a feed containing more than 136 kg/day (300

TABLE 2.6. Additions to Activated Sludge Systems (adapted from Reference [12]).

	Effluent Concentration, mg/L Minimum	Maximum	Median	Mean	% Removal Efficiencies Minimum	Maximum	Median	Mean
BOD₅	4	54	13	17	<90	>99	96	96
COD	33	563	98	160	60	98	91	87
TOC	9	387	38	67	64	97	90	86
TSS	17	83	54	52	0	96	0	24
Oil and grease	11	57	13	23	8	96	54	53
Total Phenol	<0.010	0.058	0.013	<0.023	99	>99	>99	>99
TKN				28				96

Limited data indicate 40 to greater than 97% removals of toxic pollutants such as chromium, hexavalent chromium, copper, cyanide, mercury, zinc, bis(2-chloroethyl) ether, Bis(2-ethylhexyl) phthalate, 2-chlorophenol, phenol, benzene, ethylbenzene, toluene, naphthalene, 1,2-dichloroethane, 1,2-dichloropropane, acrolein, and isophorane.

TABLE 2.7. Required Design Data.

Critical pilot plant treatability data specific to the waste
1. Design temperature
2. Service time
3. Hydraulic loading
4. Organic loading
5. Achievable effluent quality
6. Minimum bed depth
7. Carbon capacity

Selected operating characteristics
8. Total bed depth
9. Carbon disposal or regeneration
10. Pretreatment requirements

lb/day) organics. This corresponds to waste influent concentrations as follows:

Flow		Influent Concentrations for a	
m³/d	(MGD)	Once-Through	Regeneration
1893	(0.5)	9.6 mg/L	71.9 mg/L
3785	(1.0)	4.8 mg/L	36.0 mg/L
7570	(2.0)	2.4 mg/L	18.0 mg/L

(2) For substrate with poor adsorbability, assuming 5% carbon adsorption capacity, a once-through disposable system could accommodate 4 to 5 kg/day (10 lb/day) of organics, and require regeneration with a feed containing more than 34 kg/day (75 lb/day) of organics. The corresponding influent concentration would be:

Flow		Influent Concentrations for a	
m³/d	MGD	Once-Through	Regeneration
1893	(0.5)	2.4 mg/L	18.0 mg/L
3785	(1.0)	1.2 mg/L	9.0 mg/L
7570	(1.5)	0.6 mg/L	4.5 mg/L

A factor often overlooked is that MTZ height increases with increasing concentration, requiring a deeper overall bed depth to satisfy both MTZ and carbon inventory requirements; thereby increasing system capital costs.

Influent Variability

Theoretically, feed *Variability* should not influence the system, although such effects are subtle and could impact effluent quality. Sudden surges can physically upset the bed, resulting in adsorbent losses. In addition, peak effluent concentrations can occur as a result of increased leakage as a system equilibrium shifts to satisfy surge conditions, attempting to adjust to new MTZ limits, service time, and operating conditions. Because many units are operated on programmed cycles, based on either periodical test results or time sequence, unexpected and prolonged surges may not be detected by the operator, resulting in apparent unaccount-

able effluent deterioration. As with most waste treatment operations, equalization must be carefully evaluated to stabilize influent conditions.

Temperature and Alkalinity

Decreasing pH and increasing temperature increase the adsorptive characteristic of activated carbon. Carbon adsorption is effective at a range between 6.5 to 9, and in some cases slightly outside this range [12]. Elevations of pH, even within acceptable limits, could cause conditions favoring desorption.

PRETREATMENT REQUIREMENTS

Some important factors to be considered in evaluating pretreatment include

(1) Reduce influent organic loading, or apply a less complicated and more economical treatment method when concentrated wastes are to be treated.
(2) Remove immiscible organics or colloidal matter that could reduce system effectiveness and result in high bed maintenance. Free oil should be maintained at less than 10 mg/L.
(3) Maintain suspended solids at less than 25 mg/L to avoid excess plugging of the bed, and excessive backwashing.
(4) Avoid waste components detrimental to the adsorbent.
(5) Maintain the adsorbent at an optimum pH.
(6) Avoid secondary treatment biomass and similar organic materials that could foul the adsorbent, become anaerobic, and emit gases such as hydrogen sulfide.

PROCESS DESIGN VARIABLES

Adsorption system design involves applying process basics to develop the operating capabilities discussed. These include the following:

(1) Process configuration
(2) Carbon adsorbability
(3) Adsorbent regeneration economics
(4) Tower loadings (required area)
(5) Column depth
(6) Fate of contaminants

System Configuration

Specific configurations are discussed in the Carbon Adsorption Process Systems section. Assessing the available alternatives involves a series of sequential evaluation steps, involving:

- Step 1. The feasibility of a batch or continuous system
- Step 2. The effects of influent solids
- Step 3. The need or advantage of multiple units
- Step 4. System head loss limitations (gravity or pressure unit)
- Step 5. Carbon management requirements
- Step 6. The need for a final blend tank

Batch or Continuous System

Batch systems have limited application in industrial waste treatment, utilized only where daily waste flow volumes are considerably less than 3785 cubic meters per day (1 MGD) making fixed or moving bed column sizes impractical; and organic loadings are low minimizing the required carbon dosage. A variation of a batch process is a modified activated sludge system where carbon is "dosed" into the aeration basin to improve overall substrate removal and effluent quality. Carbon regeneration is seldom practical in batch systems.

Continuous systems should be evaluated on the basis of required daily carbon. If adsorption is considered for high organic loadings, requiring greater than 900 kg (2000 lb) per day of carbon, the capital and operating costs should be fully investigated. A fixed bed is often employed in the influent ranges economically appropriate for industrial wastes. If a narrow MTZ is required, along with a high carbon consumption, a moving bed system may be appropriate.

Influent Suspended Solids Level

High influent solids level impacts adsorption performance because of solids "filtration," which could result in rapid deterioration of available head, and reduced service time. At influent solids concentrations greater than 25 mg/L, an upflow expanded solids system or solids pretreatment should be employed. As an alternative, or where high effluent quality is required, an upstream filter can be employed.

Number of Units

There are no set criteria dictating the number of required process units. Rather, multiple unit adsorption trains are a result of related feed and operating considerations, such as

(1) A single, fixed-bed unit can be used if (a) the wastes exhibit a broad wave length allowing gradual effluent deterioration, (b) the MTZ can be contained within the contactor depth and the depth is at least 3 times the MTZ, (c) an adequate carbon inventory can be maintained, (d) the resulting service time is reasonable, (e) the cost of an additional contactor cannot be justified, (f) opera-

tion of the unit is not critical and some shutdown is allowable, and (g) the effluent criteria is not stringent.

(2) Multiple units are used if the required carbon inventory exceeds the capacity of a single unit, the required MTZ cannot be contained in one unit, or operating flexibility is essential.

(3) Multiple units in series are used for (a) the reasons stated in (2), (b) additional units are an economical method to store carbon, (c) the wastes exhibit a wave length that results in rapid breakthrough and effluent deterioration in a single column, and backup units protect against sudden noncompliance effluent discharge, (d) a high quality effluent is required, or (e) shorter units in series improve the aesthetic effects.

(4) Multiple, parallel units allow waste distribution, reducing the organic or hydraulic loading to any unit to allowable limits, and reduces the headloss to any unit and the overall system power requirement.

(5) A moving bed is used if the wastes exhibit a narrow wave length, or large carbon quantities are required prohibiting frequent regeneration of single, fixed-bed units.

Although process parameters discussed usually determine the number of contactors selected, practical construction limitations may influence the selected unit capacity. As an example, shop fabricated units are a maximum 3.7 meters (12 feet) in diameter, restricted by transportation limitations. Based on a hydraulic loading range of 1 to 7 L/s/sq meter (2 to 10 gpm/sf), each tower has a maximum capacity of approximately 1200 to 6000 cubic meters per day (325,000 to 1,600,000 gpd). Field erected concrete gravity contactor capacity could be higher.

Shop fabricated units can be transported at heights up to 18 meter (60 feet), which should provide an adequate height for MTZ and carbon inventory purposes. In fact, each shop fabricated vessel of 3.7 meters in diameter and 18 meters in total length could store approximately 64,000 kg (141,000 pounds) of carbon; assuming a carbon density of 400 kg/cubic meter (25 pounds per cubic foot) and 15 meters (50 feet) carbon bed.

Once the unit tower capacity has been established, the number of units installed should be selected to assure operating flexibility, and a high on-line operating time. For large plants this is not a problem, since multiple units are a process requirement. For small capacity systems of less than 7570 cubic meters per day (2 MGD), multiple units will have to be considered for operating flexibility, even if not a process requirement.

Gravity or Pressure System

Whether a gravity or pressure system is employed depends on the system hydraulic profile, and the available gravity head. Pressure systems allow greater head losses but are usually more expensive to construct.

Carbon Inventory

Carbon use rate and corresponding inventory requirements must be established to determine whether (1) adsorption is economically viable, (2) the quantity of carbon used warrants on-site regeneration, and (3) off-site disposal of small spent carbon quantities is feasible.

Final Blend Tank

Final effluent blending may be required to meet strict effluent quality requirements allowing equalizing significant discharge variations. Generally, final blend tanks or basins are not required if upstream equalization is provided, minimizing effluent variances by reducing influent variations to acceptable limits.

Carbon Adsorptivity

Activated carbon effectiveness can only be measured by laboratory testing, particularly for industrial wastes which rarely contain pure or single constituents, and therefore are impossible to theoretically evaluate. Adsorptivity depends on the adsorbent and the contaminants involved, but some general guidelines can be cited.

(1) Adsorption of organics increases with increasing molecular weight, decreasing solubility, decreasing ionization capability, increasing hydrolysis and the ability to form an adsorbable acid or base, and decreasing polarity [15].

(2) Molecular structure can affect adsorptivity, with branched chains being more adsorbable than straight chain organics and increased length decreasing solubility and improving adsorptivity [10]. Group substitution affects adsorbability, generally as indicated in Table 2.8 [11]. For straight carbons, the affinity to carbon adsorption is summarized in Table 2.9 [15]; structural effects are summarized in Table 2.10 [10].

TABLE 2.8. Effects of Substitution on Adsorptivity (adapted from Reference [11]).

Substitution	Adsorbability Usually
Hydroxyl	Reduced
Amino	Reduced
Carbonyl	Varies with compound
Double bonds	Varies with compound
Halogens	Varies with compound
Sulfonic	Reduced
Nitro	Increases
Aromatic rings	Increases significantly

TABLE 2.9. Straight Chain Effects on Adsorptivity (adapted from Reference [15]).

Compound	Affinity
Undissociated organic acids	Highest
Aldehydes	
Alcohols (more four carbons)	
Esters	
Ketones	
Alcohols (less four carbons)	
Glycols	Lowest

Wastewater Adsorptivity

Single compound adsorptivity information is useful but seldom applicable to wastewaters, containing various forms of pure or interacting manufacturing products, by-products, and raw materials. Table 2.11 illustrates the breakdown of the components of a municipal treatment plant effluent, and their relative adsorptivity [10].

Regeneration

Regeneration options are economically driven, with on-site regeneration dependent on carbon usage. The EPA design manual specifies 90 kg/day and 680 kg/day (200 to 1500 lb/day) carbon usage as "benchmark" guides in evaluating regeneration feasibility, based on a 137 cm (54 inch) multiple hearth gas fired furnace [12]. Below 90 kg/day the cost of recovering carbon equals or exceeds raw carbon costs. In such cases a "once-through throw away" basis is the best choice, while returning spent carbon to a central recovery facility may be advantageous, if one is available.

When daily carbon usage exceeds 680 kg/day regeneration is an economical necessity. For daily carbon uses between 90 and 680 kg/day the cost of on-site regeneration, off-site central system regeneration, or alternative treatment methods (chemical, steam stripping, incineration, etc.) must be evaluated on a case-by-case basis. Although costs utilized in this early EPA study have changed considerably, these boundary carbon usage rates are probably still valid as a preliminary design guide.

A "once-through throw-away" carbon system suggests an inexpensive dumping of spent carbon, although hazardous

TABLE 2.10. Structural Effects on Adsorptivity (adapted from Reference [10]).

Compound	Affinity
Aromatic acids	Highest
Aldehydes	
Aliphatic acids and alcohols	
Aniline	
Aliphatic amines	
Phenols	Lowest

TABLE 2.11 Municipal Effluent Adsorbability (adapted from Reference [11]).

Molecular Weight	% Adsorbed	Species
Less than 100	57	Polar organic compound
100 to 500	82	Fulvic acid material
500 to 10000	90	
Greater 50000	50	Humic carbohydrate material

waste regulatory controls may apply. Because carbon is commonly employed as advanced (tertiary) treatment to remove priority pollutants, the spent carbon saturated with organics may be classified hazardous. In such cases, regulatory restrictions may limit available disposal sites, require pretreatment prior to disposal, or both. If the supplier is willing to receive the spent carbon, with or without credit, this may be the most economical alternative.

Another word of caution, depending on regulatory interpretation, the regeneration facilities may be more stringently regulated than the adsorption unit. If the organics treated are classified as hazardous, the regeneration furnace may be classified and regulated as a hazardous device. This is expensive to operate, and considering the type of incinerator commonly used for regeneration (multiple hearth), emissions and performance standards may be difficult to achieve without installing considerable pollution control equipment. In addition, fugitive emissions from regeneration facilities may add high quantities to the total facility emissions, which would have to be included in regulatory reporting.

On-site regeneration is generally an integral part of large scale adsorption systems, greatly influencing treatment plant costs. Regeneration allows the reuse of spent carbon at a fraction of the virgin carbon cost, limiting makeup carbon to system losses of approximately 10% of the gross amount, or less. However, carbon regeneration is a complex oxidation process involving destruction of the captured organics; as well as a multistep thermal oxidation process that includes drying, baking, and activating the carbon. The thermal oxidation is commonly conducted in a multiple hearth furnace, in a controlled oxygen and steam atmosphere.

Detailed regeneration design is not the responsibility of the Process Engineer, requiring expertise available from the carbon and adsorption system suppliers. However, some general regenerative system information is required to initiate the work of other disciplines; including the required furnace capacity, a rough estimate of its size, and energy requirements. Theoretically, the carbon requirements or the furnace size is dependent on the required carbon usage, which is a direct product of the carbon dosage per volume of waste times the daily waste volume. EPA design procedures recommend that this theoretical capacity be adjusted for the furnace on-line time, or a 40% downtime factor [12]. The

size of a multiple-hearth furnace required for regeneration is estimated at about 0.09 square meters (one square foot) of hearth area per 18 kilograms (40 pounds) of carbon to be regenerated per day [12]. Approximately one pound of steam is required for each pound of dry carbon regenerated. The total regeneration fuel requirements are estimated to be approximately 2361 kcal/kg (4250 BTU per pound) of carbon regenerated, 694 kcal/kg (1250 BTU per pound) of carbon to generate the required steam, and 1667 kcal/kg (3000 BTU per pound) of carbon for the furnace operation. Fuel required for the furnace afterburner must be added to the energy requirements.

Unfortunately, not much attention is given to the regeneration system in the design development stage, with its success assumed a forgone conclusion. Some process considerations include the following:

(1) Frequently, the regeneration operation is more complex and less effective than the adsorption system.
(2) The apparent simplicity of regeneration is deceiving, since regeneration is not simply a matter of burning spent carbon, but of activating carbon at the required usage rate. Any incinerator downtime increases the required regeneration rate or virgin carbon requirements, or reduces the available adsorption operation time.
(3) The properties of the regeneration carbon will have a pronounced effect on adsorption efficiency. Changes in carbon adsorption quality and capacity are not always easy to predetermine, until either a change in service time or effluent quality is apparent.
(4) Effective carbon transport is an important consideration in the entire adsorption process.
(5) The economic impact of regeneration is often underestimated in process engineering evaluations.

The facilities should include provision for a virgin carbon makeup rate equivalent to at least 10% of the total required carbon usage.

Contact Time

Adsorption vessels are designed to meet the required MTZ height, provide a sufficient carbon inventory, and assure adequate contact time to meet effluent requirements. Contact times for municipal tertiary systems can range from 15 to 35 minutes, and explicit requirements for industrial wastes must be determined [14]. Although, if a substrate has significant adsorbability characteristics, MTZ height and contact time should usually be readily achieved by meeting service time and carbon inventory requirements. However, this must be verified for specific wastes since the waste composite adsorbability cannot be assumed to be equivalent to that of a specific component. Contact time is established by determining the system service time, which includes all the dominant operating variables.

Column Loading

Column loading is generally stated in terms of hydraulic loading, expressed as flow rate per unit area; and organic loading, expressed as weight of substrate per unit carbon weight or volume. However, available head loss must be considered when selecting a hydraulic loading, and substrate adsorptive limitations and service time must be considered when selecting an organic loading.

Head Loss and Hydraulic Loading

Hydraulic loading rates range from 1 to 7 $L/s/m^2$ (2 to 10 gpm/sf); with upflow units operating at 3 to 7 (4 to 10), and downflow at 2 to 3.5 $L/s/m^2$ (3 to 5 gpm/sf) [12,14]. Adsorber column head losses are directly related to hydraulic loading and bed height. These loadings may have to be reduced for gravity systems. Selecting a hydraulic loading involves evaluating the resulting pressure drop and service time, and must be consistent with the other critical design considerations. Waste viscosity, bed depth, carbon diameter, carbon particle size, and flow rate affect headloss.

The head loss at the completion of an adsorption cycle, when the carbon characteristics are different and the void space reduced, can only be determined by actual field testing. When the waste contains significant solids or fouling components the available pressure drop is critical because it limits the service time for gravity systems, allows processing up to the available pump head in pressure systems, and requires downtime for bed maintenance. When the available head is depleted the process terminates, whether that involves one hour or days of operation; making a desktop process design a potential disaster! Therefore, it is essential that an adequate pressure drop be provided to assure an effective operating system.

Some design considerations to minimize pressure drop include

(1) Upstream filtration minimizes the adsorber bed pressure drop depletion rate, allowing cycle times consistent with the adsorbent service time and not its filtering capacity.

(2) Downflow beds commonly utilize 8 × 30 mesh carbon because of lower head losses [12].

(3) Upflow beds, with allowance for approximately 10% expansion, can be operated with 12 × 40 mesh carbon [12].

(4) Where headloss problems are anticipated an upflow configuration may be a viable alternative.

(5) Head loss limitations may restrict the use of gravity conditions.

(6) Pressure systems provide the most operating flexibility.

Organic Loading and Service Time

Organic loading is a characteristic specifically related to waste substrate adsorbability and to system service time. The selected organic loading must be based on laboratory study results to assure optimum system service time, as discussed under Service Time in the Basic Concepts section, and illustrated in Case Study 27.

Column Depth

Total tower column heights depend on the system configuration and the minimum depth required. Total fixed bed depth is based on the minimum height required, plus that required for carbon inventory and for proper waste distribution. Contactor total unit depths typically range from 3 to 12 meters (10 to 40 feet), with 4.5 to 6 meters (15 to 20 feet) typical. As with loading, a measure of bed depth effectiveness is achievable operating service time. In a moving bed system only a small working depth beyond that required for the MTZ is required because fresh carbon is constantly being replenished and spent carbon removed.

Mechanical design aspects, minimizing short-circuiting and "dead spots," reduce the required fluid distribution height. This involves not only the inlet and outlet collection system design, but the tower geometry, expressed as the bed aspect ratio. The aspect ratio is defined as the bed depth-to-diameter ratio, generally greater than 4 to 1. The aspect ratio is critical if the feed and discharge are not adequately designed to assure proper internal distribution, and conversely relaxed if adequate distribution provisions are included.

Backwashing Requirements

Backwashing requirements should be thoroughly evaluated as part of the process design, considering backwash volume generated and suitable disposal. Specific requirements depend on influent suspended solids content, upflow or downflow configuration, and carbon size characteristics. Frequent backwashing will significantly limit adsorption as a viable treatment alternative by reducing on-line time and net water production. Some considerations applicable to backwashing design include

(1) Wastewaters with more than 25 ppm of suspended solids should include solids removal as a pretreatment. If downflow units are used for solids removal, consideration must be given to backwash effects, with the system becoming inoperable (uneconomical) with increasing

high suspended solids content. At some point, the use of a prefilter offsets the cost of reduced adsorption efficiency and capacity.

(2) Fixed beds employ backwashing rates ranging from 8 to 14 L/s/m^2 (12 to 20 gpm/sf), for 10 to 15 minutes, operating at approximately 10 to 15% bed expansion [12,14].

(3) Upflow expended beds result in reduced headloss and minimal backwashing if turbidity is limited. Where required, backwash rates range from 7 to 8 L/s/m^2 (10 to 12 gpm/sf) for 10 to 15 minutes [12,14].

(4) Backwash volume should be less than 5% of the filter effluent, or 10 to 20% for deep filters [14].

FATE OF CONTAMINANTS

Treated effluent is the major system product. Secondary washwater is generated from column backwashing, carbon transport to and from the contactors, and transfer to the regeneration system. Major polluted wastewaters could be produced from a regeneration incinerator as a result of air pollution control scrubbers. Adsorbent is the significant solid produced, which is generally regenerated. In small facilities spent adsorbent may not be economically regenerated and must be disposed. Generally, adsorption system wastes are secondary treatment plant effluents, already "stripped," and therefore generating minimal air emissions unless anaerobic conditions develop in the bed. However, emissions from regeneration furnaces must be controlled, and could present major regulatory problems.

GENERAL ENGINEERING CRITERIA

Figure 2.7 illustrates an Adsorption System Preliminary Concept Flowsheet. A complete adsorption system could involve the following process components:

Equalization (optional)	Chapter I-4
Waste feed system	Chapter I-5
Pretreatment	Chapter I-6 to I-9
Carbon storage and transport	
Carbon contactors	
Regeneration	

The Process Engineer has the specific responsibility of establishing the general adsorption system process requirements, as well as the minimum design criteria. Allowing the supplier to set them, and developing a mechanical design to suit, are the *first* major mistakes inviting poor system performance. Attempts by the Process Engineer to independently develop mechanical design criteria for such specialized equipment is the *second* major mistake contributing to process design failure and ineffective operations.

The *activated carbon system* is commonly purchased as a packaged unit, including detail design involving fabrication of the units, piping and transfer system, mechanical transfer systems, chemical storage system, etc. These design elements will incorporate all of the suppliers operating experience, defining equipment performance, required personnel protection, equipment safety, and incorporating necessary process operating safety considerations. This is especially significant for activated carbon system components such as (1) carbon storage and transfer, (2) the carbon contactors, and (3) the regeneration system. The design criteria should incorporate the suppliers recommendations for accessory equipment, consistent with the major units supplied. The information discussed in this section is intended as a guide for the engineer to work effectively with the supplier, and to understand the engineering details proposed.

ADSORPTION COLUMN DESIGN CONSIDERATIONS

Contactor

Contactors can be constructed of carbon steel, stainless steel, concrete, or fiberglass; although steel or concrete are the most common shell materials. The primary design considerations are size, pressure requirements, distribution, column maintenance, and corrosion protection. Basically, there is nothing unique about the contactor shell design, which generally adheres to pressure vessel standards common in industrial practice, or open concrete basin construction common in water and waste treatment. Contactor internal design closely adheres to criteria commonly applied to sand filters.

The shell *size* is affected by fabrication restrictions, construction economics, the number of units, and internal distribution considerations. Shop fabricated vessel sizes are restricted by practical transportation considerations, which generally result in units with 4 meters (12 feet) maximum diameter and 18 meters (60 feet) in length. Larger units can be field constructed. In many cases larger units are required to accommodate gravity flow and are constructed as common-wall, multiple-unit concrete basins. Theoretically, field fabrication allows construction of any size contactor, except for process considerations. Too large a single unit complicates fluid distribution, minimizes process reliability, and limits operator flexibility.

The available hydraulic head establishes whether a gravity or pressure system is employed, and thereby the shell *pressure requirements*. Although gravity systems are the most economical to construct, pressure systems provide a more flexible operation. Pressure units allow increased pumping head and a wider range of controlled hydraulic loading rates.

Internal fluid *distribution* is critical in the shell design; involving either a portion of the bed dedicated to distribution, or mechanical design of an adequate feed and collection system. Where the bed is used for fluid distribution, the available depth is critical because a portion of the bed (the

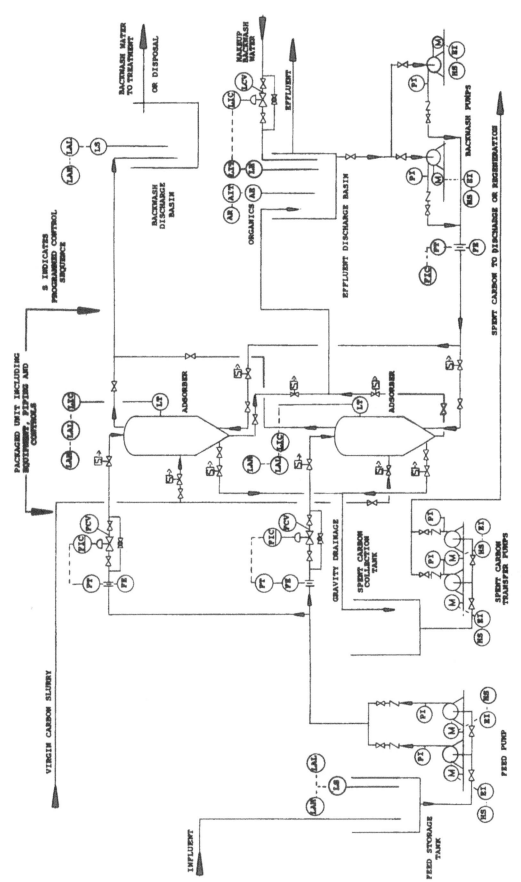

Figure 2.7 Adsorption system preliminary concept flowsheet.

79

distribution section) is ineffective for adsorption, and therefore not available for treatment. Distribution considerations require that a minimum bed depth to diameter of 4 to 1 be considered. However, proprietary contactor designs usually provide multiport entrance and internal piping to assure excellent fluid distribution and collection, with a minimum sacrifice of bed depth. Upflow configurations employ "flooding" conditions to further enhance carbon-waste contact. Cone angle designs further improve distribution and transport from the system.

Other important considerations in contactor design include

(1) Inclusion of a stainless steel well screen for inlet and outlet service protection
(2) Pressure vessels must include provision for pressure and vacuum relief valves, i.e., air pressure relief for the filling cycle, and vacuum relief for the tank draining cycle
(3) Provision for fresh carbon loading and spent carbon removal
(4) Corrosion and abrasive control

Bed maintenance involves providing backwash capabilities to remove accumulated solids, which may require both air scouring and water wash. This is especially a concern for high solids content wastes, and when the contactor serves as both a filter and adsorber. Another major maintenance concern for industrial waste systems is controlling bacterial activity within the bed, resulting from bacterial rich influent with organic content. An associated problem is hydrogen sulfide generation which is corrosive, can alter the process pH, and can create an odor problem. The practices for bacteria control are similar to sand filter operation, which include

(1) Chemical cleaning
(2) Frequent backwashing
(3) Maintaining aerobic conditions to reduce the potential for hydrogen sulfide generation
(4) Adequate secondary treatment to reduce organic concentrations and solids filtration as an adsorption pretreatment

Carbon Inventory Control

Carbon management design includes maintaining an adequate on-site *inventory*, provisions to *transport* the carbon where needed, and on-site *regeneration* (where applicable).

Inventory

Provision for ample inventory is an important consideration regardless of the process configuration, requiring an adequate supply of regenerated and virgin material to avoid process interruption. For fixed bed systems the inventory

must be adequate to replenish at least one contactor, while for a continuous system the carbon must be available at the usage rate (or greater) at all times. Inventory design includes bulk storage of regenerated carbon and virgin carbon storage. For a fixed bed system a stand-by contactor can be used for storing inventory.

Carbon Transport

Carbon transport includes pneumatic facilities to unload from trucks or tankers to bulk storage, and hydraulic transport of a slurry to and from a contactor. Bulk unloading requires coordination of the on-site transport equipment, and tie-in connections, with that of the carrier's. Bulk storage facility design must be in strict compliance with the suppliers standard criteria, and the unloading equipment must adhere to suppliers specifications.

Some commonly employed loading design guidelines include [12]

(1) Hydraulic transport is achieved by injecting, and mixing carbon with water, pumping a carbon slurry.
(2) The water piping systems are those commonly employed in water and wastewater treatment facilities; while the slurry piping design is more complex, similar to that encountered in lime slurry piping systems.
(3) A maximum slurry concentration of 0.36 kg/L (3 pounds of carbon per gallon) of water is recommended; with a 0.12 kg/L (one pound per gallon) of water commonly employed.
(4) The slurry system should not use less than 5 cm (2 inch) pipe; and include provisions for quick disconnect, flushing, and line cleaning in case flow is interrupted.
(5) Transport velocities should be adequate to prevent settling; but not excessive, to prevent abrasion and minimize energy transport costs. Velocities in the 1 to 1.5 meter per second (3.5 to 5 fps) range are recommended.
(6) Slurry flow control is generally achieved by controlling the injected water flow, direct control of the slurry flow should be avoided. Where valves are utilized in the slurry line they should be for on-off control, using rotary valves (ball, plug, or pinch) for proper seating and positive shut-off.
(7) Provision should be made to prevent slurry backup into the water systems, which is best accomplished with a break tank, although where local plumbing codes permit, double check valves may be used.
(8) Where check valves are required in a slurry line, they must be carefully selected and routinely inspected, since they tend to erode, develop seating problems, and malfunction.
(9) Piping design should involve layouts that avoid the potential for solids "hang-up," commonly occurring

with abrupt changes in flow direction, sudden restrictions, or obstructions in the piping system.

(10) Piping can be constructed of carbon steel, reinforced, or of extra strength at areas subjected to abrasive wear. More expensive materials are seldom justified unless corrosive conditions from residual waste are anticipated.

(11) The piping system should be designed so that when transport is complete the piping slope and adequate flushing allow the piping to be empty and relatively clean. Piping containing nonflowing slurry "pockets" will deposit solids, and, under some conditions, plug the line. Piping must be designed for quick disconnect for when this occurs, since attempts to "pressure" clean a plugged line may compact the solids and further aggravate the problem.

(12) Pumps should be selected with adequate clearances to prevent solids "straining," and with an abrasion protective lining. Pump speeds should be selected to avoid carbon destruction. At low slurry concentrations, open or closed impeller pumps with adequate solids clearance and speeds no greater than 800 to 900 rpm can be applied. Positive displacement pumps with no obstruction are suitable but may be more expensive. Eductors or pressure transport (blow tanks) can also be considered for this service. At the higher range slurry concentrations, diaphragm or similar positive displacement pumps with no flow obstruction internals are recommended.

Regeneration

Regeneration is commonly performed in multiple hearths, and in some cases infrared furnaces [12]. Applicable process considerations have been discussed in the Process Design Engineering section, while general incineration design criteria are discussed in Chapter III-1. However, carbon regeneration is not solely a thermal oxidation process, but also a chemical process to "recondition" carbon to suitable adsorption quality. The regeneration process is estimated to take 30 minutes: 15 minutes of drying, 5 minutes to initially remove the adsorbed material, and a final 10-minute period for complete oxidation of the remaining adsorbed material and "conditioning" the carbon to an activated state [12].

CORROSION AND EROSION PROTECTION

Corrosion is not a concern for carbon slurry, but a major concern for wet or partially dewatered carbon, which is extremely corrosive. Since most of the storage, contactor, and transport system will be subjected to wet carbon, all must be designed for corrosion control. Some guidelines include [12]

(1) Storage vessels, contactors, bins, and containers are commonly constructed of carbon steel, with a protective coating such as coal-tar epoxy applied. The more sophisticated the protective coating the more advantageous to have the vessel shop fabricated and coated. Care must be taken not to puncture the coating, during construction or operation, since this will excel the corrosion in the exposed area. As an alternate, 304 or 316 stainless steel or fiberglass vessels can be considered. Cost may exclude stainless steel, but fiberglass is a suitable alternative for some services.

(2) The mechanical conveyor system can be constructed of stainless steel.

(3) As discussed above, black steel pipe is suitable for slurry piping.

Abrasion is generally a problem only in transport lines, but experience indicates that a minimum service life of 10 years can be expected with carbon steel, making more expensive materials of construction impractical [12]. However, reinforced or extra strength sections are recommended for areas subject to high abrasion conditions.

SAFETY

Although activated carbon is considered a safe compound, prudent safety considerations should be incorporated into the design [12]. Some of these include

(1) Eliminate potential explosive conditions resulting from a direct spark or static conditions in contact with carbon *dust* emitted or accumulated in the system.

(2) Provide for personnel protection from carbon dust.

(3) Ground all treatment system components.

(4) Eliminate contact of carbon with any waste components that could cause dangerous reactive conditions, either as a result of concentration or chemical nature.

More extensive safety considerations should be obtained from the carbon supplier. In addition, the facility should be constructed in adherence with safety and health requirements as expressed by the corporate Safety and Health Officer, facility insurance underwriters, and any applicable local codes.

PROCESS CONTROL

System monitoring is essential for proper process control. Suggested process instrumentation includes

(1) Flow control of influent, effluent, and the carbon slurry feed mixture. Critical operating flows should be recorded.

(2) Quality control parameters such as influent and effluent organic concentration are usually not automatically monitored (in-line) and recorded. However, provision

should be made for automatic influent and effluent composite sampling to analyze critical characteristics.

(3) Head loss through each column should be continuously monitored and recorded.

(4) Levels in critical process tanks should be continuously monitored.

(5) Critical process variables should be monitored and alarmed to alert the operator of malfunction. This includes
  a. High influent and effluent pH concentration
  b. Low carbon supply
  c. High contactor pressure drop
  d. Influent, effluent, and carbon feed pump failure

COMMON ACTIVATED CARBON DESIGN DEFICIENCIES

Some design deficiencies which can occur in adsorption system design include

(1) As a result of poor carbon adsorption, required effluent quality is not achieved because waste components are not significantly adsorbable.

(2) Fixed carbon bed service time is too short, carbon inventory is rapidly depleted. This can occur because the organic loading is too high, or the system is under designed.

(3) The system head loss at the operating loadings is too high for the available gravity head or the pumping capacity.

(4) Carbon adsorbability quality is poor because of poor regeneration.

(5) The system pH is not in the proper range, resulting in poor adsorbability.

(6) Poor adsorbability results because of the waste operating temperature.

(7) The backwash volume is too high and ineffective because of high influent colloidal solids content.

(8) No method has been included to establish that bed "break-through" has occurred.

(9) Carbon transfer piping is constantly plugging, and no means is provided to disconnect and flush the lines.

(10) Pumps are not correctly selected for carbon slurries.

(11) Valves are not correctly selected for carbon slurries.

CASE STUDY NUMBER 27

Some limited data has been obtained to evaluate whether carbon adsorption is a viable alternative to treat 1 MGD of secondary effluent containing 50 mg/L organics to a level of 5 mg/L. Study results for loadings or 2, 4, and 10 gpm/ft² are tabulated below. A preliminary evaluation should be made on the basis of the following data:

(1) Carbon density: 23 pounds per cubic foot
(2) Bed expansion: 50%

(3) Recommended backwash rate: 20 gpm/ft²
(4) Recommended backwash time: 10 minutes
(5) Operating at 24 hours per day, 7 days per week
(6) Carbon losses assumed at 5%
(7) Regeneration employed at carbon usage greater than 1500 pounds per day, economically related between 200 to 1500 pounds per day

Is the adsorption a viable treatment option? Is the data adequate?

PROCESS CALCULATIONS

(1) Select applicable BDST equation

| | Service Time, Hours | | Rod Area |
	5 Foot Bed	10 Foot Bed	Ft²
2 gpm/ft²	1530	—	347
4	480	1130	174
8 gpm/ft²	150	430	87

Reviewing the data for a 5 and 10 foot bed, and comparing the resulting service times, a required bed area 4 gpm/ft² design basis will be assumed. An economic evaluation of service time versus capital costs would generate finite data to base the decision. In this case, a downflow column is being considered which is generally operated at 2 to 5 gpm/ft²; higher linear velocities could physically upset the bed, and enhance solids penetration.

The data for 4 gpm/ft² loading can be represented by two straight line curves, using Equation (2.2e). (When adequate data is obtained this is usually done graphically.)

$$480 = 5 \cdot a + b$$

$$1130 = 10 \cdot a + b$$

So that $a = 130$, $b = -170$.

$$T = 130 \cdot D - 170$$

(2) Determine minimum applicable depth
At $T = 0$, d = 170/129.8 = 1.3 feet.

(3) Select the operating depth
Evaluate a 5 and 10 feet carbon bed.

(4) Determine service time at breakthrough

Working Equations

$$T = [N_o/(C_o \cdot V)] \cdot D - [\{1/(K \cdot C_o)\} \cdot \{\ln (C_o/C) - 1\}]$$

$$T = 130 \cdot D - 170$$

10 foot bed: $T = 130 \cdot 10 \text{ ft} - 170 = 1130 \text{ hr}$

= 47 day inventory

5 foot bed: $T = 130 \cdot 5 \text{ ft} - 170 = 480 \text{ hr}$

= 20 day inventory

Theoretically, a five foot bed would be adequate. However, since the carbon bed serves to polish effluent from an activated sludge system, the column depth will be expanded to 10 feet, a portion of which may be subjected to heavy solids accumulation. An alternative is a reduced bed depth with an upstream filtration system.

(5) Establish required area

Area = gd waste/1440/hydraulic loading rate

Area = 1,000,000 gpd waste/1440/4 gpm/ft^2 = 174 sq ft

Assuming two units in parallel, each unit requires 87 sq ft, or a diameter of 10.5 ft. The column criteria can be meet with the following configuration:

Two columns, diameter each = 11 ft

Carbon bed depth = 10 ft

Expansion @ 50%

Total tower depth = 15 ft

Total flow area = 190 sq ft

(6) Carbon evaluation

(7) System carbon inventory

Lb carbon = total sq ft · depth · density carbon

Lb carbon = 190 sq ft · 10 ft · 23 lb/cu ft = 43,700 lb

or 21,850 pounds per contactor

(8) Carbon usage

Carbon usage rate = bed volume (lb)/service time (hr)

Carbon usage rate = 43,700 (lb)/1130 (hr) = 39 lb/hr

Assume 5% losses of 0.05 · 39 = 2 lb/hr

Total carbon required = 39 + 2 = 41 lb/hr

(9) Regeneration requirements

41 lb/hr · 24 hr per day = 984 lb/day = 0.5 tons per day

The cost of on-site regeneration for furnaces less than 1500 pounds per day may not be feasible, and can only be evaluated on the basis of specific process capital and operating cost.

Based on off-site regeneration, spent and fresh carbon storage facilities should be included. Service time for each unit is somewhere between 20 and 47 days, depending on how much of the bed volume is used for adsorption or filtration. Provision will be made for 30 days storage, based on a carbon use of 41 pounds per hour, approximately the equivalent of one contactor off-line at a time.

30 days storage · 41 lb/hr · 24 hr/23 lb/ft^3 = 1283 cu ft

This would require a vessel 11 feet in diameter by 13.5 feet high, which is about the size of the contactor. As an alternative, storage may be provided by including an additional unit which can serve the dual purpose of storage and an on-line spare.

(10) System performance

(11) Organic adsorbed

Feed: 1 MGD · 8.34 · 50 ppm = 417 lb/d = 17.4 lb/hr
Removed: 1 MGD · 8.34 · (50 − 5) ppm = 375.3 lb/d = 15.6 lb/hr

(12) Adsorption efficiency

15.6 lb/hr organics/41 lb/hr carbon = 0.4.

(13) Backwash generated
As an approximation, assume a daily backwash as follows:

Backwash rate 20 gpm/sq ft

Backwash = 20 gpm/ft^2 · 190 sq ft = 3800 gpm

3800 gpm · 10 minutes/day = 38,000 gallons per day

38,000/1,000,000 · 100 = 3 to 4% of the feed

Provision should be made for feeding this back to the treatment facilities.

(14) Pressure drop
The initial bed pressure drop is 2.25 inches per foot of column.

2.25 inches water · 10 feet = 22.5 inches of water
$$= 0.81 \text{ psi}$$

(15) Effects of influent variations on service time
The effects of changes in service time resulting from changes in required removal (effluent concentration), influent concentration, or flow (loading rate) are estimated by changing that variable in the *a* or *b* constants, and calculating the resulting service time.

$$T = a \cdot \text{depth} + b$$

$$T = 130 \cdot D - 170$$

First, the defining constants are established.

$$a = 130 = 1990 \, N_o/V \cdot C_o$$

since $V = 4$ gpm/ft^2 and C_o is 50 mg/L; $N_o = 13.07$.

$$b = -170 = -16,018 \cdot \ln [(C_o/C) - 1]/C_o \cdot K$$

since $C_o = 50$ and C is 5 mg/L; $K = 4.14$.

Substituting these constants and a depth of 10 feet,

$$T = \frac{260,000}{V \cdot C_o} - \frac{3,869}{C_o} \cdot \ln \left[\frac{C_o}{C_e} - 1 \right]$$

(16) Changing recovery requirements

Maintaining C_o at 50 mg/L and V at 4 gpm/ft^2, the effect of varying removal efficiency to service time is as follows:

Removal Efficiency	Effluent, mg/L	Service Time, hr
99%	0.5	945
90%	5	1130
80%	10	1193
60%	20	1269
40%	30	1332

(17) Feed concentration variations

Maintaining the effluent concentration at 5 mg/L and V at 4 gpm/ft^2, the effect of varying influent concentration to service time is as follows:

Influent, mg/L	Service Time, hr
100	536
75	731
50	1130
25	2386
10	6502

(18) Feed loading (flow) variations

Maintaining the influent at 50 mg/L and the effluent at 5, the effect of varying hydraulic loading to service time is as follows:

Flow, gpm/sf	
10	350
8	480
6	697
4	1130
2	2431

DISCUSSION

Adsorption appears a viable consideration as a tertiary treatment option. However, two concerns must be further investigated, the economics of on-site or off-site carbon regeneration, and the quality and quantity of the data used for the preliminary investigation.

First, the quantity of carbon required per day (0.5 tons) is probably not enough to warrant on-site regeneration, but a thorough investigation of the available options should be made. Once the regeneration alternative has been selected, the facility capital and operating costs should be determined.

Next, the data is suspect! The projected service time appears too optimistic. Extended testing should be conducted to verify (or alter) the assumed design criteria, using continuous, fresh, on-site wastes, investigating the waste quality variations, and establishing whether the waste batch tested is representative of the facility wastes. If the final design is based on "best data" results the equipment will be undersized, the carbon requirements underestimated, and the facility will not meet intended process (or regulatory) requirements.

REFERENCES

1. Bhattacharyya, D., Cheng, C.Y.R.: "Activated Carbon Adsorption of Heavy Metal Chelates from Single and Multicomponent Systems," *Environmental Progress*, V 6, No 2, Pg 110, May 1987.

2. Edwards, M., Benjamin, M.M.: "Adsorptive Filtration Using Coated Sand: A New Approach for Treatment of Metal-Bearing Wastes," *Journal WPCF*, V 61, No 9, Pg 1523, September, 1989.

3. Faust, S.D., and Aly, O.M.: *Adsorption Processes for Water Treatment*, Butterworth, Newton, MA, 1987.

4. Fleming, H.L.: "Application of Aluminas in Water Treatment," *Environmental Progress*, V 5, No 3, Pg 159, August, 1986.

5. Huang, C.P., Vane, L.M.: "Enhancing AS^{5+} by a Fe^{2+}-Treated Activated Carbon," *Journal WPCF*, V 61, No 9, Pg 1596, September, 1989.

6. Hutchins, R.A.: "New Method Simplifies Design of Activated Carbon Systems," *Chemical Engineering*, Pg 133, August 20, 1973.

7. James M. Montgomery Consulting Engineers: *Water Treatment Principles and Design*, John Wiley & Sons, 1985.

8. Ku, Y., Peters, R.W.: "Innovative Uses for Adsorption of Heavy Metal from Plating Wastewaters: I. Activated Carbon Polishing Treatment," *Environmental Progress*, V 6, No 2, Pg 119, May 1987.

9. Masschelein, W.J.: *Unit Processes in Drinking Water Treatment*, Marcel Dekker, Inc, New York, 1992.

10. Reimers, R.S., Englande, A.J., Miles, H.B.: "A Quick Method for Evaluating the Suitability of Activated Carbon Adsorption for Wastewaters," Paper presented at *31 Annual Industrial Waste Conference*, Purdue University, May, 1976.

11. Rubin, A.J.: *Chemistry of Wastewater Technology*, Ann Arbor Science Publishers, Chap 8, Pg 123, 1978.

12. U.S. Environmental Protection Agency: *Process Design Manual for Carbon Adsorption*. U.S. EPA, 625/1-71-002a, October, 1973.

13. U.S. Environmental Protection Agency: *Treatability Manual*, Four Volumes, EPA-600/8-80-042a, 1980.

14. WEF Manual of Practice: *Design of Municipal Wastewater Treatment Plants*, Water Environment Federation, 1992.

15. Giusti, D.M., Conway, R.A., Lawson, C.T.: "Activated Carbon Adsorption of Petrochemicals," *Journal WPCF*, V 46, No 5, Pg 947, May, 1974.

16. Lukchis, G.M.: "Part I: Design by Mass-Transfer-Zone-Concept," *Chemical Engineering*, Pg 111, June 11, 1973.

Ion Exchange

Ion exchange is employed to remove dissolved waste inorganics or similar dissolved waste components.

Ion exchange can be used to remove specific dissolved ions from plant wastes and reclaim usable chemicals or in some special cases to polish treated plant effluents to meet stringent effluent limits. Typically, wastewater flows through a fixed, vertical ion exchange column, where specific anions or cations are removed on highly active resin surfaces. After some time, the resin bed starts losing its exchange capacity and the effluent contaminant concentration increases until it is identical to the influent, indicating spent resin. Prior to this point, the exhausted unit is placed out of service and regenerated, and a second unit is put in operation.

Principles governing ion exchange are

(1) The reactions involve inorganic (sometimes organic) waste and ion exchange constituents that ionize.

(2) The exchanger must have a specific selectivity for the ions to be removed from the waste.

(3) For the treatment to be applicable the reaction must be reversible, so that the exchanger can be regenerated to its original form.

Ion exchange technology is primarily based on water treatment experience employed in specialized processes such as water softening, boiler feed treatment, and preparation of mineral free water for manufacturing. In some cases, direct application of water treatment experience to waste treatment is difficult because of the wide range of waste water characteristics usually not encountered in water treatment.

Ion exchange technology has limited waste treatment application because of several factors. First, it cannot always be applied to concentrated industrial wastes containing many components which could either overload, foul, or destroy the resin. Next, ion exchange technology does not destroy the contaminants, merely concentrates them. This is not a problem where the desired product is high quality water, and the rejected volume or quantity of contaminants of secondary consequence. However, in waste treatment the desired overall treatment effectiveness requires that the contaminant be removed to an environmental inert condition, and not merely transferred to a secondary stream. The volume of concentrated waste generated must be a small percent of the initial waste volume, and that volume disposable or recoverable, to make ion exchange a viable treatment scheme.

BASIC CONCEPTS

The principal characteristic of ion exchange is that the resin ion *selectively* remove specific inorganics. As applied in water treatment its primary application is removing cations such as calcium, magnesium, manganese, and iron and anions such as sulfate, nitrate, and chloride. Softening involves replacing calcium and magnesium (the significant hardness parameters) with resin sodium. Because of the trend toward sodium restricted diets, this practice is not encouraged when the final product is drinking water. In water softening the total ion content is not reduced, whereas in deionization the total solids content is decreased.

Ion exchange is discussed as an applicable process to recover inorganics from isolated streams, a pretreatment to reduce the total load and fouling constituents to ultrafiltration membranes, or a tertiary treatment to produce water quality effluent for plant reuse. The crucial requirement in employing ion exchange is that the brine can be recovered or disposed. The Process Engineer involved in an ion exchange project should first contact resin manufacturers to establish available resins with an affinity for the specific service and then contact equipment manufactures for specific resin testing and specialized equipment design. The ion exchange basics discussed are aimed at providing an understanding of the process and operating variables and as a guide in evaluating proprietary systems.

Ion exchange entails a chemical reaction in which stoichiometric quantities of a solution electrolyte contact and react with an equivalent insoluble resin electrolyte. The physical mechanisms governing the contact are similar to activated carbon transfer characteristics. Some obvious differences between ion exchange and adsorption include

(1) Adsorption involves a physical surface reaction, whereas ion exchange involves a specific reversible chemical reaction between the substrate component and the media.
(2) Different media characteristics govern physical adsorption and exchange reactions.
(3) Spent activated carbon must be reactivated, while spent exchanger resins are chemically regenerated.

An argument can be made that adsorption resins are also selective, that adsorbent regeneration is similar to chemical regeneration of ion exchange, and that in both cases the regenerated adsorbent or resin is reused. The one significant difference in the two operations is that carbon regeneration is primarily an economic consideration, and when applied destroys the collected organic substrate in the regeneration furnace. As commonly used, ion exchange produces high quality water, the concentrated inorganic substrate byproduct being discharged as a waste. In waste treatment applications, regeneration brine may be either recovered, discarded, or further treated and discarded; but always a significant process concern!

The advantages of ion exchange are that (1) the process is highly selective and can be tailored to specific substrate removal, (2) competitive costs can be obtained because of the great number of suppliers, (3) considerable operating design and operating experience is available, (4) completely packaged manual or highly automated units are available, (5) the process has been successfully applied in a wide range of water treatment processes, and (6) because of the available full scale operating experience, simple waste specific laboratory testing can be scaled-up for plant design. Its disadvantage for industrial waste application include (1) a high potential for fouling because of organic substrate components, (2) high disposal costs for regenerated streams, (3) potentially low resin life in industrial applications, and (4) potentially high regeneration costs.

ION EXCHANGE REACTIONS

The displacement of wastewater anions or cations with insoluble exchange components can be expressed by the general equilibria equations shown in Figure 3.1.

Chemical equilibria principles can be employed to discuss exchange reaction processes, as well as evaluating specific compound selectivity. As discussed in Chapter I-5, "Chemi-

$$Substrate\text{-}C_1^+ + Resin\text{-}C_2^+ = Substrate\text{-}C_2^+ + Resin\text{-}C_1^+$$

$$Substrate\text{-}A_1^- + Resin\text{+}A_2^- = Substrate\text{-}A_2^- + Resin\text{+}A_1^-$$

C_1^+, C_2^+ are the cations of two different species

A_1^-, A_2^- are the anions of two different species

R^-, R^+ are the cationic and anionic exchange materials.

Figure 3.1 Equilibria equations.

cal Processes," reaction equilibria can be defined by a mass law constant, as indicated in Equation (3.1).

$$K = \frac{(C_2^+)(R^-C_1^+)}{(C_1^+)(R^-C_2^+)} \qquad (3.1)$$

When the reaction shifts towards the right, the numerator concentrations are greater than the denominator, the exchange selectivity coefficient (K) is greater than one, and ion exchange is an effective system. As previously discussed, ion exchange involves a selective exchange of resin components' with those in solution. Where the solution contains many candidate species the exchange resin selectively displaces the more active ions (relative to the resin), as measured by the selectivity coefficient, and affected by concentration.

The selectivity constant is *not* really a constant, being applicable within the experimental conditions obtained and a narrow concentration range, specific for the waste and resin evaluated. For this reason the value is commonly referred to as the "apparent equilibrium constant." Selectivity of specific ions, and general characteristics influencing selectivity, are discussed below.

SPECIFIC ION EXCHANGERS

Resin: Function Group

The heart of an ion exchange system is the exchanger, consisting of an organic base resin and a functional group containing exchangeable ions. They are commonly manufactured of complex copolymers as uniform spheres, with a functional group containing the exchange ion participating in the reaction. Resins are insoluble solid acids or bases which "exchange" ions with the waste, converting them into an insoluble acid, base, or salt in the exchanger.

A system can be classified as a cation, anion, or special exchanger. A cation or basic exchange system involves the displacement of one positive ion or cation, by another positive ion, which in water treatment commonly involves magnesium, calcium, sodium, hydrogen, iron, and manganese. An anion or acid exchange system involves displacement of one negative ion or anion, by another negative ion, which

TABLE 3.1. Ion-Exchange Resin Classification (adapted from References [2,8]).

Type	Active Group
	Cation Exchange Resins
Strong acid	Sulfonic acid $[SO_3] \cdot H$
Weak acid	Carboxylic acid $[COO] \cdot H$
	Anion Exchange Resins
Strong base	Quaternary ammonium $Z-N-OH$
Where Z represents three methyl groups for Type I, and two methyls and one ethyl group for Type II	
Weak base	$Z-OH$
Where Z can be a primary, secondary, tertiary or quaternary amine	

in water treatment commonly involves sulfate, chloride, nitrate, carbonate, hydroxide, and fluoride. Special exchangers are constantly being developed, involving the select removal of specific ionic organic or inorganic species. Some of these may be applicable to priority regulated pollutants. Macroreticular resins are typical of special exchangers developed for removal of nonpolar organics such as carbon tetrachloride, chloroform, and toluene. A broad classification of ion exchange resins employed in water treatment is listed in Table 3.1 by type and active exchange group [2,8]. These exchangers can be used separately for the removal of specific cations or anions, or in combination for total demineralization. The characteristics of commonly employed exchangers are summarized in the appendix. Specific information on these and specialized resins are available from their manufacturers.

ION EXCHANGE PROCESSES

As with adsorption, ion exchange systems can be configured as fixed- or continuous-bed operations. *Fixed-bed systems* have been traditionally used in water and waste water application, representing the body of installed units and operating experience.

CONTINUOUS BED SYSTEMS

Continuous ion exchangers are not commonly employed in wastewater treatment. Their configuration is based on the fluidized moving bed or pulse bed principles discussed for carbon adsorption.

FIXED-BED SYSTEMS

Fixed-bed systems apply the media process chemistry discussed in Basic Concepts. Fixed-bed regeneration can be cocurrent, in the same flow direction as the treatment flow, or countercurrent. Cocurrent operation is simplest to design

and operate, and is the most common system employed, provided the regeneration step is effective and leakage is not a major consideration. In turn, countercurrent regeneration is employed to (1) minimize leakage during the service cycle, since entrance area is effectively regenerated, (2) reduce regeneration chemical consumption, (3) reduce regeneration time, and (4) decrease regeneration wastes. A major consideration in designing and operating a countercurrent system is preventing resin movement and losses during the regeneration cycle. How the media contactors are combined is the essence of ion exchange design.

Single-Column Systems

Single-column resin systems are commonly used in simple treatment schemes such as water softening, employing a cation exchange resin.

Multiple-Column Systems

Multiple-column systems are used for deionization, requiring a combination of cation and ion exchange resins employed in separate columns, optimizing the use of strong and weak anion and cation resins to maximize treatment efficiency and minimize operating costs. These systems are designed to specific waste properties as outlined in the Configuration section, consistent with the resin characteristics described in the Appendix.

Mixed Beds

This is a unique utilization of a single column in which the bed is a combined strong-acid-cation and strong-base-anion resin mixture throughout the bed depth. The effects of a multiple column system are achieved since waste is constantly being exposed to both cation and anion resins. A high effluent purity is achieved, which cannot be obtained from a single contactor. The only ions present in the effluent are from influent leakage and incomplete exchange reaction. The resin properties are designed to permit resin to separate and form two layers during backwash, as a result of the lighter anion resin density. Regeneration is conducted with caustic contacting the upper anion layer, and acid the lower layer. After regeneration the bed is water rinsed, and air mixed prior to being put back in service. In industrial wastewater application they would most likely be employed as a final polishing step. Mixed beds are prone to poor performance resulting from fouling, resin deterioration after a number of cycles, and ineffective regeneration or classification [1].

Layered Beds

These systems utilize the same principles as a mixed bed, except that a column contains either two anion or two cation resins; the resins are "layered" with the weaker ionized resin at the top and the stronger at the bottom. Regeneration is in the upflow direction. These systems are not commonly employed in industrial waste practice.

PROCESS ENGINEERING DESIGN

Ion exchange is a comparatively simple treatment system with most operating variables related to flow rate, as indicated in Table 3.2. As a result, a stable system is relatively easy to maintain, requiring limited operator attention if the columns are effectively automated. However, achieving high removals at a reasonable cost depends on the waste characteristics and service time, over which the operator has little control. Effective performance depends on resin selectivity, with poorly exchanged influent components reducing overall removal efficiency. Total influent dissolved solids are limited to about 1000 mg/L, above which selectivity is difficult to predict, effluent quality significantly deteriorates, and regeneration requirements affect the operating costs.

ACHIEVABLE LIMITS

Where ion exchange is applicable water quality standards can be achieved.

REQUIRED PROCESS DESIGN DATA

Ion exchange systems can be evaluated using bench scale columns and experimental procedures common in water treatment evaluation [3]. Column tests are similar to those conducted for adsorption evaluation; and includes obtaining enough data to develop a breakthrough curve, tracking volume treated to the effluent quality, and establishing a service time relation to the breakthrough point. Process information obtained from pilot testing include

(1) Service time
(2) Column hydraulic loading and effects on service time
(3) Regeneration conditions and expected resin recovery
(4) Alternative exchange media

Scale-up from column tests is common. However, it is essential that upset conditions be evaluated to determine whether the data developed is effective through the wide range of influent conditions encountered. Adequate data must be collected to establish the required design data listed in Table 3.3.

PROCESS DESIGN VARIABLES

Ion exchange may be considered a special case of adsorption, with much of the basics discussed in Chapter III-2 applicable. Its major differences include resin characteristics and selection, and regeneration method. Specific ion exchange process concerns include

(1) System configuration
(2) Resin selectivity
(3) Exchanger duty
(4) Contactor depth
(5) Contactor diameter
(6) Regeneration
(7) Fate of contaminants

System Configuration

Ion exchange systems are tailored to specific waste characteristics using the multitude of resins available and being

TABLE 3.2. Operating Characteristics.

Variable	Waste Characteristics Operator Controllable	Critical
Waste generated	No	Yes
Composition	No	Yes
Concentration	No	Yes
Selectivity	No	Yes
Operating Characteristics		
Flow rate	Minimal	Yes
Hydraulic loading	Limited	Yes
Substrate loading	Limited	No
Contact time	No	Yes
Service time	No	Yes
System pH	Yes	Yes
System temperature	No	No
Distribution	Yes	Yes
Regeneration	Yes	Yes
Backwash	Yes	Yes
Disposal		
Backwash	No	Yes
Brine	No	Yes

TABLE 3.3. Required Design Data.

Complete waste characterization is essential, and must include flow, flow variation, and a complete anion-cation analysis. Analysis must include total dissolved solids, alkalinity, dissolved carbon dioxide, and pH.

Critical pilot plant treatability data specific to the waste
1. Design temperature
2. Applicable resins
3. Resin capacity
4. Service time
5. Hydraulic loading, vols/day/area contact
6. Achievable effluent quality
7. Minimum bed depth

Selected operating characteristics
8. Total bed depth
9. Regeneration requirements
10. Backwash requirements
11. Pretreatment requirements

developed, in process schemes selected to optimize treatment effectiveness and regeneration costs. Special resins are constantly being developed for specific inorganic or organic component removal, their use dependent on the specific resin characteristics, and precise application can be obtained from the suppliers developing the technology. The resins discussed in this section, whose chemical characteristics are outlined in the appendix, are those commonly employed for water treatment. They are applied as sequential unit processes selected to optimize overall process efficiency and minimize regenerating costs, based on the following guidelines:

(1) *Strong acid resins* are highly functional over a wide pH range, with low leakage and high exchange rates; but are less economical and efficient to regenerate than weak acid resins.

(2) *Weak acid resins* are limited to cation removal over a pH range from 7 to 14 and exhibit high leakage. They can be economically and efficiently regenerated, using a weak or strong acid; including that regenerated from second stage strong acid resin.

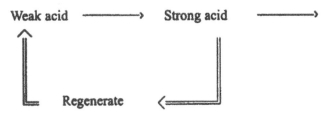

(3) *Strong base resins* are highly functional over a wide pH range and high exchange rates, but are less economical and efficient to regenerate than weak base resins.

(4) *Weak base resins* are limited to anion removal below a pH of 6, but can be economically regenerated at a high efficiency utilizing a weak or strong base; including that regenerated from the second stage strong base resin.

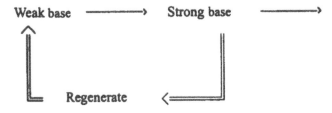

Economical ion exchange application requires optimum use of bed capacity and effective regeneration. This involves minimum chemical reagent use and combining resin beds to optimize performance and achieve the required effluent quality. Exchange application can be illustrated with alternatives available for a total deionization system.

* Level 1: Basic deionization can be accomplished with a strong-acid cation and a weak-base anion exchange system to remove all ions except silica.

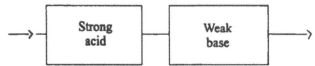

* Level 2: By replacing the weak base with a strong base, a comparable water quality can be achieved and silica can be removed.

** Level 3: A common economical deionization practice is to utilize a three-bed system: removing the cations with a strong acid exchanger, remove most of the resulting acid with a weak base, and "polish" the system with a strong base exchanger that is effective for silica removal. By regenerating the anion exchanger series, the caustic discharged from the strong-base (anion) exchanger can be used to regenerate the weak anion exchanger, reducing the regenerating costs from that of the level 2 system.

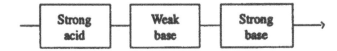

** Level 4: The same (or higher) quality water can be obtained with a four-bed system, alternating a weak ion resin with a strong ion resin, the specific configuration tailored to the waste, effluent requirements, and economics. By regenerating common exchangers in a series, a regenerating agent from a strong exchanger can be used for the weak exchanger, reducing the regenerating costs for the entire system. As an example:

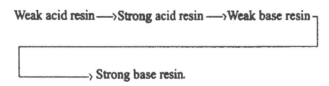

Degasifiers are commonly included after the cation exchanger where the influent contains a high carbonate level, to reduce the resulting carbon dioxide and thereby the bicarbonate loading on the anion exchanger.

Resin Selectivity

As with carbon adsorption, exchanger performance is governed by flow rate, influent concentration, bed configuration, and contact time. A major process design consideration in ion exchange is choosing an appropriate resin based on contaminant selectivity, resin capacity, regeneration efficiency, ion exchange rate, and resin stability. Significant characteristics affecting exchange are solution concentration, ionic charge (electrostatic force), and ionic size (molecular weight). These characteristics are commonly cited to describe some general guidelines governing ion selectivity for solutions containing total dissolved concentrations less than 1000 mg/L. They include [4,6,8]

(1) Exchange potential increases with increasing ion valence.
(2) Exchange potential for ions with the same valence increases with increasing atomic number (atomic weight).
(3) At TDS concentrations greater than 1000 mg/L preference may not be related to valence or atomic number, and in some cases reverse exchange may result.
(4) In some cases, large molecular weight inorganic or organic ion complexes may be excluded as a result of screening or sieving.

Table 3.4 illustrates the relative selectivity of common anions and cations, with the more active species located at the top. Some of the ion selectivity is not in strict adherence with the above cited guidelines [1,4].

These guidelines are not absolute, and exceptions are common. Resin efficiency is an integral part of ion exchange performance, and subject to many operating limitations, some of which include [1,4]

(1) The ion exchange process is not complete, and is especially limited when total dissolved solids concentrations exceed 1000 mg/L TDS.

(2) Ion exchanger selectivity can change with process conditions or waste characteristics.
(3) The process is temperature limited.
(4) The presence of certain solvents (other than water) can cause swelling.
(5) The more select a resin is to an ion, the more difficult it is to remove it.
(6) Waste components must dissociate to be exchanged, since only ions can diffuse in or out of the bed.
(7) High molecular weight ions cannot penetrate the exchanger interior.
(8) Resin beds treating industrial wastes are prone to bacterial growth which will foul the bed, reducing the service cycle. This is further complicated since some resins are chemically degraded by common bacterial growth treatment chemicals such as chlorine products.

Resin Stability

Resin stability can affect process efficiency, effluent quality, resin life, and process economics. Resin deterioration can result from physical stresses caused by excessive mechanical compression and abrasion, and temperature or process conditions enhancing shrinking and swelling. This may be a result of processing, regeneration and washing steps as well as the resin handling, transporting, and loading. Degradation can also result from direct chemical reaction between waste components and the resin material. Fouling or plugging may increase required process pressure, which could increase physical bed stresses. Each resin has specific limitations which must be carefully evaluated and matched to the waste characteristics.

Exchanger Duty

Waste is generally applied at a hydraulic loading of 1.5 to 5 L/s/m² (2 to 8 gpm/ft²), and as high as 11 (16) [5].

Waste concentrations are expressed as equivalent g/L of the cation or anion:

$$\frac{g/L\ of\ cation\ or\ anion}{(Atomic\ weight/valence)}$$

The bed capacity is expressed as gram equivalents per volume. Resin capacity must be defined in units common with the waste concentration. Resins are sometimes described in terms of capacities other than gram equivalents, such as calcium carbonate or kilograin equivalents, which can be calculated as follows:

$$g\text{-}equivalents/L \cdot 50 = g\ CaCO_3/Liter$$

$$g\text{-}equivalents/L \cdot 21.85 = Kilograins\ CaCO_3/cubic\ foot$$

TABLE 3.4. Ion Selectivity (adapted from Reference [1,4,9]).

Strong Acid	Weak Acid	Weak Base	Strong Base
Li+	H+	OH−	I−
H+	CU++	SO₄−−	Br−
Na+	CA++	CrO₄−−	HSO₄−
NH4+	Mg++	NO₃−	NO₂−
K+	K+	PO₄−−−	Cl−
Rb+	Na+	HCO₃−	HCO₃−
Cs+		I−	F−
Mg++		Br−	OH−
Zn++		Cl−	SO₄−−
Co++		F−	CO₃−
Cu++			HPO₄−
Cd++			
Ni++			
Ca++			
Sr++			
Pb++			
Ba++			

Note: The selectivity of an ion depends on the resin properties. The above listing was combined from the sources indicated, presented solely as a guide, since the position of the ions may change with application.

The exchanger *performance* is defined by the water processing rate and the quantity of contaminants that must be removed. Fixed bed processing rate must account for the regeneration down time, so that the processing rate is established as follows:

$$\text{Water processing rate} = \frac{\text{Production volume}}{\text{[Cycle} - \text{Regeneration] Time}}$$

The *Exchanger duty* is defined as follows:

Gram Equivalents removed = (WASTE Liters/time)

· (Gram equivalents/Liter) · cycle time)

Obviously, the waste concentration is critical to sizing a resin bed, and its validity depends on the accuracy of the characterization data. A complete solids material balance helps establish its accuracy. This includes balancing the total dissolved content of the waste with the waste cation and anion content, and balancing the total cation and anion content. If a balance cannot be obtained within a 10% order of magnitude, then further waste characterization may be required to avoid gross underestimating of resin requirements.

Contactor Diameter

The contactor diameter is based on the influent flow and the applied hydraulic loading, similar to any contact column; so that:

$$\text{Area} = \frac{\text{Waste Volume/time}}{\text{Loading, Volume/time/area}}$$

Resin Volume

The required resin volume is based on the total influent ion equivalents to be removed (*exchanger duty*) and the *resin capacity,* as follows:

$$\text{Volume} = \frac{\text{Exchanger Duty}}{\text{Resin Capacity}}$$

Resin capacities for water treatment service range from 0.5 gram equivalents per liter (11 kgrain $CaCO_3$ per cubic foot) to as high as 2 gram equivalents per liter (44 kgrain $CaCO_3$ per cubic foot). For industrial applications special resins with higher capacities may be available.

Resin capacity is a specific bed property, influenced by its inherent composition and construction as well as the regeneration effectiveness, waste composition, and processing conditions. Relevant process conditions include temperature, pH, waste concentration, and ion species. Resin

efficiency depends on effective regeneration, and the preparation steps after regeneration. A poorly regenerated resin produces excessive leakage, producing a poor effluent quality. Leakage is defined as the appearance of low influent ion concentration in the effluent during the initial resin operating cycle, a result of influent ions parsing through because of incomplete regeneration, poor backwashing, or both.

Contactor Depth

Based on required volume, the bed depth can be estimated using the calculated bed volume and tower area (diameter), so that

$$\text{Approximate Depth} = \frac{\text{Bed Volume}}{\text{Tower Diameter}}$$

However, the depth must be adequate to avoid immediate column breakthrough, which is a combined property of the waste and resin. This usually means that the depth will be a minimum of 0.75 meters (30 inches), and usually less than 2.5 meters (96 inches) [5]. Minimum transfer zone depths should be established for specific wastes and selected resins. The total tower depth is determined by adding an expansion depth to the bed depth; the headspace generally being 75 to 100% of the calculated bed depth [5].

Number of Contactors

Once the resin volume, diameter, and depth is established, the required number of contactors can be selected. This will depend on factors such as providing maximum operating flexibility, headloss limitations, tank construction limits, and site specific limitations, criteria similar to those cited in Chapter III-2 for adsorption.

Regeneration

The maximum degree to which a spent resin can be restored to a desired ionic form, with unlimited quantities of regenerate solution, defines the regenerative limits and its optimum effective exchange capabilities. Regeneration efficiency falls well below the optimum level because of costs and process method. Practical regenerative limits are economically driven, dependent on acceptable resin capacity and efficiency; an important consideration in evaluating ion exchange as a viable treatment alternative.

In addition, the processing method impacts the regeneration effectiveness. Frequently, the waste and regenerative solution is passed downflow into the bed, operating as a cocurrent system. However, a higher efficiency is obtained by passing the waste and regeneration solution in opposite directions, as a countercurrent operation. This allows the waste to exit the column by contacting the portion of resin most effectively treated, resulting in minimum influent leakage. Specific regeneration agents used, concentrations employed, and common required quantities are outlined in the Appendix.

Regeneration involves three general steps to restore the spent resin:

(1) Backwashing the bed to remove accumulated particles and organisms
(2) Bed regenerating to remove collected waste ions and restore the bed
(3) Final rinse to remove the regenerating chemicals

Backwash water is generally applied at a low rate of 3 to 5 liters per square meter/second (4 to 8 gpm/ft²), depending on the resin density and backwash water temperature. Total rinse volume after regeneration is generally between 7 to 13 cubic meters per cubic meter of resin (50 to 100 gallons per cubic foot).

Regenerating agents include salts, acids, or bases; with the regeneration effectiveness dependent on its strength and resin contact time. Theoretically, the regenerating agent quantities are stoichiometric relations, although in actual practice excess quantities are required to drive the regeneration process. In weak acid and base regeneration approximately 10% excess quantities are required, whereas for strong acid and bases much higher excess dosages are used. Regeneration is driven by cost effectiveness directed at optimizing chemical costs to obtain an acceptable resin capacity.

Although for the most part the brine (regeneration products) generated in an ion exchange system consists of inorganic compounds, final disposal could be subject to restrictive regulations. This should be checked early in the project. In addition, the brine developed in the regeneration steps will probably have to be thickened, and possibly dewatered, to reduce the handling and transporting costs. Municipal facilities are sometimes reluctant to accept brine because of the possibility of disrupting the biological activities in their treatment facilities.

Fate of Contaminants

Treated effluent is the major product from the system. Secondary washwater is produced as a result of column backwashing. Brine is produced in the regeneration process. Significantly, the brine will contain all the influent cations and anions, *concentrated* as a salt. Disposal can be expensive and a problem.

Frequently, ion exchange influent is a segregated plant waste stream containing specific contaminants or recoverable compounds. These streams do not usually contain appreciable quantities of volatile matter, or have been pretreated to remove these constituents.

GENERAL ENGINEERING CRITERIA

Figure 3.2 illustrates an Ion Exchange System Preliminary Concept Flowsheet. A complete exchange system could involve the following process components:

Equalization (optional)	Chapter I-4
Waste feed system	Chapter I-5
Pretreatment (optional)	
Exchanger	
Regeneration system	

An *ion exchange system* is commonly purchased as a packaged unit, including the contactors and regeneration equipment, piping, ancillary systems, and installation engineering design. The design specifications will incorporate all of the suppliers operating experience, design criteria, required personnel protection, equipment safety, and process operating safety considerations. Even where certain items such as chemical storage, chemical feed makeup or pumps are separately purchased, the design should adhere to the suppliers specifications to assure compatibility with the other equipment. The information discussed in the next sections is intended as a guide for the Process Engineer to work effectively with the supplier and to understand the engineering details proposed.

CONTACTOR

Physically, an *ion exchange* contactor has components similar to a sandfilter, with resin used instead of sand, and with all the components required for media support, influent distribution, effective bed use, and treated water collection. The big difference being that inplace regeneration is required.

Basically, an ion exchange unit must include

(1) Influent inlets and a distribution system
(2) A treated water outlet and collection system
(3) Inlet and outlet back wash nozzles
(4) Regenerate inlet and outlet nozzles
(5) Resin media
(6) Media support system
(7) Under drainage system
(8) Flow rate controls
(9) Sampling nozzles

As with sand filters, exchangers can be gravity or pressure fed, although pressure systems are commonly employed. They can also be operated as an upflow or downflow process, with downflow systems acting as a filter and exchanger. Ion exchanger vessels are commonly pressurized vertical steel vessels, lined to provide corrosion protection from the regeneration chemicals. Stainless steel and reinforced plastic can be used for some processes. Valves and piping are steel

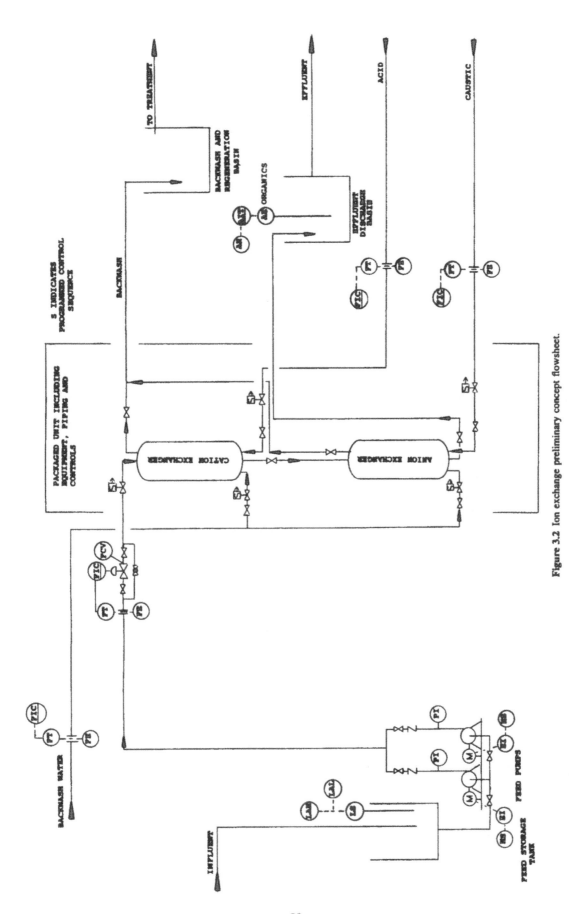

Figure 3.2 Ion exchange preliminary concept flowsheet.

93

coated, stainless steel, or PVC, depending on the regeneration chemical's corrosive properties.

The specific configuration depends on the manufacturer's proprietary design, the influent suspended solids content, and whether the system is to function both as a filter and exchanger. Influents to upflow systems should be free of suspended matter because solids will pass through to the effluent. If influent with significant suspended solids passes through a downflow system the operation may be dominated by limitations imposed by solids removal, and the length of run established by pressure drop depletion rather than exchange capacity. In such cases, the system is simplified by including a pretreatment filtration step.

Proper influent and bed distribution is critical to avoid dead spots and maintain a highly active resin bed. Influent distribution piping complies with the general concepts discussed for activated carbon systems. The exchange material must be supported to accommodate the underdrain system. The underdrain system must be capable of being completely drained to minimize effluent contamination resulting from leakage. Generally, the internals of an ion exchange contactor are similar to a filter. The reader can review the carbon and filtration design criteria for applicable design guidelines.

REGENERATION REAGENT FACILITIES

Regeneration facilities will consist of bulk storage of the concentrated reagents, and solution tanks to prepare the required regeneration volumes. The bulk storage facility's size will depend on the size of the exchanger system, the supplier location relative to the treatment facility, and shipment dependability.

BLENDING TANKS

Provision should be made for blending treated waters to obtain a suitable and consistent effluent quality, allowing for some leakage protection.

PROCESS CONTROL·

Ion exchange monitoring is essential for proper system operation. Generally, exchangers are highly automated, the sequence controlled by a predetermined time cycle, and sometimes linked to some dominant effluent quality parameter. Suggested process instrumentation includes

(1) Flow control of influent, effluent, regeneration reagents, and backwash. Critical flows should be recorded. All media regeneration steps are commonly sequentially timed, automatically activated by appropriate effluent limits, or cycle time.

(2) Quality parameters such as influent and effluent turbidity can be automatically monitored and recorded. Effluent monitoring can be based on each contactor or a composite of all contactors. Provision should be made for automatic influent and effluent composite sampling to analyze for other discharge parameters.

(3) Headloss through each column should be continuously monitored and recorded.

(4) Levels in all operating process tanks should be continuously monitored.

(5) Critical process variables should be alarmed to alert the operator of malfunction. This includes

   a. High influent and effluent control parameters
   b. High influent and effluent pH
   c. Low chemical supply
   d. Pump failure of key components such as the influent, plant effluent discharge, and regeneration components
   e. Exchanger head loss

Provision should be made for some level of automation of the chemical transfer, feed preparation, and reagent inventory. Regeneration should be completely automated.

COMMON ION EXCHANGER DESIGN DEFICIENCIES

See Adsorption, Chapter III-2, for common contactor deficiencies.

CASE STUDY NUMBER 28

Develop an *ion exchange* system for a 100,000 gallon per day waste stream, recovering both water and chromium recovery. The waste has the following characteristics:

$Copper^{++}$, 28 mg/L
$Zinc^{++}$, 18 mg/L
$Nickel^{++}$, 23 mg/L
$Cadmium^{++}$, 30 mg/L
Chromium as CrO_3, 133 mg/L

(1) Design criteria
    The following design characteristics apply:

    Cation exchanger capacity, 1.25 eq/L, 35 eq/cu foot
    Cation regeneration using 5% sulfuric acid at a capacity of 12 lb cubic foot of resin, rinse required at 100 gallons per cubic foot
    Cation cycle length, 3.5 days*
    Cation hydraulic loading, 5 gpm/sq ft
    Anion exchanger capacity, 3.8 lb CrO_3/cu ft resin.
    Anion regeneration using 10% sodium hydroxide, with a capacity of 4.5 lbs per cubic of foot resin, rinse required at 120 gallons per cubic foot
    Anion cycle length, 3.5 days*
    Anion hydraulic loading, 5 gpm/sq ft
    Recovery exchanger capacity, 1.25 eq/L, 35 eq/cu foot
    Recovery regeneration using 5% sulfuric acid, with a capacity of 12 lb per cubic foot of resin, rinse required at 100 gallon per cubic foot
    Recovery cycle length, 3.5 days*

Recovery hydraulic loading, 5 gpm/sq ft
*The required bed volume will vary with the selected cycle length.

PROCESS CALCULATIONS

Cation Cycle Exchanger

(2) Calculate individual equivalents

	mg/L	Eq Wt = MW/Valance	mg Eq = mg/L/Eq Wt
Copper	28	63.5/2 = 31.8	28/31.8 = 0.88
Zinc	18	65.4/2 = 32.7	18/32.7 = 0.55
Nickel	23	58.7/2 = 29.4	23/29.4 = 0.78
Cadmium	30	112.4/2 = 56.2	30/56.2 = 0.53
			2.74 mg eq/L

(3) Calculate total equivalents

2.74 mg eq/L · g eq/1000 mg eq · 100,000 gpd ·

$$3.78 \text{ L/gal} = 1036 \text{ eq/day}$$

(4) Calculate resin requirements

Basis: 35 eq/cu ft resin capacity
      3.5 days per cycle
1036 eq/day · 3.5 days/35 eq/cu ft resin = 104 cu ft resin

(5) Calculate diameter
(6) Calculate hydraulic related area

Basis: hydraulic load of 5 gpm/sq ft
      Maximum diameter, 12 feet
Area = 100,000 gpd/1440/5 gpm/ft^2 = 13.9 square feet
Estimated diameter = 4.2 feet
Select 5 feet diameter
Area equals 19.6 square feet
Hydraulic loading equals 3.5 gpm/sq ft

(7) Calculate depth
(8) Establish minimum bed depth

104 cubic feet volume/19.6 square feet = 5.3 feet
Allow 50% expansion, 1.5 · 5.3 feet = 7.9 feet
Use *one* unit, 5 feet diameter, 8 feet high

(9) Calculate regenerant requirement

Basis: 5% solution sulfuric acid
      12 lb sulfuric acid per cubic foot of resin
12 lb/cu ft · 104 cu ft = 1248 lb acid per cycle
1248/0.05 = 25,000 pounds of 5% acid solution

(10) Calculate rinse water requirements

Basis: 100 gal/cubic foot resin
100 · 104 cu ft = 10,400 gallons per cycle

Anion Cycle Exchanger

(11) Calculate total equivalents

133 mg/L CrO_3

133 mg/L · 100,000 gpd · 3.78 l/gal · 1/1000 g/mg

· 1/454 lb/g = 111 lb/d

(12) Calculate resin requirements

Basis: 3.8 lb CrO_3/cu ft resin capacity
      3.5 days per cycle
111 lb/day · 3.5 days/3.8 = 102 cu ft resin

(13) Calculate diameter
(14) Calculate hydraulic related area

Basis: hydraulic load of 5 gpm/sq ft
      Maximum diameter, 12 feet
Area = 100,000 gpd/1440/5 gpd/ft^2 = 13.9 square feet
Estimated diameter = 4.2 feet
Select 5 feet diameter
Area equals 19.6 square feet
Hydraulic loading equals 3.5 gpm/sq ft

(15) Calculate depth
(16) Establish minimum bed depth

102 cubic feet volume/19.6 square feet = 5.2 feet
Allow 50% expansion, 1.5 · 5.2 feet = 7.8 feet
Use *one* unit, 5 feet diameter, 8 feet high

(17) Calculate regenerant requirement

Basis: 10% solution sodium hydroxide
      4.5 lb sodium hydroxide per cubic foot of resin
4.5 lb/cu ft · 102 cu ft = 459 lb sodium hydroxide/cycle
459/0.10 = 4590 pounds of 10% sodium hydroxide solution

(18) Calculate rinse water requirements

Basis: 120 gal/cubic foot resin
120 · 102 cu ft = 12,240 gallons per cycle

Recovery Cycle (Cation) Exchanger

(19) Calculate individual equivalents
The sodium hydroxide used in the anion regeneration equals 459 pounds per cycle, with NaOH having an equivalent weight of 40, so that,

459 NaOH · 454 g/lb/40 = 5,210 gram equivalents per cycle

(20) Calculate resin requirements

Basis: 35 gram eq/cu ft resin capacity
5210 g eq/day/35 g eq/ft^3 resin = 149 cu ft resin

(21) Calculate diameter

(22) Calculate hydraulic related area

Basis: hydraulic load of 5 gpm/sq ft
Maximum diameter, 12 feet
Area = 100,000 gpd/1440/5 gpd/ft^2 = 13.9 square feet
Estimated diameter = 4.2 feet
Select 5 feet diameter
Area equals 19.6 square feet
Hydraulic loading equals 3.5 gpm/sq ft

(23) Calculate depth

(24) Establish minimum bed depth

149 cubic feet volume/19.6 square feet = 7.6 feet
Allow 50% expansion, 1.5 · 7.6 feet = 11.4 feet
Use *one* unit, 5 feet diameter, 12 feet high

(25) Calculate regenerant requirement

Basis: 5% solution sulfuric acid
        12 lb sulfuric acid per cubic foot of resin
12 lb/cu ft · 149 cu ft = 1788 lbs acid per cycle
1788/0.05 = 35,760 pounds of 5% acid solution

(26) Calculate rinse water requirements

Basis: 100 gal/cubic foot resin
100 · 149 cu ft = 14,900 gallons per cycle

Note: Resin capacities, regeneration requirements, and rinsing
requirements are specific to the resins selected, and should
be obtained from specific evaluations based on supplier experience. If possible, all units should be the same configuration
and size to simplify piping, layout, and construction costs.

APPENDIX

EXCHANGER CHARACTERISTICS [2,6,4,7]

Strong Acid (Cation) Exchangers

R-SO3 – H   R-SO3 – Na

Function and Use
Can split *strong* or *weak* salts
Softening and demineralization

Water Treatment General Criteria
Maximum *hydraulic* capacity: 30 cu m/sq m/hour
                              12 gpm/sq ft
Minimum *hydraulic* capacity: 3 cu m/sq m/hour
                              1 gpm/sq ft
Minimum *bed depth* 0.6 meters
                    24 inches
Maximum operating *temperature*: 120°C
                                : 248°F
resin capacity, gram-eq/L
                              (kilograin CaCO$_3$/ cubic ft)
(NaCl) regeneration: 0.8 to 1.5 (17–33)
(H$_2$SO$_4$) regeneration: 0.5 to 1 (11–22)
(HCl) regeneration: 0.7 to 1.4 (15–31)

Regenerant, grams/L, (lb/cuft) of resin
(NaCl) regeneration: 80 to 250 (5–16)
(66°Be H$_2$SO$_4$) regeneration: 35 to 200 (2.2–13)
(20°Be HCl) regeneration: 80 to 500 (5–31)

Typical Exchange Reactions

The functional group is commonly derived from the sulfonic
group HSO$_3$, which will be represented by the symbol R. With the
acid (H) as the functional component, the resin can remove all
cations, including sodium, and can be represented by the following
typical equations.

$$2 \, RH + Ca(HCO_3)_2 = CaR_2 + 2 \, H_2O + 2 \, CO_2$$

$$2 \, RH + MgSO_4 = MgR_2 + H_2SO_4$$

$$RH + NaCl = NaR + HCl$$

Typical Regeneration Reactions

$$MgR_2 + H_2SO_4 = 2 \, HR + MgSO_2$$

Operating Characteristics

(1) Applicable pH range: wide range
(2) Regeneration: excess strong acid, with typical efficiencies rang-
    ing 25 to 45% in concurrent operation
(3) Leakage: low
(4) Capacity: low relative to weak acid exchanger, see above
    and below
(5) Stability: High, can last years without loss capacity

Weak Acid (Cation) Exchangers

R-COO – H

Function and Use
Cannot split *strong* salts, without
buffering solution. Can remove
only those cations equivalent to
the amount of alkalinity present
in water; most efficient removal
calcium and magnesium hardness.
Because of high removal efficiency
effectively used as first of two
cation exchangers demineralizers

Water Treatment General Criteria
Maximum *hydraulic* capacity: 20 cu m/sq m/hour
                              8 gpm/sq ft
Minimum *hydraulic* capacity: 3 cu m/sq m/hour
                              1 gpm/sq ft
Minimum *bed depth*: 0.6 meters
                     24 inches
Maximum operating *temperature*: 120°C
                                : 248°F

Usable *resin* capacity, gram-eq/L (kilograin CaCO₃/cu ft)
                        : 0.5 to 2.0
                        : (11 to 44)

Regenerant, g/L, (lb/cu ft)
            HCl or H_2SO_4: 110% theoretical

Typical Exchange Reactions

$$2\ RH + Ca(HCO_3)_2 = CaR_2 + 2\ H_2CO_3$$

$$2\ RH + Mg(HCO_3)_2 = MgR_2 + 2\ H_2CO_3$$

$$2\ RH + Na_2HCO_3 = 2\ NaR + H_2CO_3$$

Typical Regeneration Reactions

Regeneration can be accomplished utilizing a strong or weak acid.

Regeneration utilizing sulfuric acid can be represented as follows:

$$CaR_2 + H_2SO_4 = 2\ HR + CaSO_4$$

Operating Characteristics

(1) Applicable pH range: above 7
(2) Regeneration: strong or weak acid with high efficiency, usually >90%, excess 110% adequate, can be regenerated with waste acid from second strong acid exchanger
(3) Leakage: high leakage of sodium, low of calcium
(4) Capacity: high relative to strong acid exchanger
(5) Stability: more resistant to oxidants such as chlorine than strong acid exchanger

Strong Base Anion Exchangers

$$Z - N^+ OH$$

Type I: Z represents three methyl groups; Type II: Z represents two methyl and one ethyl group

Water Treatment General Criteria
Maximum *hydraulic* capacity: 17 cu m/sq m/hour
                        7 gpm/sq ft
Minimum *hydraulic* capacity: 3 cu m/sq m/hour
                        1 gpm/sq ft
Minimum *bed depth*: 0.75 meters
                        30 inches
Maximum operating *temperature*: 50°C
                        : 122°F
Usable *resin* capacity, gram-eq/L (kilograin CaCO₃/cu ft)
                        : 0.35 to 0.7
                        : (8 to 15)

Regenerant, g/L, (lb/cu ft)
NaOH regeneration: 70 to 140
                        (4 to 9)

Function and Use
Can remove *strong* or *weak* acids
Type I for maximum silica removal
Type II less effective silica, can
remove other weak ions

Typical Regeneration Reactions

The resin without the functional group DERIVATIVE will be represented by the symbol Z. The exchange reactions can be represented by the following equations.

$$2\ ZOH + H_2SO_4 = Z_2 \cdot SO_4 + 2\ H_2O$$

$$ZOH + HCl = Z \cdot Cl + H_2O$$

$$ZOH + HNO_3 = Z \cdot NO_3 + H_2O$$

$$ZOH + H_2CO_3 = Z \cdot HCO_3 + H_2O$$

$$ZOH + H_2SiO_3 = Z \cdot HSiO_3 + H_2O$$

Typical Regeneration Reactions

Regeneration utilizing caustic can be represented as follows:

$$Z \cdot Cl + NaOH = ZOH + NaCl$$

Operating Characteristics

(1) Applicable pH range: wide range
(2) Regeneration: excess strong sodium hydroxide, with typical efficiencies ranging 18 to 33%, Type I more difficult to regenerate than Type II.
(3) Capacity: somewhat less than comparable acid exchanger
(4) Stability: prone to fouling, less stable than cation resins

Weak Base Anion Exchangers

$$Z - N$$

N an be primary, secondary, tertiary or quaternary amine

Water Treatment General Criteria
Maximum *hydraulic* capacity: 17 cu m/sq m/hour
                        7 gpm/sq ft
Minimum *hydraulic* capacity: 3 cu m/sq m/hour
                        1 gpm/sq ft
Minimum *bed depth*: 0.75 meters
                        30 inches
Maximum operating *temperature*: 40°C
                        : 104°F
Usable *resin* capacity, gram-eq/L (kilograin CaCO₃/cu ft)
                        : 0.8 to 1.4
                        : (17 to 31)

Regenerant, g/L, (lb/cu ft)
NaOH regeneration: 35 to 70
                        (2 to 4)

Function and Use
Removes *strong* ionized acids,
but not weak acids
Useful preceding a strong base
exchanger to save cost regenerate
chemicals and to protect from
fouling organic chemicals

Typical Exchange Reactions

$$2\ ZN + H_2SO_4 = (ZN)_2 \cdot H_2SO_4$$

$$ZN\ \ \ + HCl = ZN \cdot HCl$$

$$ZN\ + HNO_3 = ZN \cdot HNO_3$$

Typical Regeneration Reactions

Regeneration can be accomplished utilizing a caustic, soda ash, or ammonium hydroxide.

Regeneration utilizing soda ash, sodium carbonate, can be represented as follows:

$$(ZN)_2 \cdot H_2SO_4 + Na_2CO_3 = 2\ ZN + Na_2SO_4 + CO_2 + H_2O$$

Operating Characteristics

(1) Applicable pH range: below 6

(2) Regeneration: near stoichiometric amount of sodium hydroxide at efficiency near 90%

(3) Capacity: somewhat lower than comparable acid exchanger

(4) Swelling: about 12% from OH− to salt form

(5) Stability: resistant to organic fouling

REFERENCES

1. C. Calmon, and H. Gold: *Ion Exchange for Pollution Control,* Volumes I and II, CRC Press, 1979.
2. Cheremisinoff, N.P., Cheremisinoff, P.N.: *Water Treatment and Waste Recovery—Advanced Technology and Applications,* PTR-Prentice Hall, 1993.
3. Eckenfelder, W.W. Jr.: *Industrial Water Pollution Control,* Second Edition, McGraw-Hill, 1989.
4. James M. Montgomery, Consulting Engineers: *Water Treatment Principles and Design,* John Wiley and Sons, 1985.
5. McKetta, J.J. and Cunningham, W.A.: *Encyclopedia of Chemical Processing and Design,* Marcel Dekker, Inc, New York, 1998.
6. Nordell, E.: *Water Treatment for Industrial and Other Uses,* Second Edition, Reinhold, 1961.
7. Perry, R.H. and Green, D.: *Perry's Chemical Engineers' Handbook, Sixth Edition, McGraw-Hill, 1984.*
8. Sanks, R.L.: *Water Treatment Plant Design,* Ann Arbor Science, 1978, Chap 25.
9. Haas, C.N. and Vamos, R.J.: *Hazardous and Industrial Waste Treatment,* Prentice Hall, 1995.

Wastewater Stripping

Stripping is employed to remove volatile dissolved organics.

STRIPPING is used to remove volatile dissolved materials from a waste stream by continuously contacting the waste with an inert gas (air or steam), increasing the waste temperature, or reducing the system pressure. Stripping removes all components which are more volatile than water, discharging them with the stripping gas and some water vapor.

Figure 4.2 illustrates a typical air stripper configured so that air is flowing into the system through the tower bottom and waste flowing from the top. Air flowing upward becomes enriched with volatile materials stripped from its countercurrent contact with the waste. The effluent leaving the tower bottom is disposed or further treated. The air stream leaving the tower, containing the stripped waste contaminants, is either sent to an air pollution device or exhausted to the atmosphere at "controlled concentrations."

BASIC CONCEPTS

Absorption, and related stripping theory, have been well developed in Chemical Engineering practice, and its principles are readily adoptable to waste treatment. Effective waste stripping is dependent on a liquid component's vapor pressure and relative volatility.

PARTIAL AND VAPOR PRESSURE

The pressure exerted by a liquid component under equilibrium conditions is referred to as its partial pressure. The more volatile a component the higher its partial pressure. As a frame of reference, when the total pressure exerted by the liquid components (the sum of the individual component partial pressures) exceeds atmospheric pressure, the liquid will boil. The *more volatile* a compound the higher its *pure vapor pressure*. The pressure exerted by a *pure* compound is its *vapor pressure* at 100% concentration, whereas the pressure exerted by the same compound in solution is a fraction of the vapor pressure.

EQUILIBRIUM DATA

The partial pressure of individual components in an ideal mixture is expressed by Raoult's Law, stating that at equilibrium the partial pressure (p) of a component in solution is equal to the product of its vapor pressure (P^*) and its mole fraction (x) in solution. According to Dalton's Law the partial pressure of a component in an ideal gas is equal to its mole fraction in the gas times the total pressure.

$$p_i = x_i \cdot P^* \text{ in an ideal solution} \qquad (4.1a)$$

$$p_i = y_i \cdot \pi \text{ in an ideal gas} \qquad (4.1b)$$

where P^* is the vapor pressure of component i, p_i is the partial pressure of component i, x_i is the mole fraction of i in solution, y_i is the mole fraction of i in the gas phase, and π is the total system pressure.

An ideal solution is one in which its characteristics are a composite of the individual components, such that the total volume of the solution is equal to the sum of the volume of the individual components, the total pressure exerted by a solution is equal to the sum of the partial pressures of the components, etc. In dilute waste treatment applications, the system is assumed to have these characteristics.

However, ideal conditions cannot be assumed for all separation processes, instead conditions between liquid and vapor phases are commonly stated in terms of fugacity. The thermodynamic details for fugacity are beyond the scope of this book, but it is sufficient to state that at equilibrium, when expressed in terms of fugacity, the concentrations in the two phases are equal.

$$fg = fl \qquad (4.1c)$$

In essence, the vapor phase concentrations, as expressed in terms of liquid concentration or vapor concentration, are equal. In ideal conditions,

$$yi \cdot \pi = xi \cdot P^* \qquad (4.1d)$$

In simple terms, corrections for nonideal conditions are made by applying an activity coefficient (σ, σ'); as indicated in Equation (4.1e).

$$\sigma' \cdot yi \cdot \pi = \sigma \cdot xi \cdot P^* \qquad (4.1e)$$

Assuming that ideal conditions occur in the gaseous phase, compensating for any nonideal conditions by correcting the corresponding liquid phase partial pressure,

$$yi \cdot \pi = \sigma \cdot xi \cdot P^* \qquad (4.1f)$$

(The interested reader should review a basic Chemical Engineering thermodynamics text for a more technical and complete explanation of fugacity, standard fugacity, and activities.)

Based on experimental data, the term $\sigma \cdot P^*$ is combined, expressed as a constant (H), forming the basis for Henry's Law,

$$p = H \cdot x \qquad (4.1g)$$

where H is referred to as the Henry's Law constant. Henry's Law constants are frequently used, and many times misused, to define a system's equilibrium curve. Some important factors governing this constant include

(1) The equilibrium curve defined by Henry's Law is applicable over a limited range because the activity coefficient is both temperature and concentration dependent, while the vapor pressure is temperature dependent.

(2) The straight line relationship of either Henry's or Raoult's Law is generally limited to low pressures and concentrations, which happen to be common in waste treatment stripping systems.

(3) When Henry's Law constants are cited, an assumption is made that activity coefficient corrections are made to the equilibrium curve as defined by ideal conditions. This may not be the case! These constants are many times derived by theoretical means, and the accuracy is dependent on the estimating method.

(4) Since many tower "simplified" design methods assume a Henry's Law constant, the Process Engineer is "driven"

to find a "number" which satisfies the estimating method.

(5) Henry's Law constants are difficult to find, where available they are limited to a single or narrow temperature range, and some times an approximated value.

(6) The reference temperature for a constant is especially critical if used for steam stripping, which operates at elevated temperatures.

These constants are often reported as a function of temperature, defined by the relation [3]

$$\log H = \frac{-dHv}{R \cdot T} + K \qquad (4.2)$$

where H is Henry's constant, atm $\cdot$ cubic meter $\cdot$ °Kelvin, dHv is the molar heat absorbed in evaporation of a component, kcal/kmole, dHv/R is sometimes reported as the constant "a," R is 1.987 kcal/kmole $\cdot$ °Kelvin, K is a constant, sometimes reported as "b," and T is the Kelvin temperature.

RELATIVE VOLATILITY

The relative volatility of a single component in low pressure conditions can be estimated using vapor pressure data and assuming Raoult's Law applies; so that,

$$ya/xa = Pa^*/\pi \qquad (4.3a)$$

where xa, ya is the molar concentration of a component in the liquid and gaseous phase, Pa^* is the vapor pressure of the component, and π is the system total pressure.

Equation (4.3b) defines the relative volatility (Φ) of component a to component b in terms of molar concentration. Equation (4.3c) expresses it in terms of partial pressure, as defined by Raoult's Law. Equation (4.3d) summarizes the definitions for relative volatility.

$$\Phi = [ya/xa]/[yb/xb] \qquad (4.3b)$$

$$\Phi = Pa^*/Pb^* \qquad (4.3c)$$

$$\Phi = [ya/xa]/[yb/xb] = Ha/Hb = Pa^*/Pb^* \qquad (4.3d)$$

In waste treatment all volatile components are related to water, as expressed by Equation (4.3e).

$$\Phi = Ha/H \text{ for water} = Pa^*/\text{vapor pressure for water} \qquad (4.3e)$$

Significantly, if Φ is greater than 1, component "a" has the potential to be stripped, and the greater the value the more feasible is the stripping process.

STRIPPING BASICS

A waste component's stripping potential depends on its volatility relative to water. The rate at which it can be separated depends on the available driving force. An analogy between other transfer operations illustrates the similar dependence on a driving force,

$dQ/dt \propto [T - t]$ [temperature difference]: heat transfer
$dC/dt \propto [C - c]$ [concentration difference]: mass transfer
$dF/dt \propto [P - p]$ [pressure difference]: filtration

In stripping processes the driving force is defined by the difference in a component's exerted partial pressure and its maximum partial pressure. A component's maximum vapor phase pressure being an equilibrium value; so that

(1) The difference between a component's exerted partial pressure and maximum vapor pressure is a measure of the stripping driving force.
(2) The maximum quantity of volatiles that can be transferred to a stripping gas is the quantity at equilibrium conditions (for that component). This is equivalent to the exit gas molar component concentration in equilibrium with the waste feed concentration, as illustrated in Figure 4.1.
(3) The equilibrium vapor pressure of a component, and therefore the stripping potential, increases with temperature.

Generally, industrial waste strippers treat a continuous waste flow in either countercurrent multistage *tray* towers or continuous contact *packed* columns. In multistage operations there is an intermittent stage-wise contact between the stripping gas and the wastewater, while in continuous contact columns the two fluids are in constant contact. In packed towers sequential differential packing heights act as minute stages, the tower performance a result of the integrated results of each minute section. Because the same mass transfer principles define the stripping potential in either system,

stage-wise contact basics will be discussed and applied to packing.

Figure 4.1 illustrates an *ideal contact stage* in multistage units such as tray towers, the maximum separation possible in one contact unit with infinite residence time. The waste component is stripped from the influent at concentration Xo, resulting in an effluent concentration Xi. The absorbing medium being an inert gas at an initial component concentration Yo, absorbs all the stripped waste component. The resulting exit gas concentration Y^*, in equilibrium with the waste concentration Xo, represents the theoretical maximum (saturated) concentration for that component. In practice, the achievable concentration will be Yi, less than the theoretical concentration Y^*, because an infinite contact time would be required to reach ideal conditions. Corrections to a theoretical stage are made by applying a contact stage efficiency. The concentration Yi and the stripping efficiency increase with temperature.

As previously stated, the significant driving force is defined by a component's actual partial pressure (p) and its maximum partial pressure (P^*) at equilibrium conditions. So that, assuming Henry's Law applies, the maximum partial pressure is expressed by Equation (4.4a).

$$P^* = H \cdot x \qquad (4.4a)$$

Assuming an operating (total) pressure of π atmospheres, then

$$P^*/\pi = H \cdot x/\pi \qquad (4.4b)$$

$$y^* = H \cdot x/\pi \qquad (4.4c)$$

When the total system pressure π is one atmosphere,

$$y^* = H \cdot x \qquad (4.4d)$$

The driving force throughout the stripper is represented by the difference between the component equilibrium pressure (P^*) and its gas phase partial pressure (p),

$$P^* - P \qquad (4.5a)$$

or its equivalent vapor concentration difference,

$$y^* - y \qquad (4.5b)$$

Stripping efficiency is affected by volatility, which is temperature dependent, both of which define the available driving force. Therefore, any process conditions increasing the stripping driving force, will increase process efficiency. Realizing that a volatile waste component is a dissolved gas, factors influencing its solubility will affect its removal. This includes (1) increasing the liquid temperature, (2) reducing

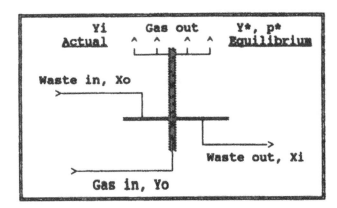

Figure 4.1 Ideal contact stage.

the system pressure, or (3) providing an inert gas (air) to transport the contaminant from the system.

Transfer from the liquid and gas phases can occur with gaseous components being *absorbed* by the liquid (such as an air pollution scrubber), or liquid components being *desorbed* (or stripped) by a gas stream. Stripping is the waste treatment unit operation discussed in this chapter. Available waste stripping equipment include

(1) A tower filled with a packing material, commonly referred to as a *packed tower*
(2) A *spray tower* in which the waste liquid is sprayed into an otherwise empty tower
(3) A *tray tower* containing bubble caps, sieve trays, or valve trays
(4) Stirred vessels, used for batch or semicontinuous operation

Specific process design procedures for packed and tray towers are discussed in the Stripper Process Evaluations section.

STRIPPER PROCESS EVALUATIONS

Note: in order to avoid any confusion the following symbols are used in this section to define flow and concentration: L, G are mass loading, mass/area − time; L_m, G_m are molar flows, all concentrations are expressed as mole fractions.

Waste treatment of volatile materials can be achieved with three basic stripping configurations:

(1) Air stripping with direct exhaust
(2) Air stripping and treating the exhausted volatiles
(3) Steam stripping

Any of these basic functions can be performed as a batch or continuous operation, in a packed or tray tower, to remove volatile organic compounds or ammonia.

Continuous systems can operate as a single-stage flash unit or multiple contact tower, employing air or steam as the stripping media, under a range of temperatures or pressures. A batch system is simply an agitated reactor containing a specific waste volume, into which steam or air is injected for a period of time until the desired effluent quality is achieved. This system is seldom used for waste treatment, and where employed is limited to small waste volumes. However, the principle is sometimes applied by intermittently aerating pretreatment or polishing ponds to remove volatile constituents.

Flash stripping is basically a one-stage stripper, with waste continuously fed to an agitated and aerated vessel or basin. These strippers are ineffective for low volatile concentrations unless very high retention times are employed, and low removal efficiencies are acceptable. Removal can be enhanced by using steam as the stripping media, a costly process when applied to large waste volumes. For the most part

these strippers are costly, ineffective, produce uncontrollable fugitive emissions, and are seldom used for waste treatment. They are sometimes used as an "expeditious" remedy to a reoccurring problem. Frequently, flash strippers are mechanically aerated basins, many times the secondary treatment basin itself.

Waste stripping is almost always performed in a tower as a continuous process, using ambient air as the stripping media. The simplest air stripper is an open structure or tower, with the contaminants exhausted at low levels to meet applicable air emission requirements [Figure 4.2(a)]. Air emission restrictions could limit the use of open towers (or basins) unless the volatile compounds do not represent a health problem and are not odorous. In which case, the basic system may have to be enhanced to reduce atmospheric emissions; using exhaust treatment or steam stripping, symbolically represented in Figures 4.2(b) and 4.2(c).

STRIPPING-PACKED COLUMN

Process design of a packed column involves the following sequential steps:

(1) Select the packing
(2) Use appropriate liquid-vapor data to establish the equilibrium line
(3) Use material balance data to establish an operating line and the liquid to gas ratio
(4) Determine the air flow rate
(5) Calculate the exit waste concentration
(6) Determine the tower diameter
(7) Evaluate the column hydraulic effectiveness
(8) Estimate the number of transfer units
(9) Estimate the height of a transfer unit
(10) Estimate the tower height
(11) Estimate the column pressure drop

Packing Characteristics

Packing size and *properties* are important design considerations affecting tower hydraulics, removal efficiency, and potential fouling. The primary packing characteristic is the wetted area per volume. The greater the area per volume, the greater and more effective the mass contact between the stripping gas and the waste. The balance is between small packing with high mass transfer area, increased plugging potential, and high pressure drop or large packing with the opposite qualities. Characteristics of typical packing are cited in Table 4.1 [3].

Equilibrium Line

As discussed under stripping basics, the equilibrium line sets the limiting or maximum concentrations that can be

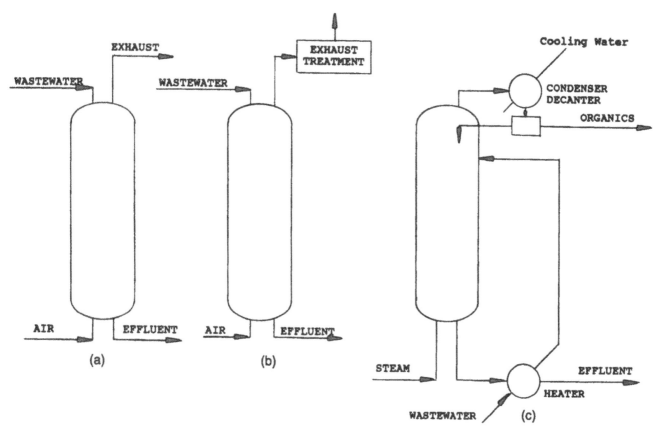

Figure 4.2 Continuous stripping processes.

achieved by contacting the gaseous and waste streams. The "ideal contact stage" being the basis for measuring stripping performance, as discussed in the Basic Concepts section. As illustrated in Figure 4.1, xi and the corresponding equilibrium value Y^* define equilibrium (limiting capacity) values in an ideal contact. In a continuous system the difference between the bulk concentration and equilibrium values defines the substrate component driving force between the two phases, as illustrated in Figure 4.3.

Therefore, selecting representative equilibrium data, and the reliability of the data, are consequential factors in developing stripper process requirements. In the dilute wastes frequently encountered in waste treatment, Henry's Law constants are usually used to define the system equilibrium line. As previously discussed, the use of a Henry's Law assumes that (1) the constant does not significantly change within the range of tower conditions encountered, and (2) the constant is a product of the vapor pressure, P^*, and the activity coefficient, σ.

Liquid-to-Gas Ratio (Operating Line)

The liquid-to-gas ratio is based on a stripper material balance, which for a single component is illustrated in Figure

4.4 and expressed as Equation (4.6a), based on mole concentrations and flows.

$$Lo \cdot Xo - Le \cdot Xe = Go \cdot Ye - Ge \cdot Yo \quad (4.6a)$$

If the molar flow rates do not significantly change, so that $Li = Lo$ and $Gi = Go$, Equation (4.6a) can be simplified to Equations (4.6b) and (4.6c).

$$L_m \cdot (Xo - Xe) = G_m \cdot (Ye - Yo) \quad (4.6b)$$

$$G_m/L_m = (Xo - Xe)/(Ye - Yo) \quad (4.6c)$$

The minimum gas capacity, the theoretical volume required to absorb an equilibrium quantity of the stripped component equivalent to Xo, can be stated in terms of the minimum gas-to-liquid mole ratio. This is graphically depicted in Figure 4.5, which relates the operating line to the equilibrium line. The minimum gas-to-liquid ratio is represented in Figure 4.5(a), when the operating line at initial conditions of Xe and Yo is tangent to the equilibrium line. This represents the condition when the exit gas is at its maximum theoretical concentration (Ye).

TABLE 4.1. Dumped Packing Characteristics (adapted from Reference [3]).

Packing	Size		Surface Area a_p		% Void Space ϵ	Packing Factor, Pf	
	mm	in	sq m/cu m	sq ft/cu ft		1/m	1/ft
Berl saddles,	25	1	250	76	68	360	110
ceramic	38	1½	150	46	71	215	66
	50	2	105	32	72	150	46
Intalox saddles,	25	2	255	78	77	320	98
ceramic	38	1½	195	59	80	170	52
	50	2	118	36	79	130	40
	75	3	92	28	80	70	21
Intalox saddles,	25	1	206	63	91	105	32
polyproplene	50	2	108	33	93	69	21
	75	3	88	27	94	50	15
Pall rings,	25	1	205	62	94	157	48
metal	38	1½	130	40	95	92	28
	50	2	115	35	96	66	20
	90	3½	92	28	97	53	16
Pall rings,	25	1	205	62	90	170	52
polypropylene	38	1½	130	40	91	105	32
	50	2	100	30	92	82	25
	90	3½	85	26	92	52	16
Raschig rings,	25	1	190	58	74	510	155
ceramic	38	1½	120	37	68	310	94
	50	2	92	28	74	215	66
	75	3	62	19	75	120	37
Raschig rings,	25	1	185	56	86	450	137
steel	38	1½	130	40	90	270	82
	50	2	95	29	92	187	57
	75	3	66	20	95	105	32

In Figure 4.5(b), the equilibrium line is curved concave down, so that the maximum concentration that the gas stream can achieve is when the operating line touches the equilibrium line at the feed concentration Xo. Figure 4.5(c) depicts the special case when the equilibrium line is represented by the Henry's Law equation, as a straight line. The operating line touches the equilibrium line at the inlet feed concentration, Xo, when the gas concentration is the equilibrium value corresponding to the inlet waste concentration. When the equilibrium curve is a straight line the minimum gas-to-liquid ratio can be represented by Equation (4.6d).

$$[G_m/L_m]_{minimum} = (Xo - Xe)/(m \cdot Xo - Yo) \quad (4.6d)$$

The design gas-to-liquid mole ratio must be greater than the theoretical minimum increased by 20 to 50% of the minimum value [3]. As a "rule-of-thumb" the stripping factor can be set at 1.4 for preliminary evaluations, and optimized on the basis of an economical evaluation of the resulting gas quantity, tower diameter, and tower height. The selected gas rate must accommodate the column hydraulics, providing a column diameter which will assure effective packing wetting; or proper tray mechanics if a tray column is employed.

Air Flow Rate

The air rate can be calculated using the design liquid-to-gas ratio expressed by Equations (4.7a) and (4.7b), representing a corrected minimum ratio. Mole rates and concentrations are used in developing the balance.

$[G_m/L_m]_{actual}$

= required air flow/wastewater design rate   (4.7a)

Required air flow

= $[G_m/L_m]_{actual} \cdot$ wastewater design rate   (4.7b)

Establishing Fluid Concentrations

All subsequent calculations require a complete system material balance, fixing all entering and exit concentrations. At this point in the design the following information is available:

(1) The inlet waste composition is known.
(2) The inlet stripping gas concentration is known, usually zero.

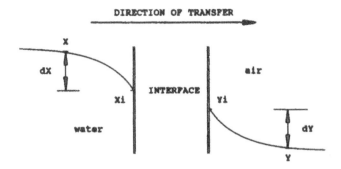

Figure 4.3 Stripper mass driving potential.

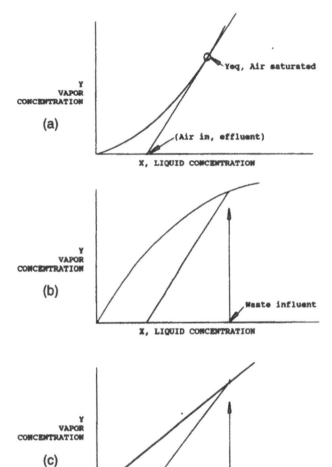

Figure 4.5 Stripper minimum gas-to-liquid ratio.

(3) The effluent waste composition is the required design composition.

(4) The liquid to gas operating line defining the material balance is known.

Based on the operating equation, the outlet stripping gas concentration can be calculated using Equation (4.8).

$$Y = \{[L_m/G_m]_{actual} \cdot (Xo - X)\} + Yo \qquad (4.8)$$

This is a complicated way of stating that all the stripped wastewater substrate is transferred to the air stream, with the inlet air concentration at zero; the exhaust gas concentration being equal to the moles of material stripped divided by the inlet air flow (moles/hour).

Packed Tower Diameter

The tower diameter establishes the system hydraulics, and its selection is predicated on balancing the fluid flows to assure effective liquid and gas contact. If the tower is not sized correctly the system will be overloaded or inadequately wetted, either of which reduces the stripping capabilities. The tower diameter is a function of the liquid and gas rates, liquid and gas physical properties, packing size, packing geometry, and packing orientation. Correlations have been developed for both dumped and stacked packings defining the pressure drop or flooding gas velocity limits to the tower loadings and packing characteristics, as illustrated in Figure 4.6 [3,6].

In this plot, the abscissa is

$$(L/G) \cdot (d_g/d_L - d_g)^{1/2} \qquad (4.9a)$$

Because d_g is relatively small, the abscissa is commonly expressed as

$$(L/G) (d_g/d_L)^{1/2} \qquad (4.9b)$$

The ordinate is

$$\frac{G^2}{g} \cdot F_p \cdot \frac{Fd}{dg \cdot dL} \cdot \mu^{0.2} \qquad (4.9c)$$

L is the liquid rate, kg/s · m^2 (lb/hr · ft^2)
G is the gas rate, kg/s · m^2 (lb/hr · ft^2)
dg is the gas density, kg/cu m (lb/ft^3)
dL is the liquid density, kg/cu m (lb/ft^3)
μ is the liquid viscosity, m · Pa · s (cp)
g is the acceleration due to gravity, 9.8 m · s^2 (4.17 × 10^8)

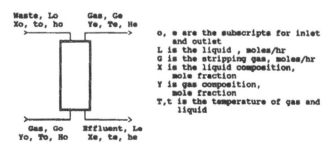

Figure 4.4 Tower material balance.

o, e are the subscripts for inlet and outlet
L is the liquid , moles/hr
G is the stripping gas, moles/hr
X is the liquid composition, mole fraction
Y is gas composition, mole fraction
T,t is the temperature of gas and liquid

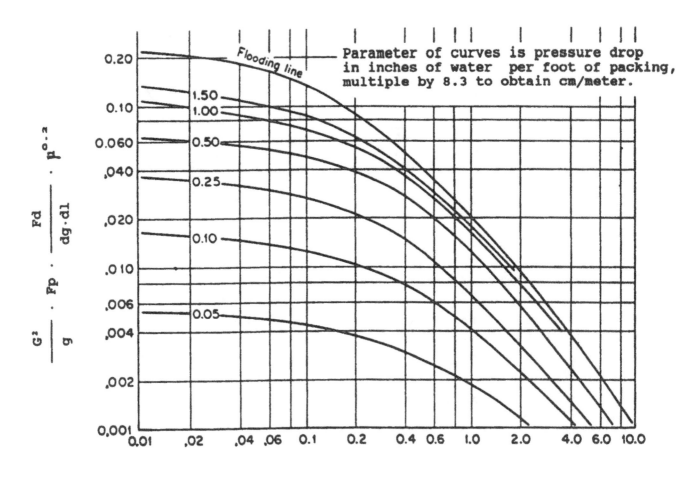

Figure 4.6 Stripper flooding curve. (Eckert, J.S.: "Selecting the Proper Distillation Column Packing," *Chemical Engineering Progress,* V 66, No 3, Pg 40, March, 1970. Reproduced with permission of the American Institute of Chemical Engineers. Copyright © 1970 AIChE. All rights reserved.)

F_p is the packing factor, a product of the a_p and ϵ factors, obtained experimentally (see Table 4.1)

Fd is a liquid-density correction factor, defined as the ratio of the density of water to the density of the liquid. For all practical purposes, this factor is 1 for wastewater.

An abscissa value can be calculated using the design L/G mass ratio, and the liquid, gas, and packing properties. In turn, a corresponding ordinate value can be obtained from the flooding curve, and the flooding gas mass loading (G) calculated. The packed tower gas flooding rate should be multiplied by a factor of 0.62 (0.5 to 0.75 range) to obtain a velocity that minimizes the potential for flooding conditions to occur during normal operation [3]. The mass gas loading can be converted to the allowable gas velocity as indicated by Equation (4.10a). Based on the selected design gas loading rate, the required tower area (A_t), and corresponding diameter, can be calculated.

G, mass rate/density of air = velocity, m/sec (fps)   (4.10a)

$$kg/sec\ (lb/sec)/kg/cu\ m\ (lb/ft^3) = m/sec\ (fps)$$

$$A_t = G_V/V_t \qquad (4.10b)$$

where G_V is the tower gas flow rate, cu m/s (ft^3/sec); V_t is the allowable tower velocity as defined by Equation (4.10a), m/sec (fps); and A_t is the tower area, sq m (ft^2).

Column Hydraulic Effectiveness

The column hydraulics should be checked to assure adequate liquid distribution, that dry spots are avoided, and the media is fully active. Although proper liquid distribution is not restricted to a single flow, hydraulics can be

TABLE 4.2. MWR Characteristics
(adapted from Reference [3]).

	sq meters/min	sq feet/minute
Ring packing, larger 0.076 m (3 in)	0.00132	0.01417
Grid with pitch larger 0.051 m (2 in)	0.00132	0.01417
Other packings	0.00201	0.02167

checked on the basis of the minimum wetting rate, defined as [3]

$$MWR = V_L/a_p \qquad (4.11)$$

where V_L is the volumetric liquid velocity in meters per minute (fpm), a_p is the external packing surface per unit volume, square meter per cubic meter (ft^2/ft^3), and MWR is the minimum wetting rate in square meters per minute (ft^2/m).

When the waste flow is too low to meet the recommended MWR values cited in Table 4.2, recirculation should be considered.

Based on the criteria cited in Table 4.2, volumetric waste loadings would range from 0.2 to 0.4 cubic meters per square meter-minute (5 to 10 gpm/ft²) with a packing factor of 150 to 200 sq m/cu m (45 to 65 ft²/ft³); or the equivalent superficial velocity of 0.2 to 0.4 meters per minute (0.7 to 1.4 feet per minute). *Tower recycle capacity* is an important operating variable to control the tower hydraulics since this may be the only expedient way of providing adequate tower flow, supplementing the waste flow.

Number of Transfer Units

The number of stripping transfer units can be defined in terms of the integrated arithmetic mean equation [3]

$$M = \frac{1}{2} \cdot \ln\left[\frac{1 - Xo}{1 - Xi}\right] + \int_{Xo}^{Xi} \frac{dx}{X - X^*} \qquad (4.12)$$

where Xi is the inlet contaminant mole concentration, Xo is the outlet contaminant mole concentration, and X^* is the equilibrium contaminant mole concentration.

It should be noted that the concentration is expressed in terms of mole fraction, which is applicable for the dilute wastes commonly encountered. Some procedures use mole *ratio* as a better approximation to represent operating conditions as a straight line.

Concentrated Wastes

For concentrated solutions, the number of transfer units can be estimated by adding the integrals within the waste operating concentrations to the logarithmic term.

Dilute Wastes

For dilute solutions, the number of transfer units (Nl) for a stripping operation can be estimated using the *log mean driving force* equation, simplified for conditions commonly encountered in waste treatment. For dilute solutions adhering to Henry's Law, in which both operating and equilibrium lines are straight, and heat effects negligible, the *log mean* equation can be simplified, using the stripping factor (S).

$$Nl = \frac{S}{S-1} \cdot \ln\left[\frac{Cr \cdot (S-1) + 1}{S}\right] \qquad (4.13a)$$

$$S = \frac{H \cdot G_m}{L \cdot P_t} \qquad (4.13b)$$

$$S = \frac{HG_m}{L_m}, \text{ when } P_t \equiv 1 \text{ atmosphere.} \qquad (4.13c)$$

These procedures may not apply at significantly reduced or elevated pressures, and when the H constant is not applicable at the process temperature and pressure conditions.

Cr is the concentration factor defined as

$$Cr = \frac{Xi - Yi/H}{Xo - Yi/H} \qquad (4.13d)$$

S is the stripping factor, dimensionless
H is Henry's constant for the compound to be removed, atm
L_m is the liquid loading rate, kg moles/hr-m²
G_m is the air loading rate, kg moles/hr-m²
Xi is influent concentration of the compound to be removed, mole fraction
Xo is the effluent concentration of the compound to be removed, mole fraction
Yi is the inlet gas concentration, mole fraction
P_t is the ambient pressure, atm

A further simplification can be made, assuming Y/H is a small number, or as is usually the case the inlet air concentration is zero, so that

$$Nl = \frac{S}{S-1} \cdot \ln\left[\frac{(Xi/Xo) \cdot (S-1) + 1}{S}\right] \qquad (4.14)$$

Values for this relationship are commonly presented as plots, such as Figure 4.7 [3].

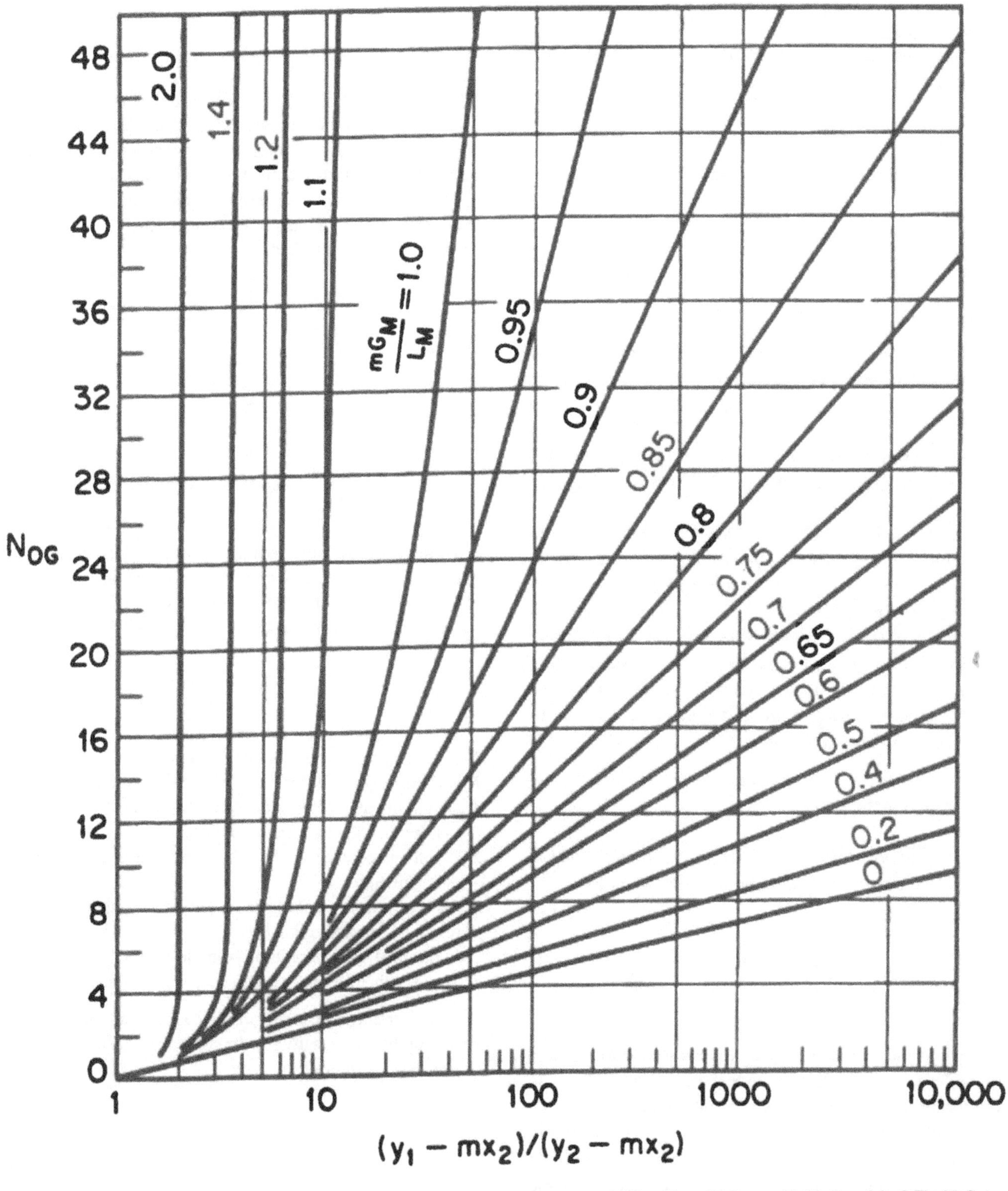

48
44
40
36
32
28
24
20
16
12
8
4
0
N_{OG}
1
10
100
1000
10,000
$(y_1 - mx_2)/(y_2 - mx_2)$
2.0
1.4
1.2
1.1
$\frac{mG_M}{L_M} = 1.0$
0.95
0.9
0.85
0.8
0.75
0.7
0.65
0.6
0.5
0.4
0.2
0

Figure 4.7 Stripper NTU plot for dilute wastes. (Sherwood and Pigford, *Mass Transfer,* Third Edition, Figure 14-9, page 14–20. Copyright © The McGraw-Hill Companies, Inc. Reproduced with permission. All rights reserved.)

Height of a Transfer Unit

The height of a transfer unit is a measure of the resistance of a component through the gas (Rg) and liquid (Rl) phases, the total resistance expressed as

$$R = Rg + Rl \qquad (4.15a)$$

Based on Equation (4.15a), the total resistance can be defined in terms of the individual mass transfer coefficients (k) and the slope of the straight line equilibrium curve (m).

$$\frac{1}{K} = \frac{1}{kg} + \frac{m}{kl} \qquad (4.15b)$$

Note: m is used in place of the Henry constant (H) to avoid confusion with the transfer height (H)

The transfer unit (H) is proportional to the reciprocal of the mass transfer coefficient (K or k). Therefore, the height of a transfer unit can be expressed in terms of the gas phase as

$$Hog = Hg + Hl \cdot \frac{m \cdot G_m}{L_m} \qquad (4.15c)$$

or in terms of the liquid phase as:

$$Hol = Hl + Hg \cdot \frac{L}{m \cdot G} \qquad (4.15d)$$

Process Parameters Related to HTU

The parameters that contribute to the individual HTU values are the gas (G) and liquid (L) mass transfer loadings and the waste characteristics. The significant waste characteristics include its temperature, viscosity, density, the diffusivity of its components, and the solubility of the components being stripped.

The relation of the liquid and gas loadings to phase transfer can be summarized as follows [3]:

(1) The gas transfer coefficient is proportional to the individual mass loading rates as

$$kg = G^{0.7} \cdot L^{0.3} \qquad (4.16a)$$

The corresponding height of a transfer unit can be expressed as

$$Hg = [G/L]^{0.3} \qquad (4.16b)$$

Note: G, L in items (1) and (2) are mass loadings.

(2) The liquid transfer coefficient is proportional to the liquid mass loading rates as

$$kl = L^{0.5} \qquad (4.16c)$$

The corresponding height of a transfer unit can be expressed as

$$Hl = L^{0.3} \qquad (4.16d)$$

(3) When a gas is relatively insoluble in water, denoted by a large Henry's Constant, stripping is viable and the liquid phase resistance usually dominates. In such cases the overall resistance is assumed to be in the liquid phase and the height of a transfer unit can be approximated using the liquid phase resistance, neglecting the gas phase resistance.

(4) The converse is true if the gas is relatively soluble in water.

(5) The gas or liquid loadings employed could negate the absolutes implied in (3) and (4), and the individual transfer heights increase or decrease in proportion to the applied rates as indicated in (1) and (2).

Application of Basics to Waste Stripping

The basics discussed are important when establishing the height of packing required for a treatment system, "testing" common presuppositions on the basis of the specific operating parameters and waste characteristics. Assumptions commonly applied to waste stripping evaluations include:

(1) Because stripping involves desorption of volatile components from wastewaters, it is convenient to express the height of a transfer equation in terms of the liquid phase, as defined by Equation (4.15d).

(2) Since stripping is usually applied to wastes containing volatile and therefore relatively insoluble components, the liquid phase resistance controls, the gas phase HTU can be neglected, and the overall HTU is approximately equal to that of the liquid phase.

$$Hol = hl, \qquad Hg \cdot \frac{L}{m \cdot G} \text{ is negligible} \qquad (4.17)$$

This is not always true, stripping is applied to wastewaters containing ammonia and similar compounds which are soluble and therefore could have significant gas and liquid phase resistance.

(3) It is commonly assumed that a reasonable estimate can be made from available theoretical coefficient equations. This may be true if the relations were developed for the specific waste components being considered, which is almost never the case. Most relations were developed from laboratory tests using specific gases, under limited conditions deliberately detailed by the researcher, and many times ignored by the Process Engineer. The results of theoretical evaluations are usually presented in terms of loadings, with exponents used to correlate all op-

erating and tower conditions. The difficulty in using general models is that the limited compounds tested, the volatility of the components, and the packing type and process loadings may not correspond to the conditions being evaluated. Estimated values could be a fraction (or multiples) of operating values.

Realistically, test data will not always be available or easily obtained, and estimates must be made. In such cases, the Process Engineer must consider the accuracy required and the importance of the stripping operation to the overall facility treatment system, deciding whether more accurate data must be obtained. When the stripper is a vital process in meeting effluent quality, site specific verification studies are essential. Heights of transfer units should be established for the specific wastes being treated, taking into account all the waste components and the applied tower characteristics.

Liquid Phase HTU Estimates

The height of a transfer unit (HTU) can be approximated using related reported data or theoretical equations. Chemical engineering design handbooks commonly cite the following Sherwood and Holloway [3] relations as a means of estimating the height of a liquid transfer unit.

$$Hl = \frac{L}{K_L a dl} = \frac{1}{\sigma} \cdot \left[305 \cdot \frac{L}{u} \right]^n \cdot \left[\frac{u}{dl \cdot D} \right]^{1/2} \quad (4.18)$$

Hl is the height of the liquid phase unit, meters
Kl is the mass transfer coefficient, [kg moles/(s · sq meter)
(kg · moles/cu meter)]
L is the liquid rate, kg/s-m^2
u is the liquid viscosity, Pa · s
D is the diffusion coefficient of the component in water, m^2/s
dl is the liquid density, kg/cubic meter
σ and n are coefficients, as indicated in Table 4.3

This equation *must* be used with the metric units defined. The original version of this equation was presented using English units and without the 305 factor. In such cases Hl is in feet, L is in lb/square foot/hr, u is in lb/ft/hr, dl is in

lbs/cubic feet, D is in sq ft/hr, and the σ' constant must be used.

Similar relations have been reported correlating compound removal in air-water systems [3,10–13]. The use of any of these methods should include a complete understanding of their limits. Other correlations or data may prove more advantageous to specific process conditions and should therefore be used. The Sherwood and Holloway correlation may be useful in the early phases of a system evaluation.

Gas Phase HTU Estimates

Gas phase transfer coefficient or HTU correlations are limited because of the difficulty in isolating the gas controlling variables and in establishing the effective transfer area [3]. Generally, theoretical estimates summarize the combined loading effects as

$$Hg = G^a/L^b \quad (4.19)$$

In addition, this coefficient is usually assumed to be negligible in common stripping operations, although this is not always true, as illustrated in Case Study 29.

Tower Height

The installed tower height establishes the removal efficiency and the resulting achievable effluent quality. The theoretical packing bed height must be adjusted for two factors, packing efficiency and compensation for entrance and exit losses (minimum one foot). Space for additional packing height should be considered for any future requirements. The theoretical bed height (Z) is a product of the height of the transfer unit (HTU) and the number of transfer units (NTU).

$$Z = \text{HTU} \times \text{NTU} \quad (4.20)$$

Pressure Drop

Actual packed bed pressure drop data can be found in the manufacturer's literature, commonly included as part of the flooding curves. The minimum *dry tower* pressure drop can be estimated by the orifice correlation [3,4]:

$$dP = C_1 \cdot P_g \cdot U_l^2 \quad (4.21a)$$

where dP is the pressure drop, inches water/ft of packing, P_g is the gas density, lb per cubic foot, U_l is the superficial gas velocity, feet per second, C_1 is the packing constant

TABLE 4.3. Sherwood and Holloway Constants (adapted from References [3,4]).

	Packing				
	mm	in	σ	σ'	n
Raschig rings	25	1	330	100	0.22
	38	1½	295	90	0.22
	50	2	260	80	0.22
Berl saddles	25	1	560	170	0.28
	38	1½	525	160	0.28
Tile	75	3	360	110	0.28

TABLE 4.4. Pressure Drop Coefficients for Dry Packed Beds (adapted from Reference [4]).

	Size		
	in	mm	C_1
Raschig rings	1	25	0.53
	1½	38	0.32
	2	50	0.25
Intalox saddles	1	25	0.43
	1½	38	0.18
	2	50	0.14
	3	76	0.073
Intalox saddles	1	25	0.18
	1½	38	0.11

(Table 4.4) [4], and constants are for units indicated (to obtain cm water per meter of packing multiply by 8.33).

Leva [3] developed the following correlation to estimate the pressure drop under irrigated packing conditions:

$$dP = C_2 \cdot 10^{C_3 \cdot V_t} \cdot P_g \cdot U_t^2 \qquad (4.21b)$$

where V_t is the superficial liquid velocity, feet per second, and C_2, C_3 are constants presented in Table 4.5 [3].

Tower pumping power requirements can be estimated using Equations (4.22a) and (4.22b).

$$hp = Q \cdot H \cdot s/3,960 \qquad (4.22a)$$

or

$$KW = q \cdot h \cdot d/367,000 \qquad (4.22b)$$

where Q is the gpm waste flow, q is the cubic meter/hour waste flow, H is the ft liquid head, h is the meter liquid head, s is the specific gravity, and d is the water density, kg/cu m.

The blower power can be estimated using Equations (4.23a) and (4.23b).

$$hp = 0.000157 \cdot Qf \cdot P \qquad (4.23a)$$

or

$$KW = 0.0000272 \, qf \cdot p \qquad (4.23b)$$

where Qf is the fan flow, cfm; qf is the fan flow, cubic meters per hour; P is the inches of water column drop; and p is the centimeter of water column drop.

STRIPPING-TRAY TOWERS

Tray tower process design follows the basic sequence outlined for packed systems, except that contact is based on intermittent stage contact rather than the integrated contact within a packed segment. Based on this definition, a continuous stripper can be explained as a series of ideal contact stages, as illustrated in Figure 4.8. Contact stage one being the entrance, followed by stage two, and the exit being stage N. A theoretical contact stage is defined in Figure 4.1. As a result, the number of theoretical trays represent the number of contact stages required to meet effluent quality. The number of trays installed is based on the theoretical stages adjusted for tray efficiency. The product of the number of trays times the tray spacing establishes the tower height.

Process evaluations to establish the equilibrium line, the liquid to gas ratio, the air flow rate, and tower material balances are similar to those discussed for a packed column. Calculations to estimate the number of stages and column diameter are explicit to tray towers. Because of the mechanical and hydraulic considerations of tray design to optimize contact efficiency, specific tray mechanical details, and resulting tower details should be left to the manufacturer.

TABLE 4.5. Leva Pressure Drop Coefficients for Irrigated Packed Beds (adapted from Reference [3]).

	Size			
	in	mm	C_2	C_3
Raschig rings	1	25	0.80	0.0348
	1½	38	0.30	0.0320
	2	50	0.28	0.0236
Berl saddles	1	25	0.40	0.0236
	1½	38	0.20	0.0181
Intalox saddles	1	25	0.31	0.0222
	1½	38	0.14	0.0181

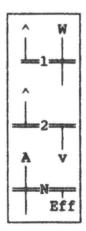

Figure 4.8 Tray tower stages.

Number of Ideal Stages

The number of ideal stages generally follows the same principles that govern the number of packing transfer units, modified for tray characteristics, tower configuration, and the waste concentration. Dilute solutions could involve air stripping and directly exhausting limited stripped pollutants, exhaust gas treatment, or a closed steam stripper system. Concentrated waste streams always involve closed systems with the substrate concentrated and recovered or further treated, as discussed for steam stripping. Direct discharge of a concentrated air stream could result in a serious air emission problem.

When the waste is dilute (concentrations are expressed as mole fraction) and the operating and equilibrium lines straight, the number of theoretical plates can be estimated in a manner similar to a packed column, using the equation [3]

$$Nl = \frac{\ln \left[(1 - A) \cdot Cr + A\right]}{\ln S} \qquad (4.24a)$$

$$S = \frac{H \cdot G_m}{L_m} = \frac{1}{A} @ P_t = 1 \text{ atmosphere} \qquad (4.24b)$$

$$Cr = \frac{Xi - Yi/H}{Xo - Yi/H} \qquad (4.24c)$$

A is the reciprocal of the stripping factor S
G_m is the molar gas phase, kmole/s-sq meter
L_m is the molar liquid phase, kmole/s-sq meter
X is the substrate molar fraction in the liquid
Y is the substrate molar fraction in the gas
i is the inlet condition
o is the outlet condition
P_t is the pressure, atmospheres
H is Henry's Constant for the compound to be removed, atm

Plate Efficiencies

Plate efficiencies are a basic property of the tray design, and best obtained from the manufacturer or from tests. Empirical methods are discussed in design manuals but beyond the scope of this text [3]. Approximate efficiencies based on an O'Connel absorption correlation are indicated in Table 4.6 [3].

Plate Spacing

Tray spacing depends on specific mechanical and hydraulic properties, the costs' effect on tower height, and maintenance space required for large diameter units. Towers treating relatively clean liquids, requiring minimum maintenance, could have spacing of approximately 30 centimeters (12 inches); with dirty liquid requiring as much as 60 centi-

TABLE 4.6. Approximate Overall Tray Efficiencies (adapted from Reference [3]).

$dl/H \cdot M \cdot \mu^*$	0.02	0.08	0.35	1.0	3.5	10
Overall plate efficiency, %	10	20	30	40	50	60

*Where dl is density, lb/cf; $H = y^*/x$, M = molecular weight, and μ is viscosity in cp.

meters (24 inches). The additional total height costs imposed by increased tray spacing may be offset by a smaller tower diameter resulting from increased tray capacity, as indicated by Equation (4.25). The approximate relation between tower diameter and spacing for bubble-cap trays is indicated in Table 4.7 [6].

These guidelines are presented with the understanding that tray spacing is more complex that selecting a convenient number, since it depends on (1) the type tray selected, (2) the individual tray pattern, (3) the tray hydraulic stability, (4) the ability to install and maintain the trays, (5) the tray efficiency, and (6) total tower *costs*. For that reason the tower internal mechanical design should be left to the manufacturer.

Tower Diameter

Tower diameter, as with packed towers, is governed by the system hydraulics, defined as the optimum gas velocity below the flooding value. Historically, Equation (4.25) has been proposed as a guide to approximate workable vapor rates and tower diameter [5].

$$V = K \cdot \left[\frac{(Dl - Dg)}{Dg}\right]^{0.5} \qquad (4.25)$$

where V is the superficial gas velocity through the tower, fps; Dl is the liquid density, lb per cubic foot; Dg is the gas density, lb per cubic foot; and K is a tower related constant ranging from 0.15 to 0.2, Table 4.8. Based on the constant

TABLE 4.7. Approximate Bubble Cap Spacing (adapted from Reference [6]).

r, feet (meters)	Bubble Cap Tray Spacing, inches (cms)	
	Minimum 6	(15)
4 or less (1)	18–20	(46–51)
4–10 (1–3)	24	(61)
10–12 (3–4)	30	(76)
12–24 (4–7)	36	(91)

TABLE 4.8. Tray Tower Capacity *K* Value (adapted from Reference [5]).

Plate Spacing, in	Liquid Seal, in		
	0.5	1	2
6	0.02–0.04		
12	0.09–0.11	0.07–0.09	0.05–0.07
24	0.185	0.17	0.16
36	0.205	0.195	0.19

in Table 4.8 and the units defined, meters/second can be obtained by multiplying by the factor 0.3048.

The tower diameter selected depends on the type of tray selected and the required seal. As a example, bubble cap trays can have a seal of 0.5 inches (1.3 cm) if operated at a vacuum, 1 inch (2.5 cm) for atmospheric conditions, and up to 3 inches (7.6 cm) if operated at a high pressure [6]. The selected seal must be optimized to minimize tower pressure drop and resulting operating costs, and therefore left to the manufacturer as part of the mechanical design.

The tower capacity for sieve, valve, and bubble plates has been defined in terms of the maximum flooding velocity [3,8]

$$V = Cb \cdot \left[\frac{\sigma}{20}\right]^{0.2} \cdot \left[\frac{(Dl - Dg)}{Dg}\right]^{0.5} \qquad (4.26)$$

where σ is the surface tension, dyne/cm or mN/meter; Cb is the capacity parameter, fps or meter/s, Table 4.9; and V is estimated as feet per second or meter per second, depending on the selected Table 4.9 Cb value, chosen as feet per second or meter per second.

Velocities ranging from 65 to 85% of the estimated flooding velocities are commonly applied for tower design. The cited references should be reviewed for tray design limits that apply to this equation [8].

STEAM STRIPPER

Figure 4.9 illustrates three typical steam strippers, each having similar process components. Feed is preheated using the hot treated effluent bottoms for heat exchange, and fed to the top of the column. The cooled effluent is discharged. Superheated steam is fed through the bottom of the tower, the waste and steam in counterflow. Within the column, the heated waste desorbs its volatile organics, whereas the steam absorbs the stripped volatiles. The vapor stream leaving the top of the tower, consisting of steam and the stripped organics, is condensed and discharged as a concentrated aqueous stream. In many cases, a reflux stream is recycled back to the top of the tower to assure a proper liquid to vapor ratio, or to act as an absorbing agent in the top enriching section.

The condensed overhead product is discharged from the system for recovery or disposal.

At this point it is important to distinguish between steam stripping and distillation. Steam stripping assumes a low influent concentration where steam can be replaced for air as an *inert* stripping agent. Distillation theory applies to a concentrated influent, where the gas stream is an active stream; the liquid and gas concentrations defined by appropriate equilibrium data, not a Henry's Law ideal straight line. In general, steam stripping can be employed to [1,9]

(1) Remove small quantities of volatile compounds; especially since many industrial wastes contain immiscible, very high boiling point organic compounds that can be easily removed by condensation after stripped.

(2) Improve waste component volatility at an increased temperature.

(3) Remove components forming low boiling azeotropes with water.

(4) Remove compounds with a boiling point less than 150°C and a Henry's Law constant greater than 0.0001 atm-cu meter/mole.

(5) Recover organic product.

(6) Replace an air stripper requiring further exhaust treatment (such as incineration) with a more economical alternative. This assumes that condensing equipment costs are less than air exhaust treatment systems.

(7) Meet low effluent requirements.

Steam stripping effectiveness depends on the solubility and volatility of the wastewater components to achieve a concentrated disposal product. Since the overhead is a condensed water-organic product, the overhead "reject" stream must be a small fraction of the feed, and significantly more concentrated, for the treatment to be viable. This usually requires that vapor concentrations greater than 10% be exhausted from the feed plate. When the vapor concentration is not adequate, an absorber must be added above the feed plate as an enriching section to increase the overheads concentration.

Applicable steam stripper design procedures depend on the feed substrates solubility limits, which could include total miscibility, total immiscibility, or partial miscibility. A totally miscible system implies complete solubility, which at the low concentrations frequently encountered in waste treatment, Henry's Law and relevant design procedures discussed for air strippers apply. When the substrate is totally immiscible, the ideal partial pressure defined by Raoult's or Henry's Law for soluble components does not apply, and tower height cannot be estimated using the equilibrium procedures discussed. When the substrate is partially immiscible the partition will be controlled by the applicable operating concentration and temperature, which are defined by a solubility curve as illustrated in Figure 4.10. Understanding the transfer mechanisms and physical characteristics for each of these solubility conditions is essential. As already men-

| T, inches | 6 | | 9 | | 12 | | 18 | | 24 | | 36 | |
| T, cm | 15 | | 23 | | 31 | | 46 | | 61 | | 91 | |
Fp	F/S	M/S	F/S	M/S	F/S	M/S	F/S	M/S	F/S	M/S	F/S	M/S
0.01	0.15	0.04	0.18	0.05	0.23	0.07	0.29	0.08	0.39	0.11	0.50	0.15
0.03	0.15	0.04	0.18	0.05	0.23	0.07	0.29	0.08	0.39	0.11	0.50	0.15
0.05	0.15	0.04	0.18	0.05	0.22	0.06	0.28	0.08	0.38	0.11	0.48	0.14
0.07	0.15	0.04	0.18	0.05	0.20	0.06	0.27	0.08	0.35	0.10	0.44	0.13
0.10	0.15	0.04	0.17	0.05	0.19	0.05	0.26	0.07	0.33	0.10	0.42	0.12
0.20	0.13	0.03	0.15	0.04	0.17	0.05	0.22	0.06	0.28	0.08	0.36	0.10
0.30	0.12	0.03	0.13	0.03	0.16	0.04	0.18	0.05	0.25	0.07	0.30	0.09
0.50	0.09	0.02	0.11	0.03	0.13	0.03	0.16	0.04	0.18	0.05	0.24	0.07
0.70	0.08	0.02	0.09	0.03	0.10	0.03	0.13	0.03	0.17	0.05	0.19	0.05
1.00	0.06	0.01	0.07	0.02	0.08	0.02	0.10	0.03	0.13	0.03	0.15	0.04
2.00	0.04	0.01	0.04	0.01	0.04	0.01	0.05	0.01	0.06	0.01	0.07	0.02

$Fp = (L/G) \cdot (d_g/d_l)^{0.5}$

L = liquid mass rate per hour
G = gas mass rate per hour
d_g = gas density, mass per volume
d_l = liquid density, mass per volume
T_{in} = tray spacing, inches
T_{cm} = tray spacing, centimeters
F/S = feet per second flooding velocity
M/S = meters per second flooding velocity

Assumptions
1. Low or nonfoaming wastes.
2. Weir height is less than 15% of the plate spacing.
3. Ratio of the bubble cap or perforated area to the total active tray area is greater than 10%. If not, following corrections applied.

Area Slot/Total Active Area	Correction
0.10	1.0
0.08	0.9
0.06	0.8

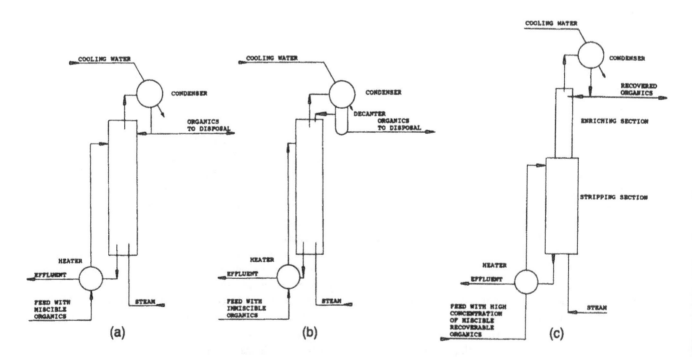

Figure 4.9 Steam stripping configurations (adapted from Reference [1]).

114

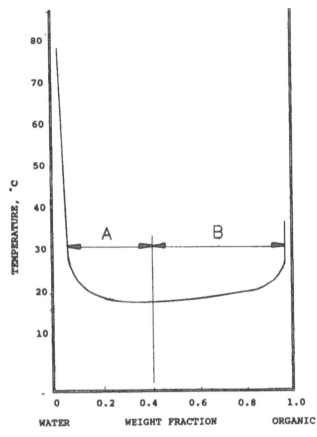

Figure 4.10 Solubility of triethylamine in water (adapted from Reference [14]).

tioned, totally miscible waste components adhere to the theoretical procedures discussed for air stripping.

Steam stripping of totally immiscible mixtures involves a unique vapor pressure behavior completely different from that defining soluble components. Each liquid phase behaves as though the other phase(s) is not present, exerting its own full vapor pressure, independent of the other phase or concentration. The total vapor pressure is the sum of the pure vapor pressure of each of the liquids, assuming insignificant solubility of one in the other. Therefore, the total pressure of a waste containing water and an immiscible impurity, being steam stripped, can be defined as

$$\pi = P_{imp} + P_{steam} \qquad (4.27)$$

where π is the total system pressure, P_{imp} is the impurity vapor pressure, and P_{steam} is the steam vapor pressure.

$$\frac{Y_{steam}}{Y_{imp}} = \frac{\pi - P_{imp}}{P_{imp}} \qquad (4.28)$$

or

$$\frac{W_{steam}}{W_{imp}} = \frac{(M \cdot P)_{steam}}{(M \cdot P)_{imp}} \qquad (4.29)$$

Where Y is the mole fraction in the vapor $[n_a/(n_a + n_b)]$, W is the mass of the component in the vapor, and M is the molecular weight of the component.

These conditions remain constant as long as the two phases exist. The driving force is controlled by the theoretical steam to component ratio in the vapor. Significantly, in steam stripping the resulting combined vapor pressure conditions reduce the boiling temperature of the waste component, as long as the two liquid phases exist, increasing the removal effectiveness. Accordingly, the tower design is not affected by vapor equilibrium concentrations of successive stages, since varying vapor conditions do not change. Instead, design is based on allowing adequate contact of the steam and the waste, and adequate time to complete stripping of the volatile phase.

A partially miscible system is described by a solubility curve, one in which temperature defines the solubility limits of a component, as illustrated by the triethylamine-water system in Figure 4.10. In that system the two liquids are completely immiscible below the minimum temperature 18.5°C; and partitioned into two liquids according to concentration and temperature above 18.5°C. As an example, a 40% waste at 30°C will be partitioned into two phases, one water phase containing about 5% triethylamine and another organic phase containing about 95% triethylamine; the liquid weight ratio of aqueous to organic being equal to the distances B:A. This misicibility characteristic, dependent on concentration and temperature, affects the products formed and the stripper configuration.

Based on the concepts discussed, the stripper configuration will depend on the solubility and volatility of the waste components. The first consideration being the concentration of the vapor exiting the feed section, the maximum being equal to its theoretical equilibrium concentration with the feed. Waste with highly soluble and volatile components will produce a low volume concentrated overhead which can be discharged for recovery or further treatment, as illustrated in Figure 4.9(a). However, as the feed plate vapor concentration decreases, because of low concentration or volatility, an enriching section above the feed is required to concentrate the vapors and reduce the overheads product volume, as illustrated in Figure 4.9(c). Partially or totally immiscible contaminants form a highly immiscible condenser product which can be decanted from the water. The liquid being recycled to the stripper for further processing, as illustrated in Figure 4.9(b). This is the case with some remediation wastewaters where the contaminant is present below its solubility limits, still presenting a water contamination problem, forming an immiscible solution upon stripping.

Stripper Design

Any of the Figure 4.9 configurations can employ packed columns or trays, treating either dilute or concentrated waste streams. Design procedures for miscible waste components are similar to those discussed for air stripping, using packed or tray columns; except that (1) steam is substituted for air, and (2) a condenser is required, steam is not directly discharged. Design procedures must be altered when the feed components are not soluble throughout the tower operating conditions.

(1) If the waste components are completely immiscible then physical separation should be employed as a primary treatment, especially at high concentrations. Components can be more economically removed since steam and condensing water are not required. Steam stripping can be used to remove the remaining trace concentrations. However, in such cases tower design must be based on required stripping residence time and adequate distribution of the steam and waste. Design cannot be estimated using the transfer unit and height of transfer procedures discussed for miscible solutions; but based on specific pilot testing and available process operating experience. However, the overheads product can be easily separated using an accumulation tank with decant capabilities.

(2) If the waste components are partially miscible, the solubility within the tower operating range establishes whether the contact procedures discussed for packed or tray towers can be used, or the residence time required to strip the immiscible components must be established. In addition, the solubility of the condenser products determines if separate, immiscible phases are formed. In some cases the influent is soluble at the waste concentration, enabling the use of Henry's Law constants to establish the tower design, whereas the condenser can be operated below the minimum solubility temperature to partition the condenser product into a recycled aqueous phase and a disposable concentrated organic layer.

Finally, some applicable steam stripping design considerations include [1,9]

(1) Suspended solids concentrations should be less than 2%.
(2) Operating pressures should be as low as possible to minimize steam use and maximize removal efficiency. Vacuum strippers operating at approximately 2 psia are effective in maximizing removal efficiency and preventing organic polymerization or precipitation. However, such systems require vacuum pumps and more complicated and expensive shell design. In addition, more noncondensables are introduced into the system in a vacuum system, reducing condenser effectiveness.
(3) Pressure systems are commonly employed in the 5 to 10 psig range.
(4) The gas (steam) to waste ratio must be maintained as near to the minimum requirements as possible, more so than in air systems, to reduce steam costs.
(5) Defining the component partitioning within the tower and at the condenser, as a result of volatility and solubility, is crucial in applying design methods and defining the treatment configuration.

Product Recovery

When product recovery is the primary consideration, three important factors affect the separation process: (1) the volatile feed concentration must be significant, (2) the gas stream is an active component of the separation containing the same components as the waste (water and the volatile components), and (3) the simple stripper becomes a more complex distillation process. The tower must be designed as two separate sections—a *stripping section* below the feed point and an *enriching section* above the feed point. Adequate packing height (or trays) must be provided above the feed point to assure that the contaminant is condensed and removed as an organic rich product (above 90 to 95%). Adequate packing must be provided below the feed point to assure that effluent quality is met. In most cases recovery quantities would involve large flow rates, and corresponding large organic quantities, making a tray tower the most likely configuration. In such cases, the system is almost always part of the manufacturing facility design, meeting required reusable material specifications, designed by the Chemical Process Engineer, applying computerized complex procedures not commonly used in waste stripping.

AMMONIA STRIPPER

Ammonia stripping chemistry, and associated equilibria relations, are indicated in Figure 4.11. It is evident that the process is pH and temperature dependent. The reaction is driven toward ammonium ion formation at a pH of 7 or lower, toward ammonia formation with increasing pH, and near complete ammonia generation at a pH range from 10.8 to 11.5 [7]. The 10.8 to 11.5 range is commonly employed for waste ammonia stripping.

EPA Evaluations

Operating criteria observed from EPA field studies are detailed in the *Process Design for Nitrogen Control* manual, and can be summarized as follows [7]:

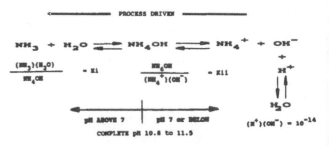

Figure 4.11 Ammonia stripping chemistry.

(1) *Types of towers commonly used:* Countercurrent and cross-flow towers, with countercurrent towers found to be more effective.

(2) *Tower packing:* Much of the packing utilized is similar to that employed for cooling tower design, consisting of redwood horizontal slats placed 2 inches (5 cm) apart, and vertically stacked to achieve the desired bed height. Other packings evaluated included 1/2 inch (13 mm) diameter PVC pipe and triangular shaped splash bars.

(3) *Hydraulic loadings:* The hydraulic loadings reported are related to both the bed depth and removal efficiency. Loadings greater than 3 gpm/ft² (2 L/sq m/s) resulted in rapidly declining efficiencies for 12, 20, and 24 feet (4 to 7 meters) tower heights, and less than 50% removals at loading in 6 gpm/ft² (4 L/sq m/s) range. At 24 foot heights removal efficiencies greater than 90% were observed at loadings from 1 to 3 gpm/ft² (0.7 to 2 L/sq m/s). Equivalent efficiencies were observed at a 20 foot height (6 meters) and loadings ranging from 1 to 2 gpm/ft² (0.7 to 1.4 L/sq m/s), and 70% removals were recorded at a 12 foot (4 meters) packing height and a 1 gpm/ft² (0.7 L/sq m/s) loading.

(4) *Air loadings:* Warm weather air loadings of 300 cubic feet per gallon of waste (2245 cu m/cu m) resulted in 90% removals, and 95% removal at 500 (3741) loadings. Cold weather (4°C) temperatures required loadings up to approximately 800 cubic feet per waste gallon (5986) to achieve 90% removal.

(5) *Temperature:* As would be expected, temperature greatly influenced the system performance because ammonia solubility increases with decreasing temperature resulting in less effective stripping, and the waste rapidly reaches ambient conditions near the tower top.

(6) *Alkalinity:* As indicated above, a pH range from 10.8 to 11.5 is recommended for maximum ammonia removals. EPA cited facilities used lime for pH adjustment.

(7) *Scaling:* Lime treatment increased calcium carbonate deposits, requiring lime control measures. This included maximizing calcium carbonate formation in the lime preparation step, using scale control chemicals, or using other alkaline neutralizing agents.

(8) *Air exhaust quality:* Ammonia tower air exhausts will contain 95% or more of the wastewater ammonia content, which could present an air pollution problem unless the quantity is below allowable emission levels, or it is removed.

It should be pointed out that the EPA studies cited are observations from a limited data base. More sophisticated techniques, applying the packed or tray column techniques discussed, are frequently applied for industrial ammonia stripping processes.

An acceptable emission level requires that either the exhaust concentration is not a problem, or that the mass rate is low enough to dilute to acceptable ground level concentra-tions. An acceptable ground level concentration may be defined by regulatory requirements, or as a guide compared to the following criteria:

(1) Ammonia odor threshold is at 35 mg per cubic meter.

(2) Eye, nose, and throat irritations are reported at concentrations ranging from 280 to 490 mg per cubic meter.

Actual regulatory limits could be a fraction of these values based on site specific conditions, existing air quality, and the specific population surrounding the site. Available removal technologies include two proprietary processes involving (1) ammonia recovery as a concentrated ammonium or ammonium sulfate fertilizer product in a low pH absorption tower, or (2) a nitrification-denitrification biological process converting ammonia to a nitrate, and reducing the nitrate to nitrogen gas. Ammonia recovery is only practical if a product market is available, while exhaust treatment to nitrogen could economically impact using ammonia stripping as a viable treatment method.

EPA reported ammonia stripping operating limitations include:

(1) Limited packing sizes can be used because of potential calcium carbonate scaling.

(2) Tower temperature reductions result because of the large air volumes needed.

(3) Imposed tower emission limits.

The design of an ammonia stripping tower is further explored in Case Study 29.

PROCESS ENGINEERING DESIGN

Wastewater stripping is a comparatively simple treatment system, with most operating variables related to air flow rate or wastewater control, as indicated in Table 4.10. Air

TABLE 4.10. Operating Characteristics.

Variable	Operator Controllable	Critical
Waste Characteristics		
Waste generated	No	Yes
Composition	No	Yes
Concentration	No	Yes
Waste volatility	No	Yes
Selective volatility	No	Yes
Operating Characteristics		
Waste flow rate	Minimal	Yes
Air flow rate	Yes	Yes
Feed temperature	Preheat	Yes
Recirculation	Yes	Yes
pH Control	Yes	Yes
Pretreatment	Yes	No
Emission control	No	Yes

or steam stripping design is relatively simple, with its validity largely dependent on accurate equilibrium data representative of expected operating conditions. Performance is fixed by the bed depth and feed characteristics, with optimum conditions achieved by controlling tower hydraulics. Waste treatment effectiveness is primarily a result of the volatility of the contaminants of concern, which can only be altered by the operator if the tower temperature can be controlled. Frequently, the major concern is emission control and not effluent quality. Emission control is a design consideration, and unless adequate equipment is provided, it is not an operating variable.

PERFORMANCE DATA

Reliable air and steam stripping data are difficult to cite because (1) stripping is not a widely used (or reported) waste treatment operation, with its prominence advanced for ammonia removal; (2) stripping is many times operated at less than optimum process conditions; (3) cooling tower design, rather than chemical process technology is frequently employed; (4) air flow is used as the primary control and poorly correlated to other operating parameters; (5) the volatility of the components as a limiting parameter is not always appreciated; and (6) it is commonly considered in removal of minute quantities of priority pollutants, which is a difficult separation process. Removal efficiencies of 95 to 99% are feasible when applied to applicable wastes components, utilizing available design techniques employed in the chemical and petroleum industry, and optimizing the operating parameters.

REQUIRED PROCESS DESIGN DATA

Laboratory tests can be conducted to establish Henry's Law constants for the component(s) in the wastewater, establishing the stripping contact efficiency. Stripping performance is affected by the packing selected, tower configuration, and hydraulic conditions. Such information can be easily obtained from pilot units prototypes to accurately reflect the tower physical effects. Resulting data can be correlated to obtain (1) a suitable liquid-vapor equilibrium relation, (2) a tower operating line defining the transfer from the liquid to vapor phase, (3) optimum liquid/gas rates, (4) achievable effluent quality, (5) achievable overheads product quality, whether that be an air emission rate or a condenser reject stream. The required data is listed in Table 4.11.

WASTE EVALUATION

An initial step in stripper design involves evaluating waste characteristics, such as flow and concentration, that define the required stripping gas capacity. *Feed flow* affects tower stability, especially since a stripping tower has a low residence time and low internal equalizing capabilities. There-

TABLE 4.11. Required Design Data.

Critical laboratory data specific to the waste
1. Check the waste composition and treatment chemistry
2. Design temperature
3. Tower operating pressure, psia
4. Equilibria data

Process criteria that should be obtained from full-scale laboratory studies, but can be estimated
5. Stripping media to be employed
6. Operating liquid to gas ratio employed
7. Contact effectiveness
8. Required bed height
9. Achievable effluent quality
10. Air exhaust concentration

And where applicable, the following additional data
11. Optimum pH range
12. Quantity of neutralizing agent
13. Fouling potential
14. Overheads treatment

And if steam stripping is employed
15. Steam rate
16. Overheads decant concentration
17. Steam pressure and temperature

Selected operating characteristics
18. Packing or trays to be used
19. Product recovery is practical
20. Feed should be heated

fore, a suitable design rate must be selected to provide hydraulic stability, enhanced by upstream equalization and tower recycle capabilities to effectively control the total feed rate. Once a tower is operating outside its flow capacity performance deteriorates. Excessive flow velocities cause liquid hold-up and flooding; while low flows result in poor packing wetting, "dead" spots, and channeling.

Feed volatile content impacts removal efficiencies. In fact, most wastes contain multiple components, many of which could be volatile, all of which are seldom identified. For the sake of simplicity, strippers are frequently designed on the basis of a representative volatile component, usually the least volatile component to be stripped. However, stripping is not selective, removing mass components whose volatilities are in the range of the targeted compound, somewhat proportional to the feed concentration. Neglecting all volatile components that could be affected by the selected design conditions could result in an underdesigned system. A more precise design can be obtained by utilizing multicomponent design procedures—which is more complicated and require more precise equilibrium data—or basing the design on pilot test results.

Gas flow is an operating variable selected to stabilize the tower hydraulics and to provide adequate "carrying" capacity to transport stripped volatile components. It is critical that a suitable design waste feed flow and a concentration are selected, so that a corresponding gas flow range can be determined.

PRETREATMENT

Stripper *pretreatment requirements* are usually minimal, based on (1) removing suspended or dissolved solids that could foul the packing, (2) eliminating oil, grease, or floatable components, (3) adjusting the pH to meet process requirements and assure compatibility with the tower or packing materials, and (4) assure that the feed is at the required operating temperature. It is prudent to install a coarse waste feed filter to remove solids that could foul either a preheater (if included) or the packing.

PROCESS DESIGN VARIABLES

Waste stripper design involves applying the process basics discussed to meet the required process requirements. These include the following items:

(1) Quality of design data
(2) Operating pressure
(3) Operating temperature
(4) Process selection
(5) Packing versus trays
(6) Energy balance

Quality of Design Data

Because stripper design is so often based on theoretical models, many times simplified for a specific application, the quality of the design data and the model assumptions should be scrutinized (or at least understood). Some common design assumptions include

(1) Most waste models assume the waste contains a single, well-defined volatile compound. However, wastes seldom contain single or necessarily identifiable components. If the most volatile component is assumed as the design basis, and it is not proportionally the predominant compound, the resulting tower design may not adequately meet process requirements.
(2) Stripping involves dilute solutions in which the molar fractions (x and y) are small, the molar flows are constant, and the operating conditions defined by a straight line.
(3) The equilibrium curve is a straight line, accurately represented by Henry's Law.
(4) The height of a transfer unit is a constant value for each stage.
(5) In steam stripping calculations constant molar flows, along with isothermal conditions, simplify the calculation.
(6) Air stripping calculations are based on constant molar flows and adiabatic conditions, and any temperature change either insignificant or accounted for in the de-

sign. Isothermal conditions are not valid assumptions because humidification can reduce the waste temperature.

(7) The number of theoretical stages or transfer units can be converted to the required tower height based on available efficiency values.
(8) The gas–liquid contact does not involve a chemical reaction.

It cannot be overly emphasized that the assumptions of the selected design procedures should be reviewed, understood, and consistent with the required design accuracy.

Operating Pressure

High pressure operating conditions should be avoided because waste volatility decreases with increased pressure. In some special cases it may be desirable to operate the system under slightly vacuum conditions to improve stripping conditions, although the additional equipment and operating (vacuum) costs are seldom warranted for waste treatment. Most waste treatment strippers operate at *atmospheric pressure*, or slightly above, to accommodate any downstream pressure requirements.

A special problem for steam stripping is minimizing air leakage into the tower, resulting in excessive noncondensables in the condenser, affecting condenser performance. In turn, "bleeding" noncondensables from the system can be a problem since organics will discharge with the gases, the condition further heightened if the components are odorous. When steam stripping is applied, the operating pressure should be as low as possible to simplify the design and reduce fouling potential.

Operating Temperature

Operating temperature is a critical design parameter since it impacts stripper performance. Air strippers are commonly designed for atmospheric conditions, ideally at operating temperatures of 16°C (60°F) or higher. The tower temperature is a product of the waste temperature, ambient conditions, and (minimizing) system heat losses. The primary consideration in evaluating air stripping is the component volatility at ambient conditions, preheating is seldom applied unless an excessive waste heat is "begging to be recovered." This necessitates that the component be very volatile for effective year around operation, because an open system is a cooling tower functioning at peak heat transfer capacity in the winter months. Steam stripping is generally considered when stripping is the most viable option, increased tower temperature is beneficial, air stripping will not meet effluent requirements, direct contaminated air exhaust is not practical, or product recovery is economically feasible.

Besides the required tower temperature, the temperature change in the tower is a significant factor. Most applied design procedures assume either an adiabatic or isothermal condition, although more sophisticated techniques could be applied to integrate varying tower conditions.

In an isothermal system the temperature is assumed constant throughout the tower depth, resulting in constant thermodynamic conditions. This is the design assumption for most strippers treating low concentration wastes.

When large waste concentrations are encountered, simple stripper assumptions are not applicable. The primary design limitation is that a direct exhaust of volatiles is not possible, a closed system is required, and condenser product is recovered. Depending on the feed contaminant(s) volatility, the tower bottoms may require heat input using a reboiler or direct steam to meet effluent quality. Under these conditions, design procedures applied for basic stripper design are no longer valid. The major deviation from stripper basics is that the temperature conditions change within the tower length; the bottom being near the boiling point of water and the top near the feed composition boiling point or that of the overheads product. In such cases, Henry's Constants are not applicable, equilibrium data must be used, and *distillation theory* applied. Design of distillation and recovery columns is beyond the scope of this text, but detailed in Chemical Engineering texts [3–6].

Design assumptions of either adiabatic or isothermal conditions should be understood. These include

(1) The validity of simplifying assumptions that the equilibrium is a straight line function, or that the molar flows are relatively constant, depends on a small tower temperature gradient.

(2) If heat losses from a system are large, the tower operating temperature will change.

(3) Temperature losses in an air stripper are primarily from evaporative cooling, which could be reduced if the air stream were saturated prior to entering the tower. This is not a common practice, but not difficult to accomplish using a simple spray section. Other tower losses from relatively low bulk fluid temperatures are negligible, except in extreme winter conditions. Where severe conditions are expected, losses can be minimized by insulating.

(4) In steam stripping the first consideration is the feed concentration and its boiling point, since the temperature can range from 212°F at the hot end to the feed boiling point. For low concentrations the temperature span is low since the feed boiling point is close to water. As the volatile feed concentration increases the resulting boiling point defines the tower operating temperature range, and a significant temperature difference can occur between the inlet and outlet conditions. In such cases a simple Henry's law constant may not define the entire tower conditions and more complex design procedures will have to be employed. In addition, constant molar flow rates are an important design assumption, based on using a straight line to define operating conditions. Adiabatic conditions must be assured to obtain these conditions.

Any deviations from adiabatic or isothermal conditions does not exclude stripping as a viable treatment system but may exclude simplified design techniques.

Process Selection

The basic air stripper, employing an open tower, is a relatively simple process. However, process conditions could warrant equipment enhancements, such as (1) using steam instead of air, (2) employing a reboiler instead of direct steam, (3) required downstream exhaust treatment, or (4) product recovery.

Air or Steam Stripping

The use of air or steam as the stripping agent depends on the waste characteristics. Specific design applications are discussed in detail in the Air and Steam Stripping sections. Some additional considerations include

(1) Once process criteria are established, capital and operating costs are a strong driving force favoring air or steam.

(2) Air stripping is applied to low volatile and low waste concentrations. Low waste concentrations are necessary to minimize air missions to acceptable levels. However, low volatility is required for effective separation at a low concentration.

(3) Steam is applicable over a wide range of influent concentrations, and could be cost effective at concentrations down to 2 ppm [1].

(4) Low waste volatility at ambient conditions will necessitate improving stripping conditions by either directly heating the waste or using steam (or a reboiler).

(5) Steam may be economical if the required amount is at least eight times less than air [8].

(6) Steam must be available, installing a new boiler for common waste flows could negate many of its advantages.

(7) Steam may enhance fouling and foaming problems.

(8) Steam is more expensive than air.

(9) A steam system could be more expensive and complex than a basic air stripper, although downstream exhaust treatment equipment for air strippers could rapidly change the economics.

(10) Leakage into a steam system injects noncondensables, which could complicate condenser design and reduce performance, and when vented allow organic or odorous emissions.

(11) More complex controls may be required for a steam system.

Steam stripping commonly implies live steam injection, although a reboiler can be used as the bottom separation stage. A reboiler requires additional capital costs, justified only when large steam quantities are to be used, and condensate recycle is an economical advantage.

Air Emission Control

Exhaust control is an important consideration when evaluating an air stripping system. Direct tower exhausts are only permissible when feed concentrations are low, and allowable exhausted odor or compound limits high. In reality, no matter how low the emissions, or high the tolerance levels, the potential for future liability exists "when new knowledge is developed," at which time it is expensive to prove "what you did not cause."

The options for controlling or treating stripper exhausts include

(1) Incinerate the overhead exhaust, ideally as makeup air to a large waste incinerator, or if necessary a separate incinerator. Whether this is less expensive than directly incinerating the waste will have to be evaluated.

(2) Another approach is to run the gas stream through an *adsorber*, assuming that the volatile compounds are easier to adsorb from the gas stream than the original waste stream. This is frequently the case, especially when the waste stream contains considerable quantities of other nonvolatile components detrimental to carbon or the chosen adsorbent.

(3) Use steam stripping as a means of avoiding a direct polluted exhaust, collecting the volatile as a concentrated organic which can be more readily disposed, incinerated, or recovered.

(4) Recover the exhaust components as a usable product, or convert them to a usable product. Proprietary processes for ammonia recovery are available, and were discussed under *ammonia stripping*. This assumes that there is a market for recovered products warranting process recovery equipment.

Although air stripping of volatiles may not appear practical if emission control equipment must be included, the same problem may occur in other treatment processes. Many of the other alternatives (such as biological systems) air strip the same volatiles, discarding them as *fugitive emissions* from one or more of the system components. In fact, as toxic air emission control (direct or fugitive emissions) becomes highly regulated, controlled removal of volatile organic material as a pretreatment system may be a prerequisite to applying any of the more conventional treatments. In many cases stripping is a more economical alternative than attempting to collect and destroy fugitive emissions from open basins or each of the transmission points. If fugitive emissions from other treatment systems with similar wastes are not a problem, then direct discharge of stripped air emissions is probably not a problem.

Product Recovery

Waste recovery is usually employed when the feed contaminant concentrations are high. This frequently requires a distillation system employing an enriching system to concentrate the product and either direct steam or a reboiler in the stripping section to meet effluent quality. System configurations are discussed in the Steam Stripping section.

Packing versus Trays

Packed columns are frequently used for the waste volumes common in industrial facilities. As a general rule, packing is cheaper than plates for towers less than 0.6 meters (2 feet) in diameter, in some cases up to 1.2 meters (4 feet), unless some special alloy packing is used [3]. Basically, the reasons for using packed columns in waste treatment systems are [3,8]

(1) Smaller diameter towers associated with packed beds are economical for waste flows commonly encountered.

(2) Packing is more economical than trays for corrosive wastes.

(3) Tray bubble caps, sieves, and valves are more difficult to maintain than packing.

(4) The system can operate at a large turn-down ratio, allowing a large range of waste flows for a specific tower configuration.

(5) They generally require a lower pressure drop than the tray tower.

(6) Construction is relatively simple.

(7) Media can be easily handled and replaced. Packing changes to alter system performance parameters such as pressure drop, waste characteristic change, or treatment efficiency are relatively simple.

(8) They are preferred for wastes with foaming characteristics.

However, trays have some distinct advantages that make them a viable selection for specific treatments [3,8]:

(1) High liquid rates are more economically handled with trays.

(2) They may be more effective for difficult separations, requiring a large number of transfer units to achieve stringent effluent limits.

(3) More effective hydraulic contact is possible, with less chance for liquid channeling.

(4) Individual plates are more easily designed for cleaning than packing, and could be less of a concern for dirty wastes.

(5) They are less affected by expansion and compression forces resulting from high influent temperature variations that could damage packing.

Energy Balance

A system energy balance is important to verify some basic design conditions:

(1) That Henry's Law assumptions are valid. The system temperature change is insignificant, and the Henry Constant does not significantly change throughout the tower.

(2) That the gas and liquid flows are reasonably constant throughout the tower.

Any large heat release, resulting in significant temperature increases, indicates that system flows or equilibrium relationships will not remain constant throughout the operating range. A tower energy balance is defined by Equation (4.30).

$$Li*hi + Gi*Hi = Lo*ho + Go*Ho + Q_d \quad (4.30)$$

where L is the liquid mass rate; G is the gas mass rate; H is the gas enthalpy, heat content/pound; H is the waste enthalpy, heat content/pound; i refers to the inlet conditions; o refers to the outlet conditions; and Q_d is the heat of desorption.

The assumptions of constant flow rates, isothermal conditions, and applicability of a single Henry's Law constant throughout the column will not be valid if heat losses are high.

Large Heat Losses

If the heat of desorption (Q_d) is significant, an appreciable temperature gradient within the tower could result. Equally significant are large tower surfaces, which when combined with poor insulation, result in large ambient losses in winter conditions. In either of these conditions the ideal process conditions assumed for simple stripping evaluations may not be valid. However, large surface losses can be minimized with insulation, while large temperature changes from desorption would require significant waste concentrations.

Steam Stripping

Isothermal and adiabatic conditions are commonly assumed for steam stripping involving liquid and gas concentrations and low tower pressure drops. However, if the waste feed temperature is low, the cold waste could result in appreciable steam condensation, a gas flow decrease, and an effluent flow increase. This can be reduced by preheating the waste influent with the heated effluent, reducing the influent heating requirements.

Air Stripping

Depending on ambient conditions, especially the inlet air humidity, there can be a significant change in the water temperature, and an appreciable loss in liquid volume due to humidification. Assuming adiabatic conditions, the wastewater temperature could approach the wet bulb temperature. An estimate can be made of the air humidification and corresponding liquid temperature change using procedures common to cooling tower design [3,8].

FATE OF THE CONTAMINANTS

Treated effluent will be discharged from the tower bottoms. If air stripping without exhaust control is employed, *all* the waste volatile components removed will be discharged from the tower. In some cases the air streams cannot be exhausted without further treatment. Even if normal operating emissions are minor, upset conditions can result in neighboring nuisance problems, and resulting regulatory response. For air or steam stripping systems, an evaluation must be made of any possible detrimental effects of such emissions, including whether (1) the components emitted are hazardous or toxic, (2) the emission level (treated or uncontrolled) is significant, and (3) the estimated ground level concentrations are a potential hazard. The quick answer is that stripping is commonly utilized for removal of small quantities of volatiles in a waste stream, resulting in insignificant quantities of air emissions. That is usually the case, but it should be pointed out that in such cases stripping is frequently not effective.

Three streams can be discharged from a *steam stripper:*

(1) An inert (bleed) gas stream containing noncondensables and a minimum amount of organics

(2) A highly concentrated organic stream from the condenser, which can usually be incinerated or recovered (Any water separated in an immiscible separator can be recycled to the tower.)

(3) Treated effluent containing a minimal amount of organics

GENERAL ENGINEERING CRITERIA

Figures 4.12 and 4.13 illustrate typical steam and air stripping Preliminary Concept Flowsheets. A waste stripper includes the following basic process components:

Equalization (optional)	Chapter I-4
Waste feed pH or pretreatment	Chapter I-8
Waste feed system	

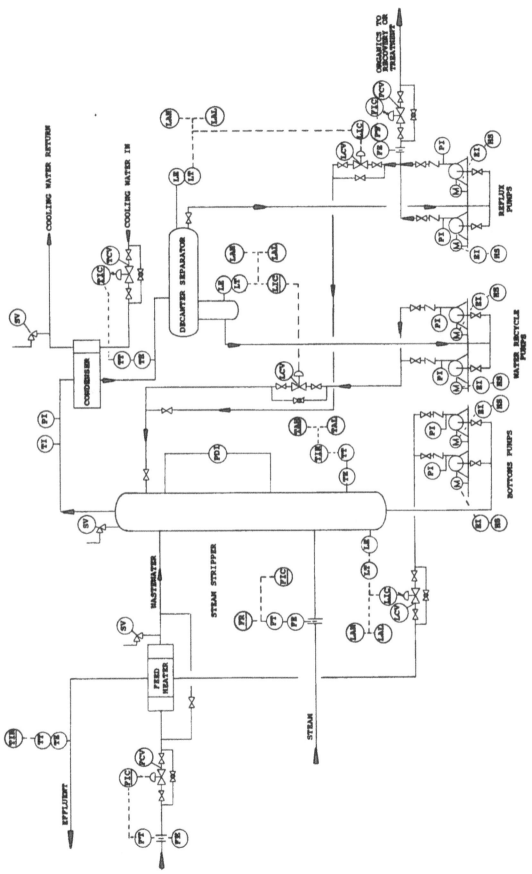

Figure 4.12 Steam stripping preliminary concept flowsheet.

Air blower
Liquid recirculation pump
Tower, internals, packing
Feed heat exchanger (optional)
Air pollution control (optional)

If the tower is designed as a steam stripper an overheads condenser and an optional reboiler would be required. As indicated, some of these components are discussed in separate chapters, with specific criteria detailed.

The Process Engineer is responsible for defining critical waste stripper process criteria. Allowing the equipment manufacturer to define these criteria, and develop a mechanical design to suit, is a major mistake inviting poor system performance. However, equipment details are usually proprietary designs, and the responsibility of a mechanical engineer associated with the equipment manufacturer.

The *stripper* is commonly purchased as a packaged design; with the tower, tower internals, and packing or trays an integral design. Heat exchangers are part of the packaged system. Shell and internals may be fabricated at different subcontracted shops; with pumps, blowers, controls, piping, and heat exchangers purchase as separate items. It is critical that fabrication, delivery, assembly, and installation responsibility be clearly identified. Some basic tower considerations are discussed as a guide in evaluating proprietary designs.

STRIPPER COLUMN DESIGN CONSIDERATIONS

Packing and tray towers contain similar elements, the major difference being the specialized internal design. Packed column elements will be discussed, detail tray design information can be obtained from the manufacturer. Steam strippers also require heat exchanger equipment using mechanical design features commonly employed for continuous chemical distillation processes. Figure 4.14 illustrates a typical packed tower, the major components being the packing, liquid distribution, gas distribution, packing hold-down section, and entrainment separators.

Packing Characteristics

Packed columns are commonly described as *random*, in which media is poured into the shell, or *oriented*, in which media is stacked in a deliberate pattern. Oriented installation is more labor intense, restricted to three inch or larger cylindrical shapes, and employed to achieve a lower pressure drop and better liquid distribution. Oriented packing is seldom used in waste treatment, should almost never be considered in design evaluations, and it should be understood that oriented packing could become random packing sometime during the life of the installation.

Feed Liquid Distribution

Effective stripping requires appropriate gas and liquid contact, highly dependent on effective media wetting, a result of distributing the influent over a multitude of points. This can be accomplished by (1) a dedicated section of packing, (2) surface spraying, (3) a feed plate with many openings, or (4) a perforated ring. Influent distribution types are illustrated in Figure 4.15.

Packing is not an effective distribution method and is seldom employed. Spray nozzles can be effective, but involve increased pumping costs, are subject to nozzle plugging, and must include mist eliminators to prevent heavy exit entrainment. An orifice plate can be effective, providing openings are adequate for countercurrent gas and liquid flow, and sensitivity to flow variations is not a concern. The weir type and perforated ring distributors eliminate many of these problems and are the most frequently used.

No influent distributor is perfect, and liquid channeling is possible as the liquid flows into the packing depths. A means is commonly employed to "police" the flow, and redirect any liquid channeling to the walls. The tower hydraulics must be evaluated and the internals designed to prevent any restrictions resulting in poor packing liquid distribution. Generally, redistribution is considered every 3 meters (10 feet), or every five tower diameters, whichever is smaller. Redistribution usually involves an internal device to collect the liquid migrating toward the walls, directing the flow through the packing area.

Feed Gas Distribution

Gas distribution at the tower bottom is as important as liquid distribution at the top. This is commonly accomplished with a coarse mesh plate, allowing gas to enter and distribute through the packing, with a layer of larger packing on top of the plate to support the smaller stripper packing.

Packing Holddown Plates

The packing is held in place with a holddown plate to assure that it does not move or shift out of position. Such plates can consist of wire mesh on weighted supports.

Entrainment Separators

Liquid entrainment is not a problem if adequate head space is provided and low velocities are employed, unless (1) sprays are used as inlet distributors, (2) the liquid is corrosive and any minute exhaust discharge can create a problem, or (3) extreme protection of the environment or surrounding property is a paramount concern, especially in time of system malfunction. Entrainment equipment can consist of a layer of packing in the head space, a mesh plate,

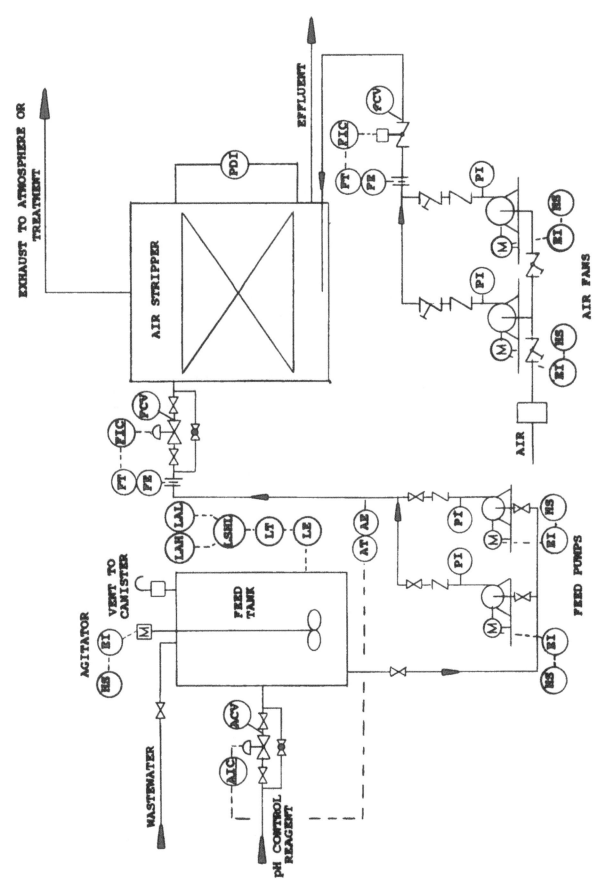

Figure 4.13 Air stripping preliminary concept flowsheet.

125

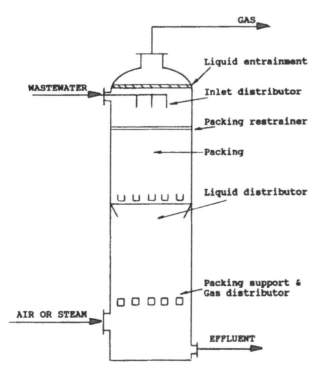

Figure 4.14 Packed tower components.

or special designed equipment that provides for impingement or abrupt flow direction to disengage any collected liquid.

Tray Columns

Figure 4.16 illustrates a typical tray column in which liquid enters the tower onto the tray, providing intimate gas and liquid contact, containing bubble caps to control inlet gas flow. A down-spout directs liquid flow to the next tray, with a liquid seal maintained to provide tray stability. The tray could contain a perforated (sieve) plate instead of bubble

caps, or any other proprietary design to allow high efficiency contact between the liquid and gas.

Tower Shells

The tower shell can be made of a variety of materials ranging from steel, steel coated, plastic and even wood. Packed columns can be constructed of any of these materials, whereas tray columns must be constructed to be able to handle a heavy tray hydraulic load and to accommodate a more sophisticated design. Towers must contain manhole and handholes to install and maintain the packing or trays.

PROCESS CONTROL

Stripper controls are basically aimed at manual system operation, and therefore installed for monitoring critical process variables. Tower controls should be installed for:

(1) Flow control (and recording) of waste influent, effluent, air, liquid recirculation, and steam flows.

(2) Provision for sampling waste influent, at a minimum as a "grab sample," preferably by automatically collecting composite samples over an operator controlled time period.

(3) Provision for automatically monitoring influent pH, and alarming for any conditions outside the specified limits.

(4) Provision to adjust steam flow, temperature, and pressure. These parameters should be continuously monitored and recorded. Stream flow should be totalized.

(5) Provision for continuously monitoring and recording tower wall, tower liquid, liquid feed, liquid exit, air inlet, and tower exhaust temperatures.

(6) Provision to monitor and record tower pressure and pressure drop.

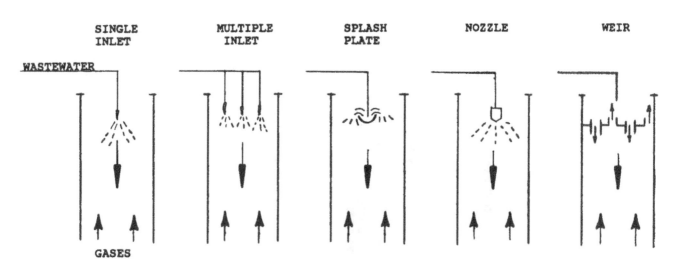

Figure 4.15 Packed tower distributor types.

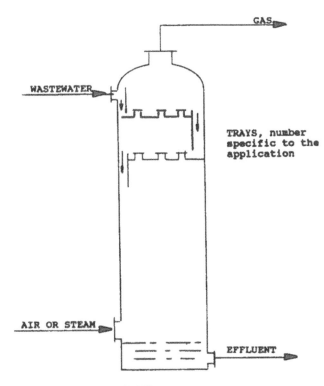

GAS

WASTEWATER

TRAYS, number
specific to the
application

AIR OR STEAM

EFFLUENT

Figure 4.16 Tray tower components.

(7) Provision for sampling waste effluent, at a minimum as a "grab sample," preferably by automatically collecting composite samples over an operator controlled time period.

(8) Provision for automatically monitoring effluent pH, and alarming for any conditions outside the specified limits.

(9) Running lights should be provided for all major mechanical equipment.

In addition, the system should contain adequate tower site ports at critical distribution sections.

COMMON STRIPPER DESIGN DEFICIENCIES

General

(1) Poor stripping efficiencies because of low volatile wastes.

(2) Cannot meet effluent quality because of insufficient packing height (or number of trays).

(3) Poor system design resulting from inadequate equilibrium data.

(4) Tower hydraulically overloaded because of excessive flow variation and a poor design basis, a result of inadequate characterization data.

(5) Inadequate controls provided to monitor waste flow into the tower.

(6) Inadequate controls provided to monitor air or steam flows.

(7) High liquid temperature losses in winter conditions from the air stripper because of low humidity and cold inlet air.

(8) High liquid temperature losses in winter conditions from a steam stripper because the tower is not insulated.

(9) The tower is not sheltered, and difficult to monitor or maintain in inclement or winter weather conditions.

(10) Heavy entrainment from the tower results because a mist eliminator was not provided.

(11) Lines freeze during winter planned or emergency shut downs because critical lines are not self draining, or heat traced and insulated.

(12) No means was provided to monitor tower temperature or pressure.

PACKED TOWER

(13) Poor tower loading results in inadequate packing wetness. No means is provided for effluent recycling to adjust influent flow.

(14) Poor packing wetness results from heavy sidewall flows and a lack of redistribution plates.

(15) Packing is constantly plugging because of heavy solids and tar in the feed.

(16) High packing breakage because of steam stripping thermal stresses.

(17) The blower air capacity is inadequate.

Tray Tower

(18) The tray seals are inadequate.

(19) Heavy foaming occurs on the trays.

(20) Trays are heavily corroded because of waste characteristics.

Steam Stripper

(21) Heavy fouling occurs in the steam stripper as a result of waste components and elevated temperatures.

(22) The steam capacity provided is inadequate.

(23) The steam flow cannot be controlled.

(24) The overhead product is dilute, an enriching section is required.

(25) The decanter is inadequate to separate the immiscible phase.

Ammonia Stripper

(26) The pH cannot be adequately controlled.

(27) Heavy scaling results from lime use.

CASE STUDY NUMBER 29

Conduct a preliminary evaluation of an *ammonia air* stripper for the following operating and process characteristics:

Waste flow, gpd: 500,000
Influent ammonia nitrogen, mg/L: 60
Required effluent ammonia nitrogen, mg/L: 5
Wastewater temperature, °C: 20 (68°F)
Contaminant MW: 17
Atmospheric pressure

PROCESS DATA

(1) u, liquid viscosity, 2.44 lb/ft-hr = 1 cp
(2) Dl, ammonia diffusion coefficient, 0.00006 sq ft/hr (estimated)
(3) d_L, liquid density, 62.3 lb/cu ft
(4) d_g, air density, 0.0753 lb/cu ft
(5) Stage efficiency, %
(6) Equilibrium data
(7) Molecular weight of ammonia, 17

Specific waste performance data is not available, use available literature data to determine whether stripping is feasible.

PROCESS CALCULATIONS

Note: These calculations should be performed using mole fractions. However, since all critical calculations involve proportional values, concentrations are expressed as molar ppm, and conversion to mole fractions are not necessary.

Selecting the Equilibrium Data

Henry's Law constant will be estimated from two sources. First, the solubility of ammonia at 20°C is 2 wt/100 wt of water, with a partial pressure of 12 mg Hg [3].

$$P = 12/760 = 0.0158 \text{ atmospheres}$$

$$[2/100] \cdot [MW \ 18/MW \ 17] = 0.0212 \text{ mole fraction ammonia}$$

$$m = y/x = 0.0158/0.0212 = 0.75$$

For dilute solutions equilibrium data can be estimated using the correlation: [15]

$$\ln \frac{p}{m'} = \frac{-4425}{T} + 10.82$$

where p is the ammonia partial pressure, atm, m' is the moles of ammonia per 1000 grams of water; and T is absolute temperature, °K.

°F	°C	H, atm
40	4	0.33
68	20	0.77
80	27	1.07

For the design conditions indicated, a Henry's Constant of 0.75

atm will be used. The sensitivity of the design to the Henry's Constant will be examined.

General Tower Material Balance

(1) Convert the waste influent, effluent and inlet air ammonia concentration to molar values

Influent waste concentration, 60 mg/L
60 · MW water 18/MW ammonia 17 = 63.5 *molar* ppm
Effluent waste concentration, 5 mg/L
5 · MW water 18/MW ammonia 17 = 5.3 *molar* ppm
Inlet air stream, 0 *molar* ppm

(2) Convert waste flow to moles per hour

500,000 gpd · 8.34 lb/gal/24 = 173,750 lb/hour
173,750 lb/hr/MW 18 = 9653 moles/hour

(3) Calculate the theoretical molar exit gas concentration. The maximum exit air concentration is that in equilibrium with the inlet waste water concentration.

$$Ye = H \cdot x$$

$$Ye = 63.5 \text{ mg/L} \cdot 0.75 \text{ (Henry } K) = 47.6 \text{ mg/L}$$

(4) Calculate the theoretical (minimum) G/L molar ratio

$$\text{Minimum } G/L = \frac{(X_{in} - X_{out})}{(Ye' - Y_{in})} = \frac{(63.5 - 5.3)}{(47.6 - 0)} = 1.22$$

(5) Select an operating G/L molar ratio. Actual G/L should be at least 1.5 times minimum G/L, select a ratio of 2 moles gas/mole liquid; which is equivalent to 3.2 lb gas per pound of waste liquid.

(6) Calculate the gas flow as moles/hour and pounds/hour

Gas moles/hr = G/L · L = 2 · 9653 moles/hour = 19,306 moles/hr
19,306 moles per hour · 29 = 559,874 lb per hour

(7) Calculate the resulting volumetric gas flow

19,306 m/h · 359 scf @ 32°F/mole · (528°R/492°R)/60 = 124,000 cfm
124,000/(500,000 gpd/1440 mins) = 357 air cfm/gpm waste

(8) Complete a system material balance and calculate the exit air molar ammonia concentration

0.5 MGD · 8.34 · (25 − 5)/24 = 9.56 lb ammonia removed/hr
Quantity ammonia in inlet air = 0 lb/hr
Quantity ammonia in outlet air = 9.56 lb/hr
9.56/559,874 lb air/hr · 1,000,000 = 17.1 ppm
17.1 (mass) ppm · (29/17) = 29.2 (molar) ppm

Tower Diameter

(9) Select the packing and associate packing characteristics. The tower configuration will be sensitive to the packing characteristics. Large packing is commonly employed because ammonia stripping is sensitive to pH control, chemical additions

to maintain pH, and potential scaling. Scaling because of precipitation at the adjusted ph, or because of influent solids. Therefore, large packing, three inch or larger, will be investigated. The properties of such packing could vary, defined by its packing factor, σ factor, and "n" constant. The following packing properties will be assumed:

Packing	Packing Factor Pf, 1/feet	σ sq ft/cu ft	"n"
1	25	20	0.09
2	50	30	0.15
3	75	90	0.28
4	100	140	0.31

Get *specific* packing characteristics from the supplier!! Based on Figure 4.6, flooding limits can be established!

(10) Calculate the gas flooding rate

$$L/G \cdot (d_g/d_L)^{1/2} = (1/3.2) \cdot (0.0753/62.3)^{1/2} = 0.0108$$

From Figure 4.6, the corresponding *flooding* ordinate is 0.22.

(11) Calculate total flooding rate

$$G^2 = \cdot \left[\frac{d_g \cdot d_l}{Pf \cdot u^{0.2}} \right] \cdot \text{ordinate} \cdot g$$

where

g = gravity, 417,000,000
Pf = packing factor, 50
u = viscosity, 1 cp
d_g = gas density, 0.0753 lb/cubic foot
d_l = liquid density, 62.3 lb/cubic foot
G = the gas mass velocity in lb/hr-square foot

Based on a resulting ordinate of 0.22, related to an abscissa of 0.0108 on the flooding curve, G equals 2,935 lbs/hr-ft^2.

2935 lb/hr-ft^2/[0.0753 lb/cu ft $\cdot$ 3600 sec/hr] = 10.8 fps

(12) Establish the operating gas loading rate. Assume 60% of flooding rate, or $Gf = 0.60 \cdot 10.8 = 6.5$ fps/ft^2.

(13) Calculate the required tower area and diameter

124,000 cfm *air*/60/6.5 fps/ft^2 = 318 square feet
Diameter = 20.1 feet $\approx$ 20 feet = 314 sq ft

Results using the other packing characteristics can be summarized as follows:

Packing factor	25	50	75	100
Theoretical velocity, fps	15.3	10.8	8.8	7.7
Design velocity, fps	9.2	6.5	5.3	4.6
Area, square feet	225	318	389	450
Diameter, feet	17	20	22	24

Tower Height

(14) Calculate the system stripping factor

$$R = (\text{Henry } K/P \text{ atm}) \cdot G/L = 0.75 \cdot 2 = 1.5$$

(15) Calculate the concentration factor

$$Cr = Xo/Xe = 63.5/5.3 = 12$$

(16) Calculate the number of transfer units

$$NI = \frac{R}{R-1} \cdot \ln \left[\frac{Cr \cdot (R-1) + 1}{R} \right]$$

$$= \frac{1.5}{1.5-1} \cdot \ln \left[\frac{12 \cdot (1.5-1) + 1}{1.5} \right] = 4.6 \text{ units}$$

(17) Determine the height of a transfer unit

L = [500,000 gpd $\cdot$ 8.34/24]/227 = 765 lb/hr/sq ft
Pf = packing factor, 50
σ = 30 sq feet/cu feet
n = 0.15
u = 2.42 lb/ft $\cdot$ hr
p = 62.3 lb/cubic foot
Dl = 0.00008 sq ft/hr

$$HI = \frac{1}{\sigma} \cdot \left[\frac{L}{u} \right]^n \cdot \left[\frac{u}{p \cdot Dl} \right]^{1/2}$$

$$HI = \frac{1}{30} \cdot \left[\frac{546}{2.42} \right]^{0.15} \cdot \left[\frac{2.42}{62.3 \cdot 0.00006} \right]^{1/2} = 1.9 \text{ feet/unit}$$

The solubility of ammonia and the high gas loading make the assumption that the gas phase unit transfer is negligible is questionable. The magnitude of the gas phase transfer unit will be estimated assuming the following relation [3]:

$$Hg = 1.01[G]^{0.31}/[L]^{0.3} = [G/L]^{0.3}$$

G = 19,305 moles/hr $\cdot$ 29 = 559,874 pounds/hr
L = 173,750 pounds/hr

Mass loadings (pounds/hr/ft^2) should be used, but in this equation mass ratio is relevent.

$$Hg = [559,874/173,750]^{0.3} \approx 1.4 \text{ feet}$$

$$Hol = HI + Hg \cdot \frac{L}{m \cdot G}$$

$$Hol = 1.9 + 1.4 \cdot \frac{1}{0.75 \cdot 2} = 2.8$$

(18) Calculate the packing height

Packing height = 4.6 units $\cdot$ 2.8 = 13 feet

After applying a safety factor a 15 foot height will be used.

Results using the other packing characteristics can be summarized as follows:

Packing factor	25	50	75	100
σ, sq ft/cu ft	20	30	90	140
"n"	0.09	0.15	0.28	0.31
Total packing height, ft	14	13	10	8
Diameter, feet (above)	17	20	22	24

Power Requirements

(19) Estimate the pumping horsepower

$$hp = gpm \cdot height \cdot specific\ gravity/3960$$

$$= 500,000/1440 \cdot 12 \cdot 1/3960 = 1\ hp$$

(20) Estimate the fan horsepower

$$hp = cfm \cdot in\ drop/ft \cdot height \cdot 0.000157$$

60% of the flooding value was selected. In Figure 4.6 a 0.22 flooding ordinate was used. Correcting the ordinate,

$$(0.6)^2 \cdot 0.22 = 0.08$$

equal to a drop of approximately 0.5 in/ft at an abscissa of 0.0108.

$$hp = 124,000 \cdot 0.5\ in/ft \cdot 15\ ft \cdot 0.000157 = 150\ hp$$

DISCUSSION

(1) Sensitivity of the design to selected parameters: In attempting to design a stripper using published constants the Process Engineer should be aware that the configuration can be radically affected by most of the selected constants.

a. The correct Henry Constant is critical to the stripper design, and must be accurate for the design temperature selected and waste characteristics encountered. The required air-waste ratio will depend on the selected Henry Constant, increasing with decreasing constant value. Specific values are difficult to obtain, those reported are frequently difficult to validate, sometimes significantly varying from different reported studies.

The effects of the Henry Constant to the estimated packed column design for this Case Study can be summarized as follows:

H, atm	G/L mole/ mole 1.5 times Theoretical	cfm/ gpm	Area RQD (Diameter, ft)	Theoretical Height, ft
0.4	3.4	614	26	15
0.5	2.8	491	24	15
0.6	2.3	409	22	15
0.75	1.8	327	19	15
0.9	1.5	273	18	15

b. The selected operating gas-to-liquid ratio will affect the tower configuration and costs. A large ratio decreases the tower height, but increases the diameter.

H, atm	G/L mole/ mole	cfm/ gpm	Area RQD (Diameter, ft)	Theoretical Height, feet	
0.75	1.8	321	19	15	
0.75	2	357	20	13	<— Design
0.75	3	536	25	9	
0.75	4	714	28	7	

In any design evaluation a final adjustment may have to be made to the estimated tower diameter to reduce pressure drop, balancing capital costs (tower) against operating costs (fan horsepower).

c. The packing selected, and the corresponding properties (Pf, σ and n) will dramatically affect the tower height and diameter, as demonstrated above; and must therefore be specifically established for the operating conditions under consideration.

The Process Engineer must be aware of these significant factors in comparing studies or designs involving stripping of the same volatile components, with apparent significant differences in tower diameter or height. Most differences can be attributed to (1) the selected Henry Constant, (2) the selected air-to-waste ratio, (3) the selected packing (or chosen packing constants), (4) the influent and effluent concentration, or (5) the operating temperature.

(2) Influent/exhaust concentrations: The influent concentration of 60 mg/L is higher than that commonly reported for municipal effluents (25 mg/L range). This affects the required tower height which is sensitive to the concentration ratio of the influent to effluent. In addition, the higher the influent concentration the greater the potential for exhausting prohibitive emission quantities. As an example, at the design conditions the following quantities are estimated.

Ammonia released: 9.56 lbs/hr = 72,300 mg/minute
   Exhaust: 124,000 cfm = 3512 cu meter/minute
Concentration: 72,300/3512 = 21 mg/cubic meter.

The potential problem with uncontrolled ammonia exhausts is evident when compared to the following reported limiting concentrations [7]:

Odor threshold: 35 mg/cu meter
Eye, nose, throat irritations: 280 − 490 mg/cu meter
Toxicity levels: 1700 − 4500 mg/cu meter

CASE STUDY NUMBER 30

Conduct a preliminary evaluation of a *volatile organic air* stripper for the following operating and process characteristics:

Waste flow, gpd: 500,000
Influent *benzene*, mg/L: 60
Required effluent *benzene*, mg/L: 5
Wastewater temperature, °C: 20 (68°F)

Contaminant MW: 78
Atmospheric pressure

REQUIRED DATA

(1) u , Liquid viscosity, 2.42 lb/ft-hr
(2) Dl , Liquid diffusion coefficient, 0.00004 sq ft/hr (estimated)
(3) d_L , liquid density, 62.3 lb/cu ft
(4) d_g , Air density, 0.0753 lb/cu ft
(5) Stage efficiency, %
(6) H , Henry's Law constant, 240 atm
(7) Molecular weight of *benzene*, 78

Specific waste performance data is not available, use available literature data to determine whether stripping is feasible.

PROCESS CALCULATIONS

Note: These calculations should be performed using mole ratio. However, since all calculations involve proportional values, concentrations are expressed as ppm, and conversion to fractions are not necessary.

General Tower Material Balance

(1) Convert the waste influent, effluent and inlet air ammonia concentration to molar values

Influent waste concentration, 60 mg/L
60 · MW water 18/MW benzene 78 = 13.85 *molar* ppm
Effluent waste concentration, 5 mg/L
5 · MW water 18/MW benzene 78 = 1.15 *molar* ppm
Inlet air stream, 0 molar ppm

(2) Convert waste flow to moles per hour

500,000 gpd · 8.34 lb/gal/24 = 173,750 lb/hour
173,750 lb/hr/MW 18 = 9653 moles/hour

(3) Calculate the theoretical molar exit gas concentration. The maximum exit air concentration is that in equilibrium with the inlet waste water concentration.

$$Ye = H \cdot x$$

Ye = 13.85 mg/L · 240 (Henry K) = 3324 mg/L

(4) Calculate the theoretical (minimum) *G/L* molar ratio

$$\text{Minimum } G/L = \frac{(X_{in} - X_{out})}{(Ye' - Y_{in})} = \frac{(13.85 - 1.15)}{(3324 - 0)} = 0.0038$$

(5) Select an operating *G/L* molar ratio. Actual *G/L* should be at least 1.5 times minimum *G/L*, select a ratio of 0.007 moles gas/mole liquid; which is equivalent to 0.0113 lb gas per pound of waste liquid.

(6) Calculate the gas flow as moles/hour and pounds/hour

Gas moles/hr = $G/L \cdot L$ = 0.007 · 9653 moles/hour = 67.6 m/h
67.6 moles per hour · 29 = 1960 lb per hour

(7) Calculate the resulting volumetric gas flow

67.6 m/h · 359 scf @ 32°F/mole · (528°R/492°R)/60 = 434 cfm
434/(500,000 gpd/1440 min) = 1.25 *air* cfm/gpm waste

(8) Complete a system material balance and calculate the exit air molar benzene concentration

0.5 mgd · 8.34 · (60 − 5)/24 = 9.56 lb benzene removed/hr
Quantity benzene in inlet air = 0 lbs/hr
Quantity benzene in outlet air = 9.56 lbs/hr
9.56/1960 lb air/hr · 1,000,000 = 4878 ppm
4878 (mass) ppm · (29/78) = 1814 (molar) ppm

Tower Diameter

(9) Select the packing and associate packing characteristics. Assume a packing with packing factor Pf = 250, small packing of 1-1/2 inches or less in size. Based on Figure 4.6, flooding limits can be established!

(10) Calculate the gas flooding rate

$$L/G \,(d_g/d_L)^{1/2} = (1/0.0113) \cdot (0.0753/62.3)^{1/2} = 3.08$$

From Figure 4.6, the corresponding *flooding* ordinate is 0.006

(11) Calculate total flooding rate

$$G^2 = \cdot \left[\frac{d_g \cdot d_l}{Pf \cdot u} \right] \cdot \text{Ordinate} \cdot g$$

where
g = gravity, 417,000,000
Pf = packing factor, 250
u = viscosity, 1 cp
d_g = gas density, 0.0753 lb/cubic foot
d_l = liquid density, 62.3 lb/cubic foot
G = the gas mass velocity in lb/hr-square foot

Based on the ordinate being 0.006 related to an abscissa of 3.08 on the flooding curve, G equals 217 lbs/hr-ft².

217 lbs/hr-ft²/[0.0753 lb/cu ft · 3600 sec/hr] = 0.80 fps

(12) Establish the operating gas loading rate. Assume 60% of flooding rate, or Gf = 0.8 · 0.80 = 0.50 fps/ft².

(13) Calculate the required tower area and diameter

434 cfm *air*/60/0.50 fps/ft² = 14.5 square feet
Diameter = 4.3 feet ≈ 4.5 feet (15.9 sq ft)

Tower Height

(14) Calculate the system stripping factor

$$R = (\text{Henry } K/P \text{ atm}) \cdot G/L = 240 \cdot 0.007 = 1.68$$

(15) Calculate the concentration factor

$$Cr = Xo/Xe = 13.85/1.15 = 12$$

(16) Calculate the number of transfer units

$$Nl = \frac{R}{R-1} \cdot \ln\left[\frac{Cr \cdot (R-1) + 1}{R}\right]$$

$$= \frac{1.68}{1.68 - 1} \cdot \ln\left[\frac{12 \cdot (1.68 - 1) + 1}{1.68}\right] = 4.2 \text{ units}$$

(17) Determine the height of a transfer unit

$L = [500,000 \text{ gpd} \cdot 8.34/24]/15.9 = 10,928$ lb/hr/sq ft
$\sigma = 100$, sq feet/cubic feet
$n = 0.22$
$u = 2.42$ lb/ft · hr
$p = 62.3$ lb/cubic foot
$Dl = 0.00004$ sq ft/hr

$$Hl = \frac{1}{\sigma} \cdot \left[\frac{L}{u}\right]^n \cdot \left[\frac{u}{p \cdot Dl}\right]^{1/2}$$

$$Hl = \frac{1}{100} \cdot \left[\frac{10,928}{2.42}\right]^{0.22} \cdot \left[\frac{2.42}{62.3 \cdot 0.00004}\right]^{1/2} = 2.0 \text{ feet/unit}$$

A check will be made of the gas phase transfer unit [3].

$$Hg = 1.01 [G]^{0.31}/[L]^{0.33} \cong [G/L]^{0.3}$$

$G = 67.6$ moles/hr · 29 = 1960 pounds/hr
$L = 173,750$ pounds/hr

Mass loadings should be used, but in this equation mass ratio is relevent.

$$Hg = [1960/173,750]^{0.3} \cong 0.3 \text{ feet}$$

$$Hol = Hl + Hg \cdot \frac{L}{m \cdot G}$$

$$Hol = 2.0 + 0.3 \cdot \frac{1}{0.007 \cdot 240} = 2.2$$

The gas phase transfer unit is not significant.

(18) Calculate the packing height

Packing height = 4.2 units · 2.2 = 9.2
A 10 to 12 foot height will be considered

Power Requirements

(19) Estimate the pumping horsepower

$$HP = \text{gpm} \cdot \text{height} \cdot \text{specific gravity}/3960$$

$$= 500,000/1440 \cdot 12 \text{ ft} \cdot 1/3960 = 1 \text{ hp}$$

(20) Estimate the fan horsepower

$$HP = \text{cfm} \cdot \text{in drop/ft} \cdot \text{height} \cdot 0.000157$$

Sixty percent of the flooding value was selected. In Figure 4.6 a 0.007 flooding ordinate was used. Correcting the ordinate,

$$(0.6)^2 \cdot 0.007 = 0.0023,$$

equal to a drop of 0.1 in/ft at abscissa of 3.02.

$$hp = 434 \text{ cfm} \cdot 0.1 \text{ in/ft} \cdot 12 \text{ ft} \cdot 0.000157 = 0.1 \text{ hp}$$

(21) Exhaust concentration

Benzene released: 9.56 lb/hr = 72,300 mg/minute
Exhaust: 434 cfm = 12 cu meter/minute
Concentration: 72,300/12 ≡ 6025 mg/cubic meter.

DISCUSSION

The basic results of Case Study 29, involving ammonia, and this benzene stripper illustrate the difference the component volatility has on the tower size. In both cases the same flow, influent concentration, and effluent quality are considered; but, a 20 foot versus a 4.5 foot tower diameter is required, and the ammonia tower uses larger packing! A high volatility reduces the minimum gas-to-liquid ratio, requiring less air, decreasing the required tower diameter. Reference is made to Case Study 29 to illustrate the effects of the other operating parameters such as varying the G/L ratio, the concentration, and the packing characteristics, that generally have the same effect on the estimated tower diameter and height. The parameters selected for Case Studies 29 and 30 are about right, but not necessarily optimum. A detailed economic evaluation would be required to optimize the configurations.

REFERENCES

1. Bravo, J.L.: "Design Steam Strippers for Water Treatment," *Chemical Engineering Progress*, V 90, No 12, Pg 56, December, 1994.
2. Nirmalakhandan et al.: "Operation of Counter-Current Air Stripping Towers at Higher Loading Rates," *Water Research*, V 27, No 5, Pg 807, 1993.
3. Perry, R.H. and Green, D.: *Perry's Chemical Engineers' Handbook*, Sixth Edition, McGraw-Hill, 1984.
4. Perry, R.H. and Chilton, C.H.: *Chemical Engineers' Handbook*, 5th Edition, McGraw-Hill, 1973.
5. Perry, J.H.: *Perry's Chemical Engineers' Handbook*, Third Edition, McGraw-Hill, 1950.

6. Treybal, R.E.: *Mass-Transfer Operations*, McGraw-Hill, 1980.

7. U.S. Environmental Protection Agency: *Process Design for Nitrogen Control*, PB-259-149/38A, October 1975.

8. U.S. Environmental Protection Agency: *Process Design Manual for Stripping of Organics*, EPA-600/2-84-139, 1984.

9. U.S. Environmental Protection Agency: *Steam Stripping and Batch Distillation for the Removal/Recovery of Volatile Organic Compounds*, EPA-600/D-89/009, 1989.

10. Ball, W.P., Jones, M.D., and Kavanaugh, M.C.: "Mass Transfer of Volatile Organic Compounds in Packed Tower Aeration," *Journal Water Pollution Control Federation*, V 56, No 2, Pg 127, February, 1984.

11. Fang, C.S. and Khor, S.-L.: "Reduction of Organic Compounds in Aqueous Solutions through Air Stripping and Gas-Phase Adsorption," *Environmental Progress*, V 8, No 4, Pg 270, November, 1989.

12. Bravo, J.L. and Fair, J.R.: "Generalized Correlation for Mass Transfer in Packed Distillation Columns," *Ind Eng Chem Process Design Dev*, 21, Pg 162, 1982.

13. Amy, G.L. and Cooper, W.J.: "Air Stripping of Volatile Organic Compounds Using Structured Media," *Journal Environmental Engineering, Proceedings ASCE*, V 112, No 4, Pg 729, 1986.

14. Pratton, C.F. and Maron, S.H.: *Fundamental Principles of Physical Chemistry*, The Macmillan Company, 1951.

15. Badger, W.L. and Banchero: *Introduction to Chemical Engineering*, McGraw Hill Book Co., 1955.

Filtration

Filtration is commonly employed to remove low concentration suspended solids from wastewater.

FILTRATION is a "polishing" process employed to meet stringent effluent quality standards or as a pretreatment to protect more sensitive adsorption, ultrafiltration, or ion-exchange equipment. It is used to remove limited quantities of suspended solid, but it is not an alternative to sedimentation or flotation, which can more effectively treat heavy solid loadings. Filtration efficiency is affected by a large number of interdependent variables such as the waste solid properties, filter bed characteristics, wastewater properties, and process configuration. In some cases these variables are not considered in the design, instead the requirements are defined by a hydraulic loading selected to minimize equipment size and function within an allowable head loss. Whether the other operating variables are considered in the design or not, they affect performance! The operator is expected to achieve both a high effluent quality and a high treated water production rate with the equipment provided.

Waste filtration design is based on water treatment operating experience, expressed as standard design criteria, refined to include pilot data, and adopted to specific waste conditions. The basic differences between the two services include the wide variety of solids discharged into industrial sewers, greater influent variations, and unpredictable waste characteristics. Specific filtration treatment models are not readily available, or too theoretical for direct application.

BASIC CONCEPTS

As illustrated in Figure 5.1, the filtration process involves many steps, all related to two parameters which establish treatment effectiveness—effluent quality and filter run time. Waste solids characteristics such as concentration, size, and size distribution impact filtration efficiency. In simple terms, separation is a result of waste flow transporting solids through the bed; the waste suspended particles contact the

media grains, are removed, and are stored in the void space. Solids separation occurs as a result of straining, sedimentation, impaction, interception, adsorption, flocculation or adhesion mechanisms occurring between the waste particles and the media grain. The media void space functions as a solids depository.

The bed media size distribution must be compatible with the waste solid characteristics, allowing large solids storage at an acceptable head loss. Filter performance can sometimes be enhanced by chemical treatment to improve waste characteristics. Assuming effluent quality can be achieved, a significant measure of filtration performance is cycle time. Long cycle times result in the bed being saturated with collected solids and improved separation resulting from "ripening." It is postulated that the "ripening" effect improves the collection efficiency with filtration time, especially of small particles, because of additional collectors formed by captured particles [3]. Theoretically, removal efficiency increases with increased bed depth and increased waste particle size and distribution and decreases with increased hydraulic loading rate.

The capability of a media to retain solids without totally obstructing effluent passage establishes the system's solids capacity, and the resulting filter media run time. The run time is directly related to the filter column pressure drop. The filtration cycle is terminated and backwash cycle initiated at a predetermined pressure drop, or elapsed time. The backwash cycle is terminated and the filtration repeated when the captured solids are removed from the filter media. In essence, the filter pressure drop is the controlling process variable, establishing the allowable operating time without excessive hydraulic loading rate loss or poor effluent quality. Filtration operating efficiency is determined by filtration time, which is maximized; and the backwash volume, which is minimized.

System performance is affected by the inherent waste characteristics, and selected design parameters such as pretreatment, hydraulic loading, media selection, bed depth,

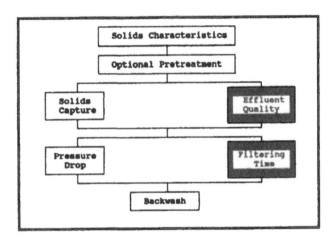

Figure 5.1 Wastewater filtration process.

allowable pressure drop, backwash method, and backwash rate. Filtering media selection is a critical design criteria, influencing all other design parameters, and establishing system performance.

HYDRAULICS MODEL

Filtration dynamics can be illustrated by assuming that the basic filter operating parameters can be related to process hydraulics, which for *initial* clear water conditions can be expressed by the Darcy-Weisbach head loss equation, Equation (5.1).

$$Hl = f\frac{L}{D}\frac{v^2}{2g} \cdot \Phi \qquad (5.1)$$

where

Hl is the head loss
f is the friction coefficient, related to the Reynold's number
L is the length or bed depth
D is the media diameter, an indirect measure of the flow area
v is the mean bed velocity
Φ is a corrective factor for nonspherical form (below)

Based on this theoretical relation, head loss increases with the bed depth, hydraulic loading (mean bed velocity), and with decreasing media size (resulting channel flow area). Unlike a pipe line, media size does not directly establish the flow area. Media size and shape determines bed void space, establishing the system's solid storage capacity and the hydraulic flow characteristics. In the Carmen-Kozeny relation for head loss through a granular filter, adjustments

to the media for nonspherical form and effective void space are expressed by the corrective factor:

$$\Phi = \frac{1}{\sigma} \cdot \frac{1 - \epsilon}{\epsilon^3}$$

where σ is particle shape factor, 1 for a sphere, 0.73 to 0.82 for common filter media, and ϵ is the porosity.

The friction coefficient (f) directly relates the filter head loss to fluid characteristics, flow velocity, and "flow conduit" area (or media void space). Initially, this can be envisioned as a strict line loss similar to a pipe line loss. Details on the Carmen-Kozeny equation, and other theoretical head loss equations such as the Rose and Hazen equations, are described in waste treatment texts [5]. Actual design head losses should be obtained from the equipment designer based on specific configurations.

In fact, the theoretical clean water head loss is not a reliable design criteria because as solids are captured and media void space depleted, the hydraulic characteristics constantly change. Head loss increases in direct relation to void space decrease, eventually restricting flow rate. Actually, after the filtration process is initiated, head losses cannot be predicted by simplified fluid flow relations, and the hydraulic profile follows that illustrated in Figure 5.2. Theoretically, the filtration system can be operated at any time cycle to accommodate waste and filter characteristics, but if the effluent quality deterioration or head loss rates are too rapid, the net production rate will be too low to be practical.

BASIC FILTRATION PROCESS

Designing a filter system requires an understanding of the operating cycle in an effort to optimize the process, and an understanding of the *limits* of each of the process variables. The basic cycle includes (1) transporting the solids to the filter media, (2) separating the solids from the liquid and to the media, and (3) removing solids from the media to "regenerate" the system. Efficient filtration depends on the selected media size, the corresponding void size, the effective void size with solids capture, and the filter bed size distribution.

The selected media size establishes the available pressure drop, the required system energy, and the pressure drop

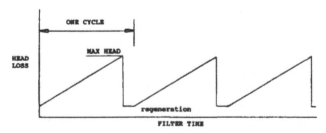

Figure 5.2 Head loss through a cycle.

depletion rate through the filtration cycle. Next, the relative media to waste solids size influences the removal rate. The finer the media size the more effective the removal rate. The media size determines the operating cycle duration by establishing the solids holding capacity and the related pressure drop. However, there is a *limit* to how fine a media can be selected. The finer the media the greater the solids capture at the subsurface and near the subsurface, the less the available solids capacity, and the sooner the allowable pressure drop is depleted. The allowable pressure drop and the depth of solids capture in the bed are critical process considerations.

Reduced void volume for solids capture *limits* the on-line process efficiency because of rapid head loss, reduced processing time, reduced filtration capacity, increased backwashing cycles, and higher volumes of backwash generated. Therefore, for a given allowable pressure drop, increased media size increases on-line efficiency. However, increased media size usually results in poorer effluent quality because larger quantities of smaller waste solids (relative to media) pass through the media. Therefore, the media selection must be balanced to optimize overall performance.

Finally, filter bed size distribution affects filtration efficiency. Ideally, the top (or inlet) of a bed should contain the largest media to capture bulk quantities of larger size solids, and the bottom (exit) section should contain smaller media acting as a polishing step. This implies a multimedia system with deliberate media sizing, to balance solids storage with polishing potential, and avoiding either too small a media size or too narrow a size distribution.

The media selection and distribution affects backwashing, and therefore regeneration of the spent bed. The smaller the media size the less effective the backwash because of the lower allowable backwash rate (velocity), which must be limited to ensure bed stability. Backwash acts as a classifying process moving the smaller particles to the bed top, the higher the backwash the more effective the solids classification. However, significant media classification causes large volumes of smaller size media to rise to the top, reducing solids capacity. This results in (1) a more rapid waste solids accumulation at the inlet area, (2) a rapid head build up, (3) ineffective use of the bed depth, and (4) reduced operating efficiency.

For all the reasons discussed, the process configuration, the bed characteristics, and operational variables must be carefully selected to optimize the process.

FILTER SYSTEMS

Filters include a wide range of devices such as strainers, screens, disposable cartridges, and leaf or tubular filters, as well as the more common granular media filters devices utilized in waste water treatment. Granular filters can be classified according to media employed, hydraulics, or configuration, as follows [2,6–8,11]:

(1) Single size or nongraded media
(2) Single, dual or multimedia
(3) Fixed or continuous moving bed

In addition, any of these beds can be constructed to operate as gravity or pressure filters, or with external or internal backwashing. These bed classifications will be further defined to facilitate selecting the proper configuration, as discussed in the Process Engineering Design section. More equipment details of proprietary designs can be obtained from manufacturers, and more general information from water treatment texts.

SINGLE MEDIA

Single media systems are usually employed as fixed or continuous moving beds, with the media nongraded; although some graded media configurations are available. No effort is made to sort the bed media, although it should be understood that, even in a single media filter, the media will not be uniform but a spread of media sizes.

Fixed Beds with Nongraded Media

Single, *nongraded*, media-fixed beds can be classified according to the applied hydraulic rate and depth of penetration as *slow sand*, *rapid sand*, and *deep coarse media*. Waste is commonly fed into the top, backwash is upflow. They are taken off-line and regenerated when the filtering cycle is complete. Slow and rapid sand filters are primarily used in water treatment systems, and are seldom subjected to the suspended solids loading encountered in waste treatment effluents. Deep coarse media filters are frequently used for wastewater treatment because of the inherent large solids storage capacity resulting from intense bed penetration into the coarse media. Resulting solids capacities can range from 16 to 160 kg/sq m (1 to 10 lb/ft^3) of bed volume per treatment cycle. Typical fixed bed, nongraded filter operating characteristics are summarized in Table 5.1 [11].

Moving Bed with Single Nongraded Media

These systems consist of single, *nongraded* media, moving beds in which the media is continuously regenerated [11]. This system is a modification of the fixed bed system, applied where difficult filtration is anticipated because of excess media plugging, making scouring and in-place, offline backwashing ineffective, and resulting in short treatment cycles. The spent bed portion is continuously removed and new sand added. For all practical purposes the entire bed is utilized; as each section moves up to the active filtration section, is spent, and removed. "Fresh" sand is always pres-

TABLE 5.1. Fixed-Bed Nongraded Filter Characteristics (adapted from Reference [11]).

	Slow Sand	Rapid Sand	Deep Coarse
Media type	Sand	Sand	1–6 mm Sand
Media depth			
mm	150–380	600	1200–1800
inches	6–15	24	48–72
Loading			
L/m2-s	0.01–0.03	1.4–4.1	4.1–5.5
gpm/sf	0.02–0.05	2–6	6–8
Application	Water trt.	Water trt.	Water trt. Wastewater

ent upstream, and "spent" sand downstream, from the working depth. Provision is made for externally cleaning spent sand.

Moving bed systems can be classified according to the wastewater flow direction relative to the sand, or to the method by which the sand is transported. In a *downflow* moving bed the waste enters the top passing through layers of sand, spent sand is withdrawn downward by the action of a regeneration pump. The operation tends to result in finer sand at the bottom. In an *upflow* moving bed the wastewater enters the bottom and flows counter to the bed movement, which is withdrawn from the bottom. In both of these cases a continuous effluent and backwash are generated, with a portion of the bed continuously removed and replaced. A *pulse-bed filter* operates semicontinuously with wastewater entering the top and passing through the bed. Once a significant pressure drop is developed as a result of a partially clogged bed, the elevated water level transmits a signal initiating a low pressure air cycle. The air is injected through diffusers above the bed, resulting in the top bed sand layer rolling and solids disengagement, after which filtration continues. At some point solids again collect on the surface, resulting in an increased pressure drop causing the liquid level to rise above the air pulsation activation level, to a backwash sensing level. At this level the effluent discharge is discontinued and a partial backwash started until the solids mat is disengaged, after which influent treatment continues. At some predetermined cycle count, the system feed is terminated and an actual full backwash cycle initiated for extensive bed regeneration. In essence, this is equivalent to a downflow system with provisions for extended solid penetration and storage by disengaging the plugged media on a regular basis.

Graded Media Fixed-Bed Operations

These systems employ a fixed bed with single *graded* media in which the bed is taken off-line and regenerated. They operate as upflow or biflow processes [11].

Upflow

In this system the waste flows upward through the vessel bottom, the bed is fluidized, and the media is hydraulically graded so that the larger sizes are at the bottom. The media grading results in large voids at the bottom for solids capacity, and finer media at the top for effluent polishing. In addition, the fluidized bed provides some natural scouring to better accommodate plugging type solids. The cleaning cycle requires stopping waste filtration, draining, low pressure air injection to disengage the plugged sand, and a water backwash. After regeneration is completed, sand reclassification restores the bed to the active condition. After a prefiltration run to rinse and collect bed leakage, the treatment cycle is started.

Biflow Filter

The Biflow Filter attempts to double the filter capacity by operating as two filters, part of the feed flowing downward, and the rest upward from the feed point. Difficulties are frequently encountered in balancing effluent quality from the two sections. The downflow filter produces an effluent quality equivalent to a rapid filter and the upflow effluent quality is limited because of the media size selected to prevent intermixing with the downflow media in the backwash cycle [2]. The system is not widely used in wastewater treatment.

MULTIMEDIA FIXED-BED FILTERS

A multimedia filter is designed to achieve media grading in reverse of that encountered in sand filters, with coarse material on top allowing solids penetration and storage, and fine media at the bottom to polish the effluent [11]. This condition can be approached with a minimum intermixing layered bed or a mixed multimedia bed.

A *layered bed* has a light coarse coal layer (sg 1.5) above a fine, but heavier, sand layer (sg 2.65). Technically, further improvement can be achieved with an additional layer(s) of lighter coarser media on top and lighter finer media at the bottom. However, available media with finite grading characteristics that will not significantly intermix in backwashing is limited. Another method to achieve this same effect is to allow intermixing, achieving a *multigraded intermixed system* with the resulting pore spaces graded from coarse at the top to fine at the bottom, resulting in an effective size of 2 mm or greater at the top and less than 0.2 mm at the bottom. Intermixing, whether desired or not, is a result of Stokes' Law applied in the upflow backwash mode, and dependent on media specific gravity and diameter differences. The unique selection of media type, number of media, and whether to allow intermixing, is a basic design consideration which must be carefully evaluated in applying these systems.

TWO-STAGE FILTERS

The two-stage filter is designed with a separate vessel containing coarse media for deep solids penetration and storage, and another vessel containing fine media for effluent quality. This avoids the problem of limited available media size, the need to carefully select the media size, and back-wash intermixing. The first (stage) vessel is designed to remove large quantities of suspended solids, and the second (stage) vessel to produce a high effluent quality [1].

PROCESS ENGINEERING DESIGN

Although tertiary filtration systems require very little operator attention, effluent quality and effective cycle time are not assured. As indicated in Table 5.2, filtration effectiveness is largely dependent on the waste characteristics, more specifically the solid properties. Where solid characteristics are not conducive to effective filtration, chemical treatment must be applied to enhance separation. Although chemical treatment can be quit effective, application is frequently unpredictable and difficult to control. This is because the required chemistry or dosage can alter with changing waste characteristics, which are not easily detected, and excessive chemical dosages can be detrimental to filtration performance.

The Process Engineer can increase operating flexibility by providing waste storage and equalization; allowing storage and testing of each batch, defining and optimizing chemical dosages, and thereby stabilizing chemical treatment for the collected inventory. Another significant operating variable is hydraulic loading, which is directly controlled by adjusting the *feed rate*. However, feed rate cannot be controlled if upstream waste storage is not provided.

Pressure drop depletion, cycle time, and net production rate are all measures of process effectiveness, none of which is directly controlled by the operator. Cycle time is a function of pressure drop depletion, a result of waste characteristics and regeneration effectiveness. If the waste contains a high solids loading, or solids which promote media plugging, the operator has little control over the cycle time. Poor performance resulting from solids depositing at the top, or depleting available head loss at low bed storage capacity, can be improved by increasing the hydraulic loading (linear velocity). The operator can improve regeneration, but if improved regeneration is too time consuming net production rate is reduced.

REPORTED PERFORMANCE DATA

Filtration systems are commonly designed using water treatment or tertiary municipal plant criteria, assuming that equivalent effluent quality can be achieved by "proper" operation. In reality, treatment performance is a product of all the influent and operating variations encountered, so that performance is more likely a range specific to the waste than the treatment plant physical characteristics. As a frame of reference, results of approximately 40 industrial facilities reported in the EPA *Treatability Manual* are summarized in Table 5.3 [10]. The reported data are for a variety of industries; such as paper mills, petrochemicals, iron and steel, hospitals, leather, textile, organic chemicals, pharmaceutical, rubber, and resin. These data are cited to demonstrate the range of performance data possible, with no assurance that the facilities reported were properly designed or operated, that these results are the best that can be achieved, or that these results can be achieved with other facilities where influent characteristics are significantly different.

Tertiary filtration can produce variable and sometimes frustrating results because of the low solid concentrations encountered. Such systems often result in unpredictable, low, and sometimes apparent negative removal efficiencies because of difficulties in obtaining representative influent and effluent samples and analytical inaccuracies encountered with low concentrations. Filter performance may be more predictable for heavy solids loading because more "target sites" and higher contact frequencies enhance agglomeration and removal. However, in some cases solids concentration may necessitate excessive backwashing, resulting in low net effluent production. In such cases, sedimentation or flotation could be a more viable treatment.

REQUIRED PROCESS DESIGN DATA

Filtration can be screened with laboratory scale columns to test media, depth effectiveness, and effluent quality. However, the data represents "order-of-magnitude" results, with no information to establish backwash requirements or achievable filter run; critical for full scale design. Scale-up

TABLE 5.2. Operating Characteristics.

Variable	Operator Controllable	Critical
Waste Characteristics		
Waste generated	No	Yes
Composition	No	Yes
Concentration	No	Yes
Solid properties	No	Yes
Filtering properties	No	Yes
Operating Characteristics		
Flow rate	Limited	Yes
Hydraulic loading	Yes	Yes
Chemicals used	Yes	Yes
Chemical dosages	Yes	Yes
Pressure	Yes	No
Head loss	Yes	No
Cycle time	Yes	Yes
Backwash program	Yes	Yes
Backwash volume	No	Yes
Net production rate	No	Yes

TABLE 5.3. Reported Performance Data (adapted from Reference [10]).

	BOD$_5$	COD	TOC	TSS	O/G
Effluent, mg/L					
Minimum	2.4	29	10	<0.01	<0.5
Maximum	23,400	260,000	25,000	7330	9940
Median	19	195	40	12	11
Mean	2280	13,400	2000	247	899
Removal, %					
Minimum	0	0	0	0	0
Maximum	51	75	49	>99	>98
Median	22	20.5	12	75	20
Mean	24	26	15	68	30

from bench scale is not recommended. Prototype piloting or data from similar operating facilities are recommended to establish cycle time and required backwashing. It is also recommended that upset conditions be evaluated to determine whether the data developed are effective through the wide range of influent conditions encountered. A word of caution regarding pilot evaluation utilizing single waste volumes. Solids characteristics can significantly change during manufacturing cycles, or during prolonged storage, so that any one sample may not be representative of daily wastes. The best pilot studies are conducted in the field utilizing continuous pilot units for extended periods. Regardless of the manner of obtaining data, the required design criteria are listed in Table 5.4.

WASTE EVALUATION

A representative design range must be selected to account for expected waste variation, because filtration performance may be affected by influent conditions, especially those "not expected." Where variation is unpredictable, provision must be made to dampen their effects and allow smooth adjustment of operating parameters. The operating effects of the feed characteristics and applicable waste pretreatment must be understood to establish a design basis.

Flow

Waste flow is the primary operating variable used to control filter bed hydraulic loading. Theoretically, with all other variables being equal, lower hydraulic loadings increase solids removal efficiency. However, the selected flow rate could be affected by the need to control solids bed penetration or pressure drop limits. Lower velocities tend to deposit solids in the upper bed layer, reducing bed capacity, rapidly depleting available pressure drop, and reducing total filtration time. Higher velocities tend to help solids penetrate deeper into the bed depth and extend bed life. However, because the head loss through a system is directly related to flow rate, excessive rates could rapidly deplete available head loss,

terminating the process in relatively short cycle times. If upstream storage is not provided and the process is sensitive to hydraulic loading, the waste generation rate establishes the influent rate, and the operator has no means of controlling the filtration process.

Filtration systems should be operated on a continuous basis, at a constant flow rate, with minimum variations. Frequent and abrupt influent variations destabilize the system, producing turbulence in the bed, causing excessive media intermixing and losses. This could be a serious problem where waste is generated intermittently, where zero flow is likely in cyclical or batch manufacturing operations, or where the waste is generated over relatively short periods. In such cases, equalization allows a continuous and consistent filtration flow, and could result in smaller units; because an average rather than peak flow can be used as the design basis.

TABLE 5.4. Required Design Data.

Critical pilot plant treatability data specific to the waste
1. Design temperature
2. Hydraulic loading rate range
3. Solids loading rate range
4. Media characteristics
5. Required chemical treatment
6. Pressure drop requirements
7. Cycle time before washing
8. Backwashing requirements
9. Effluent quality

Waste characteristics that should be
obtained from laboratory studies
10. Concentration
11. Type solids
12. Filtering characteristics
13. Chemical treatment affects on filtering characteristics
14. Media effects on filtering characteristics

Selected operating characteristics
15. Overflow rate
16. Solids loading rate
17. Media
18. Operating cycle

Solids Concentration

As previously stated, filtration is not a substitute for sedimentation or flotation, but a polishing step. Effluent quality will be affected by influent concentration and solids size distribution. Applicable concentration can range from 30 to 200 mg/L, although if high volume wastes contain more than 100 mg/L suspended solids upstream coagulation, flocculation, and removal are advisable [11]. Higher concentrations cause the unit to reach terminal capacity early, resulting in frequent backwash cycles, low net effluent production, and high backwash volumes. When employed as tertiary systems to meet stringent effluent quality of less than 10 mg/L, influent concentrations as low as 30 to 50 mg/L may be required [4,9,11]. Where waste volume and solids loadings are extremely low, disposable cartridge filters are worth considering as the most cost effective treatment method.

Solids Characteristics

Definitive waste solids sizes resulting in optimum filtration are difficult to establish. Theoretically, the transport removal mechanisms for suspended solids larger than 1 μm involves gravitational forces, whereas those less than 1 μm are removed by diffusional mechanisms [6]. Available data indicate that municipal effluents and water treatment influents tend to be bimodally distributed, equally divided between relatively low and high size distributions [9]. In a reported municipal effluent study the two modes identified were from 3 to 5 μm and from 80 to 90 μm. The particles in the lower size range were less effectively removed [9]. Because increased size distribution improves removal efficiency, coagulation and flocculation are necessary for wastes containing significant quantities of small solids. In fact, ultrafiltration may be a more viable consideration for wastes containing predominantly submicron solids.

Unlike suspended solids generated in water treatment facilities, wastewater solid characteristics are unpredictable, specific to the generating source, and therefore considerably more difficult to treat. Single-type solid species are unlikely in wastewater treatment, where a variety of solids with a large particle size distribution is common. Wastewater solids are commonly classified as hard and inactive, compressible, fibrous, colloidals, or biological [13].

Hard and inactive solids are those that do not compress, and therefore do not change shape. These solids can be easily filtered because their removal can be directly related to the selected media size, and bed regeneration could involve minimum scouring and backwash, although hardness could be an operating problem affecting media life and bed losses. These solids have characteristics similar to sand, metal particles, or grit. The filter design involves selecting a specific media size (range), which can be easily established with pilot testing.

Compressible solids such as gels, oils, greases, and plastics will change shape with pressure, presenting a special filtering design problem. Media selection is not a simple matter of size selection and orientation because these solids can be extruded through most bed sizes. They exhibit poor bed penetration through small bed media sizes, and resulting large upper layer collection. In such cases, higher pressures to increase bed penetration induce bed plugging and solids compacting, resulting in accelerated pressure drops and decreasing filtering time. Under extreme pressures a significant quantity of solids could be extruded through the media resulting in poor effluent quality. The adhesive quality of some of these solids can result in rapid solids accumulation, difficult to remove by backwashing.

Fibrous solids are not only compressible, but exhibit relatively large lengths. Their inherent "weaving" characteristics enhance surface matting and plugging, making their removal more difficult. As with compressible solids, media selection based solely on size and physical characteristics is not reliable.

Colloidal characteristics produce difficult filtration conditions, the most detrimental being a negative surface charge similar to that found in sand and granite bed media. This common surface charge results in repulsive electrostatic forces between the wastewater solids and the media, making filtration ineffective unless the colloidal solids are chemically stabilized.

Biological solids can exhibit most of the characteristics mentioned, in addition to biological activity under some conditions. In such cases, biological growth can reduce bed capacity. This problem can be minimized by chlorination, as long as the chlorination does not interfere with the effectiveness of pretreatment chemicals.

Many industrial wastewaters contain a combination of solids, requiring their aggregate effects be considered in pretreatment, process design, and media selection. Individual characteristics must be balanced to optimize effluent quality and filter run time, minimizing backwash time and volume. Major solids related considerations in establishing a treatment system include

(1) Developing a suitable floc size
(2) Minimizing adhesive or plugging characteristics
(3) Neutralizing solids electrostatic characteristics
(4) Developing a floc with adequate shear strength to withstand the shear forces encountered in the bed

For the reasons mentioned, attempting to design a wastewater treatment system without specific wastewater filtration data is chancy.

PROCESS DESIGN VARIABLES

Industrial filtration system design requires that critical system components be evaluated and specified to assure a stable process over a range of anticipated conditions. The primary process factors in filtration design include evaluation of the following:

(1) Configuration
(2) Waste preparation
(3) Bed media
(4) Loading rate
(5) Pressure drop and head loss
(6) Hydraulic control method
(7) Backwashing system
(8) Filter run time
(9) Fate of the contaminants

Configuration

Wastewater filter configuration depends on the influent solids concentration and the allowable pressure drop. Common filter media systems were discussed in the Filtration Systems section. Some selection guidelines include the following:

(1) Generally, fixed bed systems are applied in industrial waste systems, unless cycle time is so short that off-line regeneration results in a prohibitive size and low net effluent production. In which case, a continuous system should be considered.
(2) It may be difficult to reduce wastewater to required effluent quality. In addition, solids removed and "matting" in the upper filter layers may result in rapid head depletion. In such situations deep coarse filters or multimedia filters may prove advantageous to improve solids penetration and cycle times.
(3) The nature of industrial solids may make simple water backwashing ineffective, requiring either a water/surface wash or a water/air scour procedure.

Based on these basic guidelines, the Process Engineer can proceed to develop a filter configuration evaluating and selecting the *media*, evaluating *conventional* or *deep filters*, and selecting the *operating mode*. Media selection is discussed in detail in proceeding sections.

Conventional or Deep Filters

Generally, industrial waste filters utilize a combination of media types and media sizes at select depths. This could involve a choice of a deep single media or multiple media. The selection depends on the expected effluent quality, an effective cycle time, and maintaining the integrity of the bed during backwashing. Industrial filters frequently, but not always, have the following characteristics:

(1) Bed depth from 1 to 2 meters (4 to 6 feet)

(2) Use of multi- or mixed media to provide for "deep solids" storage capacity at the upper layers, and high effluent quality at the lower levels
(3) Enhanced bed regeneration procedures employing air scouring, fluidization, backwashing, and optional chemical treatment

Operating Mode

Basically, any of the configurations discussed in the Filtration Systems section can be tailored to a specific design. However, commercial filters are commonly packaged in a variety of standard configurations to meet usually encountered services. The most common filter designs are

(1) Pressure filters: These systems can be tailored to most system requirements because they eliminate pressure as a limiting variable. Because pressure is a selected design variable, operating parameters such as bed depth, media type and size, and backwash cycle can be independently evaluated.
(2) Stored backwash gravity filters: These are self contained, multiple-modular units with (effluent) backwash storage capacity. They are commonly constant rate and gravity fed units with sufficient backwash storage head to assure an adequate backwash. The backwash cycle can include air scouring, and is activated by pressure loss or a predetermined cycle time. These units are available as single or multimedia beds. Allowable head losses of 3 meters (10 feet) limits the use of these units. They are typically used for low influent concentrations to achieve high effluent quality.
(3) External backwash gravity filters: More operating flexibility can be incorporated to gravity systems by providing an external backwash. They are similar to the internal backwash filters, containing all their features, but they allow more design and operating flexibility. As a result they are applicable to a wider range of industrial wastes. They can be operated as constant or variable declining rate systems, depending on the influent feed location. They are available as modular units, with shell diameters up to 12 meters (40 feet), accommodating bed depths up to 3 meters (ten feet). Filter headloss is a design variable, measured by the difference between the feed weir and bed liquid levels. The feed weir level remains constant, the bed liquid height rises with increasing solids storage, depleting the available head with service time.

These three systems represent the basic semicontinuous filtering systems, a fourth class involves continuous filtration.

(4) Traveling bridge filter: Continuous filtration can be obtained with a traveling bridge filter, consisting of individual cells operated at head losses of less than 1 meter (2 feet), and at a variable rate. They use a moving bridge

that backwashes individual cells, utilizing clean stored effluent, while the other cells are still in the filtering cycle. The system operates at a low filtering and back-wash head, and is therefore best applied to units with nonclogging type solids at low influent solids feed rate. This unit can be considered a modification of the external backwash gravity filter with continuous filtration, since all cells are active except the row where the traveling bridge is backwashing.

Applicable controls are discussed in detail in proceeding sections.

Waste Preparation

Filtration is sensitive to waste characteristics, what should flow to the filter and how it should be conditioned, and what should be excluded from the waste. This includes some general pretreatment considerations, as well as waste chemical conditioning.

Pretreatment Guidelines

Although each waste will have to be specifically evaluated, some general pretreatment guidelines can be cited.

(1) The influent solids range should not exceed 200 mg/L to prevent short run times resulting from excessive maintenance, backwashing, and cleaning. If influent solids concentration exceed 100 mg/L pretreatment consisting of coagulation, flocculation, and sedimentation (or flotation) should be considered. A filter subjected to excessive loadings could produce regeneration volumes greater than the waste flow.

(2) Influent changes are a major cause of effluent quality deterioration. This could be a result of varying generated waste volumes, a simple concentration change, or a complex change in solid characteristics. Although chemical treatment can condition the waste and improve filtering characteristics, conditioning requirements may be affected by influent changes. Selected coagulation chemical treatment and dosages may be effective within an influent range, and equally ineffective outside that range. Unfortunately, changing conditions can only be detected, and corrections applied, as a result of laboratory jar testing. Jar testing is usually performed on random daily samples, or in response to deteriorating effluent quality. Because industrial wastes can be dumped into a drain at any time, either as an intentional manufacturing discharge or an unintentional spill, automatic corrective responses are not possible. In such cases, process design evaluation must include (a) equalization to dampen instant changes, or (b) waste storage to allow batch testing and preparation to assure a consistent feed, or (c) both.

(3) The media should not be subjected to high levels of free oil, grease, abrasive or fibrous solids; all of which are detrimental to bed life. These constituents should be removed in a pretreatment step.

(4) If chemical treatment is employed, the system pH should be adjusted to the effective conditioning range. This will be specific to the reagents used, based on supplier recommendations, and laboratory testing.

Conditioning

Wastewater chemical treatment can include solids destabilization and agglomeration as a separate unit operation upstream of the filter, or chemical addition preceding the filter. Many theories have been advanced governing chemical application to improve filtration, although most applications are evaluated on the basis of water treatment or municipal effluent experience. Establishing a definitive process chemistry requires on-line testing involving frequent jar testing. However, some general guidelines can be cited [2,6–9, 11,12].

(1) If the influent contains significant colloidal particles provision should be made for chemical destabilization, coagulation, and flocculation to produce a filterable floc.

(2) Provision should be made for addition of small doses of filter aids such as alum, cationic polymer, or both at the filter to correct any floc damage in the transport.

(3) Coating the media prior to filtration could change the media characteristics, enhancing filtration [6].

Filtration design should include capabilities to add inorganic and organic coagulants such as alum and polymers at various upstream locations, including small doses preceding the filter. Chemical dosage can range from 0.5 to 1.5 mg/L of polymer upstream, and 0.05 to 0.15 mg/L as a filter aid prior to the filter [11]. In some cases alum may be effective. With any chemical conditioner, dosage is critical and should be optimized; since too low a dosage will prove ineffective, and too high a dosage could cause rapid head loss acceleration from accumulation of excess chemicals in the media voids. In addition, where chemical combinations are used, excess alum could adversely affect polymer performance by radically changing the pH to a range outside the polymer limits. Coagulation and flocculation chemistry is discussed in Chapter I-9.

Finally, treatment to retard biological activity in the media may be required. Chlorination doses ranging from 5 to 10 mg/L can effectively reduce media biological activity [7]. Care must be taken in chlorinating systems where pH sensitive polymers are employed.

Media

Media size is commonly cited as the predominant bed property; but other properties such as density, hardness, and shape are equally important. As the media density approaches that of water, bed losses increase so that media with specific gravities less than 1.6 require increased freeboard and are prone to larger losses. Bed erosion can affect bed losses as well as reduce media size and thereby increase intermixing. Attachment of waste solids to the media necessitates heavy scouring and lengthy high velocity washes to maintain the bed; these, when combined with media hardness, can accelerate bed deterioration. Media shape affects bed performance, if its geometry limits interlocking, fluidizing, and available void space, or cannot be easily cleaned. Round shapes inherently offer the least resistance to interlocking, tend to easily fluidize, and self clean in backwashing.

Silica sand, garnet, and coal are common filter media. Their characteristics defined by effective size and the uniformity coefficient (UC). Two specific criteria utilized in the United States are the d_{10} effective size, in which 10 weight percent of the media have diameters less than d_{10}; and the d_{60} effective size, in which 60 weight percent have diameters less than d_{60}. The uniformity coefficient is defined as the ratio d_{60}/d_{10}. Commercially available media have a UC of 1.6, media with a UC ranging from 1.3 to 1.4 are available in limited supply, and media with a UC less than 1.3 are difficult to obtain. Media can also be defined by the lower and upper particle size, and allowable tolerances on these limits. This practice is not usually used in the United States.

Media Selection

As a general rule, media characteristics affect operating characteristics as indicated in Table 5.5.

As illustrated in Table 5.5, media selection alone will not optimize all the filtering variables. The configuration must be based on predominant process requirements, which could involve (1) meeting effluent quality, (2) minimizing pressure drop depletion, (3) maximizing system (flow) capacity, (4) maximizing process cycle time, and (5) maximizing backwashing effectiveness.

Media selection involves consideration of size, depth, uniformity, and the use of single or multiple components. An ideal filter system would involve total bed penetration at

TABLE 5.5. Effect of Media Size on Operating Variables.

	Effect with Media Size Increasing
Potential size classification	Decreasing
Solids storage capacity	Increasing
Allowable flow rate	Increasing
Rate pressure depletion	Decreasing
Length of cycle time	Increasing
Effect on solids removal	Decreasing
Length of backwash cycle	Decreasing

the exact time when influent breakthrough occurs, at which time the effluent concentration equals that of the influent, and the bed is totally utilized. In turn, the bed characteristics of an ideal filter would include the following [8]:

(1) The bed would have a high solid capacity.
(2) The bed would retain the floc loosely but securely.
(3) The bed could be easily regenerated by backwashing without excessive bed loss.
(4) Bed integrity with a little media classification occurs during backwashing.
(4) High effluent quality would be achieved throughout the filtration cycle.

Understanding that the ideal filtration system cannot be practically achieved, the question is how can this system criteria be best approached, by single or multimedia systems? First, the problems encountered by inappropriately selected single media beds should be understood:

(1) Single, fine-graded media will produce high effluent quality, but filtration will be limited to the surface and subsurface layers, with very little bed penetration. Unless the influent contains a very low solids concentration, rapid head loss and a short filtration cycle results.
(2) Single, coarse-graded media allows deep solids penetration and long cycle time, but produces a poor effluent quality.
(3) An optimum single grade may produce required effluent quality at a reasonable cycle time; but backwashing could produce classification resulting in media grading and floating finer media at the top, producing the same problem as single fine-graded material. The problem is compounded when extensive backwashing is required because of high solids capture or adhesive solids.

These problems can be minimized by considering the following options:

(1) Operate the filter in an upflow, taking advantage of the natural backwash media classification. This option is seldom used because of complications in design, operation, and construction [6].
(2) Select as low a UC, uniform, single-grade media which can achieve the required performance, and where backwash classification is minimized.
(3) Select a dual or multiple media such that lighter and coarse media are used as a top layer to promote solids penetration and increase storage capacity; with denser and finer size media at the bottom as a polishing step. The media (and its density) must be carefully selected so that they fluidize at approximately the same backwash velocity.
(4) Allow intermixing of select single or multimedia; achieving a multigraded, intermixed, graded system—coarse at the top to fine at the bottom.

(5) Use a two-stage filter, with the first stage containing coarse media for high solids capture, and the second stage containing finer media for effluent quality. In this system each unit can be independently backwashed at an optimum velocity.

Except for guidelines similar to those discussed, there are no "absolutes" governing media selection and configuration. The primary consideration being "whatever works," which for any specific waste must be based on prior experience or pilot evaluation. As a general guide, Table 5.6 illustrates some common media configurations [9]. As a rule, multimedia, stratified beds should be considered for high influent concentrations; a stratified system being one in which the multimedia are separated, with the coarse media on top and the fine media at the bottom. In any multimedia system size selection must include concern for intermixing during the backwashing cycles, preventing small size media migration to the top, where coarse material is desired. In fact, the degree of intermixing is a primary consideration in media selection. Intermixing for a coal and sand system can be related to the relative media diameters as follows [6]:

(1) Substantial intermixing when $d_{90\%}$coal/$d_{10\%}$sand is greater than 4.0
(2) Partial intermixing when $d_{90\%}$coal/$d_{10\%}$sand is approximately 3.0
(3) Sharp interface when $d_{90\%}$coal/$d_{10\%}$sand is less than 2.5

Media Depth

Media depth and size are interrelated as indicated in Table 5.6, indicating frequently employed media size, and com-

mon multimedia combinations. Media size can be as low as 0.35 mm and as large as 1.5 mm, rarely larger than 2 mm. Bed depths range from inches to feet, depending on the configuration selected.

Loading Rate

Generally, loading rates can range from 1.4 to 6.8 L/sq meter · s (2 to 10 gpm/sq ft); gravity filters operating at 1.4 to 4 L/sq meter · s (2 to 6 gpm/sq ft), and pressure filters up to 6.8 L/sq meter · s (10 gpm/sq ft) [11]. Significantly, unless the solids are amenable to filtration, or chemically conditioned to be filtered, even a rate of less than 1.4 L/sq meter · s will not produce an acceptable effluent. For wastes amenable to filtration some suggested guidelines applied to water and wastewater treatment include

(1) Low suspended wastes can be filtered at rates up to 6.8 L/sq meter · s (10 gpm/sq ft) without any significant deterioration of effluent quality. High influent concentrations should be filtered at rates less than 3.4 L/sq meter · s (5 gpm/ft^2).
(2) Weaker chemical flocs should be filtered at rates less than 3.4 L/sq meter · s.
(3) Municipal tertiary systems loadings are as high as 6.8 L/sq meter · s for peak wet weather design [8].
(4) Rapid, single-media (fine sand), shallow filters operate at rates lower than 3.4 L/sq meter · s [8].
(5) Applicable filtration rates increase with depth and the use of multimedia beds.
(6) Increased filtration rates promote deep solids penetration, which could result in increased operating time if

TABLE 5.6. Common Media Filtration Configurations (adapted from Reference [9]).

	Bed Configuration					
	Single	Single	Dual	Dual	Tri	Tri
Coal						
Diameter, mm			0.9	1.84	1.0–1.1	1.2–1.3
Depth, in.			36	15	17	30
cm			91	38	43	76
U.C.			<1.6	<1.1	1.6–1.8	
Sand						
Diameter, mm	1–2	2–3	0.35	0.55	0.42–0.48	0.8–0.9
Depth, in.	60	72	12	15	9	12
cm	152	183	30	38	23	30
U.C.	1.2	1.11	<1.85	<1.1	1.3–1.5	
Garnet						
Diameter, mm					0.21–0.23	0.4–0.8
Depth, in.					4	6
cm					10	15
U.C.					1.5–1.8	
Application						
Loadings	Heavy	Heavy	Moderate	Heavy	Moderate	Moderate
Floc	Strong	Strong	Weaker	Strong	Weaker	Strong

the media are selected for high solids storage capacity. Although higher loadings result in maximum solids capacity and penetration, they could impose more severe backwash requirements because of solids packing (and under extreme conditions plugging).

(7) Available pressure drop, or head for gravity systems, could significantly limit filtration rate.

(8) Excessive filtration rate variations can affect effluent quality, the extent of the problem depends on the magnitude of the rate change and the abruptness of the change.

Pressure Drop and Head Loss

Although available head is an important operating variable, increased filter pressure may not significantly increase filter capacity or solids capacity. Instead it could compress solids, resulting in rapid pressure drop buildup. Generally, filtration pressure drops range from 2 to 4 meters (8-12 feet) of water for gravity systems, above which pressurized system are commonly employed. However, this is not a rigid rule, rather selection is based on (1) the available upstream head for a gravity system, (2) a reasonable design for a pressurized system, and (3) economic benefits in constructing either system.

System process pressure drop considerations are illustrated in Figure 5.3. Processing cycle time is directly dependent on available filter pressure drop, which is equal to the available total head minus the underdrain and media losses. The water level difference between the filter inlet and outlet is the available filter head, or *head loss*. For any given filter the design loading rate is not only limited by optimum performance requirements but by available head. For pressure filters this is less of a problem, assuming the feed pumps are properly sized; but for gravity filters available head can be readily dissipated with excessive filtration rates. When pressure within the filter is less than atmosphere, due to losses and lack of an available head, a vacuum is formed. To eliminate this problem the effluent liquid level must be above the sand media surface.

General Head Loss Behavior

The head loss versus time configuration of a filtration system is directly influenced by the solids removal mecha-

nism, which could include (1) surface removal and formation of a surface mat, (2) penetration into the media, resulting in solids storage, or (3) a combination of both. The characteristic of a pressure:time (or filtrate volume) performance curve can be classified as follows [6]:

(1) A linear or straight line curve signifies a nearly direct relationship between the filtration rate and the head loss, with solids formation at the top minimal and uniform solids capture in the voids by adsorption on the media surface.

(2) A slight curvature indicates some removal by straining, and some blockage of the voids.

(3) A constant rate filtration system producing a concave upward curve signifies surface filtration occurring as a result of fine media, a significant solids loading of agglomerating solids with little opportunity for void penetration, or both. In such cases the media must be selected with a coarser surface media to allow penetration, and pretreatment included if the solids concentration or characteristics are major factors. Increased production time may be achieved with increased filtration rates, and corresponding increased linear velocities, improving solids penetration into the voids.

(4) A constant rate filtration system producing a concave downward head loss curve indicates the potential for low treatment efficiency, and poor effluent quality. This could indicate poor waste characteristics, requiring chemical conditioning.

Control Methods

An important process decision in selecting the filtration configuration is flow control, which will be either constant or variable flow, incorporating constant or variable pressure, and adjusting to system variations with a minimum effect on effluent quality. Some standard configurations are commonly employed in water treatment and wastewater treatment to meet specific performance criteria [6,11].

Constant Rate Control

As illustrated in Figure 5.4, a constant pressure is applied across the filter system, and a constant filtration rate or water

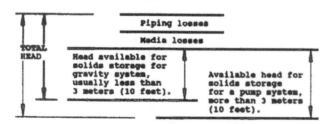

Figure 5.3 Filter head losses.

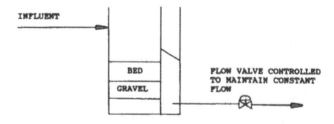

Figure 5.4 Constant rate control filtration.

level is maintained by controlling the effluent valve. Initially the effluent valve is nearly closed, with enough resistance to maintain the required effluent flow. As the bed resistance increases, the valve correspondingly opens to maintain the flow. Near the end of the cycle the valve is fully open, and the filtration run terminated. Its major disadvantages include higher initial and maintenance cost, and poorer water treatment performance and reliability because the system is prone to disruption from rapid adjustments to varying changes.

Influent Flow Splitting (Gravity Filter)

As illustrated in Figure 5.5, this method involves pumping the waste to an inlet weir box, the flow equally split and gravity fed to each filter. The level in each filter will rise to that equivalent to the filter bed resistance, so that constant flow is maintained. When a filter is out of service, the level in the other filters will gradually adjust to satisfy the additional required head. The head varies in each filter according to service time (plugging), but flow to each unit is equal. Advantages of this configuration include

(1) Constant rate is achieved without controllers.
(2) There is gradual change within a filter when another is out of service.
(3) Head loss in a filter is evident from the water level in the bed.
(4) Since the effluent weir must be located above the media to prevent accidental filter dewatering, the possibility of a negative head developing is eliminated.

A major disadvantage of this control system is that additional filter box depth is required because the effluent outlet must be raised above the media surface height.

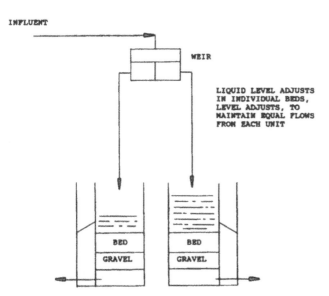

Figure 5.5 Influent flow splitting filtration.

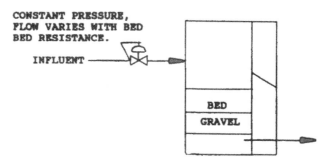

Figure 5.6 Constant pressure filtration.

Constant Pressure Filtration (Declining Rate Filtration)

As illustrated in Figure 5.6, with this control method the total available pressure is applied across the filter throughout the processing run, with the filtration rate varying according to changing bed resistance. Initially, a low resistance and a corresponding high filter rate develop. As the filter bed clogs and resistance increases, the flow rate decreases, although the pressure drop remains constant. In declining rate filtration the applied pressure is maintained constant by either applying a constant pressure (in a pressure system) or by a fixed level head tank in a gravity system.

Variable Declining Rate Filtration

As illustrated in Figure 5.7, in a variable declining rate system the feed enters each filter from a common header. The header is sized to minimize line loss and therefore the system is essentially insensitive to header loss. Any pressure loss is a direct measure of the filter bed activity. When the filter liquid level(s) is low (below the outlet trough) the system acts as a *split flow system*, all the filters receive nearly equal flows. When filter resistance increases, the liquid level rises above the outlet trough, but is equalized in the filters, resulting in a *constant pressure (liquid level) system*. The result is a variable flow through the units proportional to the bed resistance. Since the resistance is different in each filter, depending on the degree of plugging or the period into the cycle, each filter rate is correspondingly different. Although there are declining filtration rates, the

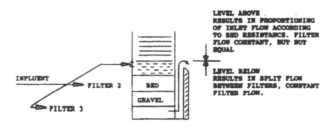

Figure 5.7 Variable declining rate filtration.

change is gradual, and effluent quality is not affected by rapid disturbances.

Declining rate is favored for gravity filters, with head losses of less than 3 meters (10 feet). Advantages of the declining rate filtration system include a constant and better effluent quality throughout the filtering cycle, with head loss decreasing with decreasing flow rates. Concerns raised about this system include selecting an appropriate design head and an apparent lack of operator flexibility.

Backwashing

Filter backwashing involves bed regeneration by removing stored solids, and preparing the void space for additional storage. Media regeneration depends on the abrasive action of the fluidized media to enhance the cleaning process. Inadequate cleaning results in a thin layer of compressible dirt or floc remaining around each grain media, gradual bed deterioration, increased pressure drop, and heavier deposits near the surface breaking and forming mudballs during backwashing.

Generally, regeneration involves

(1) Fluidized bed expansion, up to 10 to 20% of the original volume
(2) Disengaging the solids from the media
(3) Removing the solids from the system

Depending on the filter configuration and waste solid characteristics, the process can involve surface washing, air scouring, and backwashing, employed for the following reasons:

(1) Surface washing breaks up any surface solid mat formed, disengaging it from the media. Water treatment surface wash can consist of a one to two minute cycle, while two to four minutes can be employed for tertiary treatment systems [6,7]. The process employs water pressures ranging from 310 to 517 kPa (45 to 75 psig); at rates ranging from 2.5 cm (1 inch) per minute for rotary, and 8 to 15 cm (3 to 6 inches) per minute for fixed nozzle systems. The surface wash is then continued with a cocurrent backwash.
(2) Air scour disengages solids captured in the bed void spaces, minimizing backwash requirements. The air is injected into the bed bottom, commonly at 5 to 25 L/square meter · second (1 to 5 scfm/sq ft). Air scour can be used prior to the backwash, or concurrent with it. Movement of the finer bed media is of great concern, and of greater concern in concurrent wash operations. Restraining bed media is a critical mechanical design consideration [6,7].
(3) Backwash rate is a function of the collected solids characteristics, media type, number of media used, media size and size distribution, and water temperature. Backwash rates depend on the sequence employed. Wash rates of 1 to 3.5 L/square meter · second (2 to 5 gpm/ft^2) apply when air scouring is used for solids detach-

ment and fluidizing, and from 10 to 14 L/square meter · second (15 to 20 gpm/ft^2) if employed primarily for fluidization and cleaning. A greater intensity or wash duration should be employed if an aggressive cleaning sequence is required. The specific combination depends on the process conditions encountered [6,9,11].
(4) Chemical treatment may be a requirement for regenerating industrial filter treatment plants, especially if biological treatment plant effluents are treated.

A complete bed regeneration cycle can include a surface wash, air scouring, and water backwashing. The sequence can be as simple as back flushing with clean effluent or an external wash. A more complex sequence requires surface washing, air scouring to loosening the solids, and backwashing. An extremely complex sequence involves surface washing, air scouring with backwashing, and chemical treatment. Generally, for tertiary treatment water/surface wash or water/air scour backwash systems are employed [7]. Minimum fluidization is recommended for single phase media, with possible increased fluidization for some multimedia systems. Regardless of the backwash rate, the total backwash volume should not exceed 3060 to 4075 liters per cubic meter (75 to 100 gallons per square foot), since once fluidization is reached regeneration is not improved by continued washing [6,11]. However, the backwash rate and the combined regeneration sequence must be carefully selected to achieve optimum performance. The backwash rate selected depends on the media characteristics and backwash water temperature. Finally, a 2 to 20 minute flush should be provided after regeneration to allow effluent quality to be reached, and the initial effluent collected and recycled for further treatment prior to discharging [2].

Filter Run Length

Filter time is the single most significant operating variable, because it is directly related to system economics and effectiveness. In the design phase an economic trade-off must be made between the filtration rate, filter size and number of filters, and an acceptable net effluent production rate.

Net Production Rate

The effective filtration operating parameter is the net production rate, defined as the net wastewater processed over an operating cycle, expressed as Equation (5.2) [6].

$$\text{Net production rate} = \frac{\text{volume}}{\text{time}}$$

$$= \frac{(\text{treated water} - \text{backwash}) \text{ per cycle}}{(\text{treatment cycle} + \text{backwash cycle})} \quad (5.2)$$

Significantly, a reasonable production rate can be achieved

TABLE 5.7. Production Rate.

Filter Cycle, hrs	Basis: 100 gal/sf backwash, 30 minutes				
	Loading Rate, gpm/sf				
	1	2	3	5	10
	Net Production Rate, gpm/sf				
1		0.22	0.89	2.22	5.56
2	0.13	0.93	1.73	3.33	7.33
3	0.38	1.24	2.10	3.81	8.10
5	0.61	1.52	2.42	4.24	8.79
10	0.79	1.75	2.70	4.60	9.37
25	0.92	1.90	2.88	4.84	9.74
35	0.94	1.92	2.91	4.88	9.81

Note: Multiply gal/sf by 40.74 to obtain L/SqM backwash. Multiply gpm/sf by 0.68 to obtain L/s·square meter

in a relatively short cycle time, if a high filtration rate can be maintained. As illustrated in Table 5.7, a high production rate can be achieved in a 3 to 5 hour cycle if the feed rate can be maintained at 2 L/s · square meter (3 gpm/ft^2) or higher; while a 25 hr or higher cycle time is required at a feed rate of 0.7 L/s · square meter (1 gpm/ft^2) for a minimal production rate. In water treatment processes effective design criteria suggest run lengths of between 15 and 24 hours; so that, at a net filtration rate of 1.5 to 4 liters per second per square meter (2–6 gpm/ft^2) production volumes of 80 to 350 cubic meters per square meter of bed area (1800–8640 gallons per square foot) are obtained. [6] The filtration cycle is affected by the process conditions discussed, such as wastewater conditioning, media selection, and backwashing.

Bank of Filters

Although a reasonable net production rate can be obtained with a high filtration rate and a short cycle time, cycle time also affects the effective design rate. As an example, Table 5.8 illustrates the on-line and regeneration time of four filters operating in parallel, on a five hour filtration cycle, and thirty minute regeneration time. For the complete cycle illustrated, the percent of time all units are in operation are as follows:

$$\frac{4 \text{ filters} \cdot 7 \text{ "}squares\text{" } all \text{ units on}}{4 \text{ filters} \cdot 11 \text{ "}squares\text{" in a complete cycle}} \cdot 100 = 64\%$$

So that 64% of the time all four filters are in operation, and 36% of the time three filters are in operation.

Mathematically the percentage of time that all parallel filters are in operation can be estimated as indicated by Equation (5.3).

$$\frac{\text{total cycle time} - [\text{no. of filters} \cdot \text{regeneration time}]}{\text{total cycle time}} \cdot 100$$

$$(5.3)$$

The rest of the time (100%—all on line %), all parallel filters minus one (the one regenerating) are on line.

The effect of filtration run and down time on operating capacity is illustrated in Table 5.9. For a four filter system operating on a two hour filtration cycle, 20% of the time four units are operating, and 80% of the time one unit is off-line; while for a 10 hour filtration cycle, 81% of the time four units are operating, and 19% of the time one unit is off-line.

To put this in perspective, the volume that could be treated in a 10 hour cycle, assuming each filter is rated at 200 gallons per minute (757 L/m):

TABLE 5.8. Four Filter Bank Operating Cycle.

	Total 5.50 Hour Cycle										
	Hour	1			2		3		4		5
1	–	–	–	*	*	*	*	*	*	*	X
2	X	–	–	*	*	*	*	*	*	*	–
3	–	X	–	*	*	*	*	*	*	*	–
4	–	–	X	*	*	*	*	*	*	*	–

Note: Each square represents 30 minutes. Symbol "X" represents off-line of a unit; "*" represents the time when all units are in operation; "–" represents a unit in operation at a time when one unit is off-line.

TABLE 5.9. Percent of Time All Filter Are On-Line.

Run, hr	Cycles/ day*	Number of Filters in Bank			
		One	Two	Three	Four
		All Filter On, % of Time			
		(Rest of Time One Filter Down)			
2	9.60	80	60	40	20
3	6.86	86	71	57	43
5	4.36	91	82	73	64
10	2.29	95	90	86	81
15	1.55	97	94	90	87

*Assumes 30 minute off-line.

(1) Theoretical throughput of one filter at 100% is 288,000 gallons per day (1090 cu m/day); adjusted for 95% on-line time the actual process flow is 273,600 gallons per day (1036 cu m L/day).

(2) Three filters operating at 86% full flow would process 0.86 · 3 · 288,000 or 743,040 gallons per day (2812 cu m/day); while 14% of the time two filters would process 0.14 · 2 · 288,000 or 80,640 gallons per day (305 cu m/day); or a total of 823,680 gallons per day (3118 cu m/day).

(3) Four filters operating at 81% full flow would process 0.81 · 4 · 288,000 or 933,120 gallons per day (3532 cu m/day), whereas 19% of the time three filters would process 0.19 · 3 · 288,000 or 164,160 gallons per day (621 cu m/day); or a total of 1,097,280 gallons per day (4153 cu m/day).

The required design capacity to treat 240,000 gallons per day (908 cu m/day), adjusted for on-line time would be as follows:

(1) Theoretically, four filters at 60,000 gpd (227,100 liters/day), or 42 gpm (159 L/m), would be adequate.

(2) Based on a two hour filtration time, adjusting for off-line time, four filters rated at a minimum 52 gpm (197 L/m) would be required.

(3) Based on a five hour filtration time, adjusting for off-line time, four filters rated at a minimum 46 gpm (174 L/m) would be required.

(4) Based on a 10 hour filtration time, adjusting for off-line time, four filters rated at a minimum 44 gpm (167 L/m) would be required.

There are two significant points governing the capacity and selection of the number of filters in a bank. First, as illustrated, the unit operating capacity of a bank of filters increases with increased cycle time. Generally, cycle times greater than 15 and less than 24 are preferred [6]. Excessively long cycles are not desired. Next, the selection of the number of filters in a bank determines the system flexibility or ability to operate uninterrupted, with a one filter system having no process flexibility. The number of filters selected depends on (1) optimizing the capital costs, (2) assuring process flexibility, (3) the lim-

iting physical dimensions of commercially available filters, and (4) the physical limitations based on available area, equipment arrangement, and equipment access.

FATE OF CONTAMINANTS

The major discharge from a filtration system is treated effluent with low suspended solids. Backwash is generated on an intermittent basis to remove filter bed solids and regenerate the bed. In some cases the backwash cycle could include a chemical wash. This by-product stream may require separate treatment, or discharge to a select point in an existing on-site treatment system. Minimal vapors are generated, since any waste stripping would have occurred in the upstream treatment plant.

GENERAL ENGINEERING CRITERIA

Figure 5.8 illustrates a typical wastewater filtration Preliminary Concept Flowsheet. The process components specific to a *filter system* include the following:

Equalization (optional)	Chapter I-4
Waste feed system	
Chemical feed system	Chapter I-5
pH Control (optional)	Chapter I-8
Filter system	

A filtration system is usually purchased as a packaged design, with the shell, shell internals, and media integral part of a proprietary system. The system components are shipped for installation at the site. Some small units can be shipped skid mounted, with all internal piping complete, ready to be connected to on-site waste, effluent, and utility piping.

CRITICAL EQUIPMENT SELECTION

A filter can be constructed for gravity or pressure operation, the selection of which governs the internal design. However, all systems must incorporate a means of distributing the influent, a

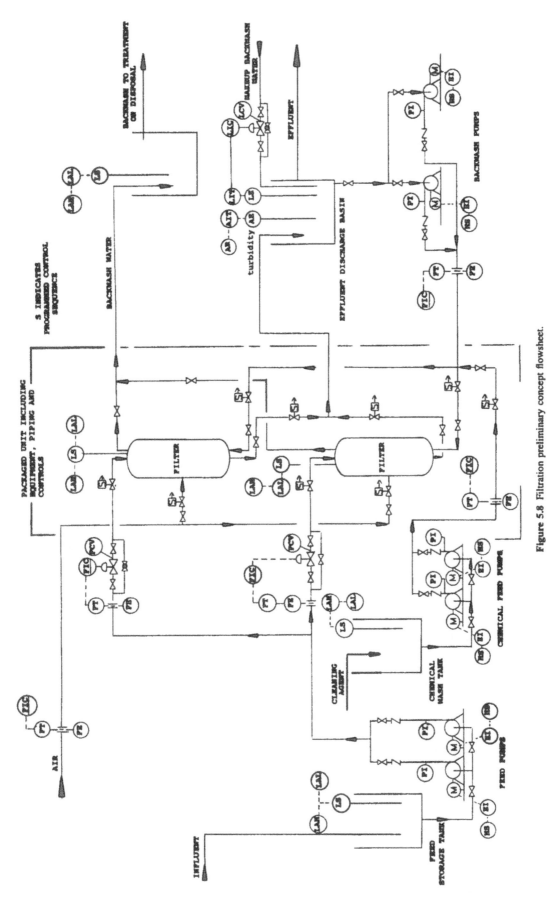

Figure 5.8 Filtration preliminary concept flowsheet.

media bed to treat the wastes, a support system for the bed, an underdrain effluent collection and backwash distribution system, and a wash water collection system. Because for the most part the mechanical construction of these elements is a standard proprietary manufacturing design, details will vary with the individual product, and are not subject to major alterations. Some common descriptions are discussed to familiarize the Process Engineer with general criteria.

Support Bed

If a conventional water backwash system is employed the filter bottom commonly contains a media supporting gravel bed, generally 30 to 45 cm (12 to 18 inches) deep depending on the filter type. The purpose of the bed is to prevent media from entering the drain system, and to provide a means of distributing the backwash water (or air) in the regeneration cycle. The support bed is graded, with 2 mm (1/16 inch) or more gravel at the top, depending on the sand bed size; and 8 cm (3 inches) or less at the bottom, depending on the underdrain system. In cases where air scouring or air-water backwashing is used, a fixed support must be used. This can include a variety of designs such as a mesh screen or a porous plate. Plugging of the orifices in these devices can be a serious problem.

Underdrain System

The underdrain system is used to collect treated water and distribute backwash. Underdrains are frequently constructed of perforated pipe, with a common header and laterals. The underflow system is designed for effluent gravity flow through the orifices, and provided with pressure restrictions to control backwash water pumped into the bed.

Wash Water Collection Troughs

A system of troughs and weirs is used at the filter top to collect the backwash, and assure uniform backwash distribution by equalizing static head. Significantly, the location of the troughs is critical, located so the trough bottom is above the expanded bed height to prevent sand losses and maintain a uniform static head. However, the height should not be too high, minimizing the residual bed backwash inventory; which could result in a high leakage rate, poor initial effluent quality as a result of residual backwash, and a resulting long start-up before required effluent quality is reached.

PROCESS CONTROL

Monitoring of the tertiary filter operation is essential for the proper operation of the system. These systems are commonly fully automated, programmed to respond to a high filter pressure loss, or in some cases a preset time cycle. This initiates the regeneration sequence, followed by a return to filtration

service. The backwashing sequences and duration can be altered by the operator to optimize the cycle. In addition, a bank of filters should be sequenced so that only one filter is off line at one time. Regardless of the degree of automation, some process on-line monitoring is required, such as

(1) Flow control of influent, effluent, backwash rate, waste, air scour rate (if used), and surface wash rate (if used). Influent and effluent flows should be recorded.
(2) Quality parameters such as influent and effluent turbidity can be automatically monitored and recorded. Effluent monitoring can be based on each filter or a composite of all filters. Provision should be made for automatic influent and effluent composite sampling to analyze for other discharge parameters.
(3) Head loss through each filter should be continuously monitored and recorded.
(4) Levels in the washwater tank, filter feed tanks, and other critical process tanks should be continuously monitored.
(5) Critical operating variables should be alarmed to alert the operator of malfunction. This includes:
    a. High influent and effluent turbidity
    b. Low chlorine, alum, polymer supply levels
    c. High chlorine supply pressure
    d. Influent, effluent, backwash, and surface wash pump failure
    e. Flash mixer and flocculator failure
    f. Air scour compressor failure
    g. Power failure
    h. Chlorine, alum, or polymer feeder malfunction

COMMON FILTER PROBLEMS

Some common problems encountered in filter operations can be summarized as follows [5,11,12]:

(1) Poor effluent quality because of poor wastewater characteristics. Generally wastes that contain a predominant amount of small solids, 10 to 20 μm or less, are poorly treated. Activated sludge and conventional trickling filters produce effluent with suspended solids containing 10 to 20 weight percent at 20 μm or smaller, of which 60 to 90% can be removed by filtration. Lagoons produce effluents with a considerable amount of solids less than 20 μm in size, and would therefore be poorly filtered.
(2) Poor effluent quality as a result of poor pretreatment, high hydraulic loadings, and high hydraulic loading variation.
(3) Poor effluent quality as a result of poor media selection.
(4) Short run times caused by high hydraulic loadings.
(5) Short run times as a result of the tendency of solids to accumulate at top few inches. This surface straining results in increased pressure drop. The problem can be corrected by increasing media size, increasing allowable

head loss, increasing filtration rate (deeper penetration), and use of multimedia carefully selected to avoid intermix.

(6) Ineffective media cleaning and mudball formation attributed to accumulated grease and the use of fine grain media. However, a predominant cause is the formation of a solid surface mat, not broken up and removed because of ineffective media cleaning. The situation becomes progressively worse if not cleaned. Chlorination of biological effluent will help prevent adhering to media, although this practice can be costly, and may not be allowed because of regulatory concerns about excess residual chlorine. Effective backwashing (see section above) is a better alternative.

(7) Buildup of emulsified grease in the media bed increasing head loss and reducing filter run. Extensive backwash procedures may have to be incorporated, or pretreatment done to remove grease.

(8) Excessive media loss as a result of stratified bed employing a high velocity backwash.

(9) Plugged media support system and media nozzles screens. Nozzles or screens with openings less than 1 mm are not recommended.

(10) Loss of media resulting from gravel support being upset because of excessive backwashing.

(11) Loss of media with washwater resulting from excessive backwashing.

CASE STUDY NUMBER 31

Develop a *filtration treatment system* to treat 1,000,000 gallons per day of waste containing 50 mg/L of suspended solids to an effluent of less than 5 mg/L. The following applicable design criteria were established by pilot studies:

(1) Surface overflow loading rate, 5 gpm/sq foot
(2) Sand losses of 5%
(3) Air scour of 5 scfm/sq ft
(4) Air scour time, 2 minutes
(5) Surface wash rate, 2 gpm/sq ft
(6) Surface wash time, 3 minutes
(7) Backwash rate of 10 gpm/sq ft
(8) Backwash time, 10 minutes
(9) Bed expansion, 10%
(10) Available head loss, 15 feet of water
(11) Operation at 7 days/week, and 24 hours per day
(12) A maximum of 1.5 mg/L of filter-aid prior to filtration unit, as a 1% solution (with a specific gravity of 1.1)

PROCESS CALCULATIONS

(1) Calculate influent contaminants

1 mgd · 8.34 · 50 mg/L = 417 lb/day of solids influent
1 mgd · 8.34 · 5 mg/L = 42 lb/day of solids effluent

(2) Calculate filter-aid required

1 mgd · 8.34 · 1.5 ppm = 12.5 lb coagulant per day
12.5/0.01 = 1250 pounds of 1% solution per day
1250/(8.34 · 1.1) = 136 gallons per day

Provide for a 250 gallon make-up and feed tank.

(3) Calculate the filter area

[1,000,000 gpd/1440]/5 gpm/ft^2 = 139 square feet

(4) Calculate the flotation area and diameter
(5) Number of units: two
(6) Calculated area

Diameter = [(4/Pi) · (139/2)]$^{1/2}$ = 9.4; select 10 ft diameter
Resulting *unit* area = 78.5 square feet
Resulting *total* area = 157 square feet
Resulting overflow rate = 1,000,000/1440/157 = 4.4 gpm/ft^2

(7) Select the filter depth: use 10 feet

Expansion 10% = 1 foot
Freeboard = 2 feet
Bottom = 2 feet
Total height = 15 feet

(8) Treatment cycle

Treatment cycle, hrs = 5.75
Total wash cycle, hrs = 0.25 (15 minutes)
Total cycle time, hrs = 6
Cycles per day = 4

Backwash generation
2 minutes air scour
3 minutes surface wash at 2 gpm/ft^2
10 minutes backwash at 10 gpm/ft^2

2 gpm · 3 minutes + 10 gpm · 10 = 56 gpm/sq ft/cycle

106 gpm/sq ft · 4 cycles/day · 2 units · 78.5 sq ft/unit

= 66,568 gallons/day

(9) Headloss evaluation

| | Reference | Maximum | |
		Design	Rate
Rate, gpm/sq ft	2.00	4.40	7.25
Available head, ft	15.00	15.00	15.00
Media loss, ft	0.50	1.10	1.82
Piping loss, ft	1.00	4.84	13.14
Operating head, ft	13.50	9.06	0

NOTE
a. 2 gpm/sq ft vendor reference data
b. Media loss = (4.40/2.00) · 0.50 = 1.10 ft
c. Piping loss = (4.40/2.00)^2 · 1.00 = 4.84 feet
d. Operating head = 15.00 − 1.10 − 4.84 = 9.06 feet

(10) Net production rate evaluation: Net production rate at operating cycle lengths of 1, 2, 5.75 (design), and 8 hours will be evaluated; loading rates of 2, 4.40 (design), and 7.25 (maximum) will be considered.

Treatment			Backwash		Net	Net
Cycle hr	gpm/ sq ft	gal/ sq ft	hr	gal/ sq ft	gal/ sq ft	gpm/ sq ft
1.00	2.00	120	0.25	106	14	0.2
1.00	4.40	264	0.25	106	158	2.1
1.00	7.25	435	0.25	106	329	4.4
2.00	2.00	240	0.25	106	134	1.0
2.00	4.40	528	0.25	106	422	3.1
2.00	7.25	870	0.25	106	764	5.7
Design Conditions						
5.75	2.00	690	0.25	106	584	1.6
5.75	4.40	1,518	0.25	106	1412	3.9
5.75	7.25	2,501	0.25	106	2398	6.7
8.00	2.00	960	0.25	106	854	1.7
8.00	4.40	2112	0.25	106	1394	2.8
8.00	7.25	3480	0.25	106	3374	6.8

Calculations

a. 1 hr · 60 minutes · 2 gpm/sq ft = 120 gal/sq ft
b. Backwash at 106 gal/sq ft (see above)
c. 120 − 106 = net 14 gal/sq ft
d. Total cycle time = (1 hr + 0.25) · 60 = 75 mins
e. 14 gal/sq ft/75 minutes = 0.2 gpm/sq ft = net production rate

Note that the cycle time is not as significant as the application rate. Maintaining a 4 gpm/ft^2 or higher rate assures a reasonable net production rate.

REFERENCES

1. Biskner, C.D., Young, J.C: "Two-Stage Filtration of Secondary Effluent," *Journal WPCF,* V 49, No 2, Pg 319, February, 1977.
2. Culp, R.L., Wesner, G.M., and Culp, G.L.: *Handbook of Advanced Wastewater Treatment,* Second Edition, Van Nostrand Reinhold, 1978.
3. Darby, J.L., et al.: "Depth Filtration of Wastewater—Particle Size and Ripening," *Research Journal WPCF,* V 63, No 3, Pg 228, May/June, 1991.
4. Givens, D.P. and Lash, L.D.: "What to Look for in Granular Media Filters," *Chemical Engineering Progress,* Pg 50, December, 1978.
5. Medcalf & Eddy, Inc.: *Wastewater Engineering-Treatment, Disposal, Reuse,* Third Edition, McGraw-Hill, 1991.
6. Sanks, R.L.: *Water Treatment Plant Design,* Ann Arbor Science, 1978.
7. Siemak, R.C.: "Tertiary Filtration, Practical Design Considerations," *Journal WPCF,* V 56, No 8, Pg 944, August, 1984.
8. Task Committee American Society Civil Engineers: "Tertiary Filtration of Wastewaters," *Journal Environmental Engineering, Proceedings ASCE,* 112, No 6, p 1008, December, 1986.
9. U.S. Environmental Protection Agency: *Process Design Manual for Suspended Solids Removal,* EPA-625/1-75-003a, January, 1975.
10. U.S. Environmental Protection Agency: *Treatability Manual,* Four Volumes, EPA-600/8-80-042a, 1980.
11. WEF Manual of Practice: *Design of Municipal Wastewater Treatment Plants,* Water Environment Federation, 1992.
12. Young, J.C.: "Operating Problems with Wastewater Filters," *Journal WPCF,* V 57, No 1, Pg 22, January, 1985.
13. Becker, C.H.: "Filtration," *Industrial Waste Engineering,* Pg 12, May, 1971.

Membrane Technology: Microfiltration, Ultrafiltration, and Reverse Osmosis

Ultrafiltration is employed as a tertiary treatment to remove low concentration suspended solids, dissolved organics and dissolved inorganics to achieve water quality limits.

BASIC CONCEPTS

MEMBRANE technology is not applicable to a wide range of raw industrial wastes or removal of large waste quantities. It is generally used as a polishing step to remove specific waste water components such as suspended solids, dissolved organics and dissolved inorganics to achieve water quality limits, produce reusable water, or to recover product from specific wastes. It has a limited application in waste treatment, being more successfully applied to single or limited species encountered in water treatment or desalination processes, and generally applied to isolated streams containing limited solutes and minimum interfering constituents. In fact, membrane filters do not destroy contaminants, merely concentrate them. This is not a problem where water is the desired product, and the reject volume of secondary consequence. However, in waste treatment contaminant removal and conversion to an environmentally inactive form is the primary objective; the concentrated reject volume produced must be small relative to the initial waste volume.

Figure 6.2 illustrates the basic continuous membrane treatment system. Wastewater is bled into a recirculation loop feeding the membrane units, maintained at an elevated pressure suitable to overcome the membrane resistance. Clean effluent and a concentrated reject stream are discharged from the filter. The unit operates in this mode until collected contaminants restrict the influent flow, at which time the unit is taken off-line and a backwash or cleaning cycle started. Waste treatment membrane technology is primarily based on water treatment experience, especially desalination processes. Direct application of water treatment design criteria to waste treatment must compensate for the significant differences in waste characteristics.

Filters can be divided into broad classifications relating molecular or particle size to average membrane restrictions, as indicated in Table 6.1 [9]. These classifications are more a matter of convention than absolute limits, and more applicable to conventional filtration and microfiltration. Particle retention in ultrafiltration and reverse osmosis processes is not solely by size or molecular weight. They have a wide pore distribution and inherent manufacturing defect holes, making average and absolute membrane size classification difficult relative to the range of small solute sizes encountered. Reverse osmosis systems involve ionic waste species, and solute rejection is a result of diffusion, liquid retained by the membrane, solubility, the liquid dialectic constant, and contaminant properties.

PROCESS MODEL

In membrane processes, liquid passes through the membrane along with some solute leakage, whereas the bulk of the solute is concentrated and rejected. The quantity of solids passing with the liquid depends on relative solute to pore size, with solids not rejected, being transported by two possible mechanisms. The transport mechanism promoting leakage shifts from pore retaining capability to diffusion, and then to predominantly diffusion at reduced pore sizes. Theoretical models developed to understand transport mechanisms for specific membrane systems are discussed below, the bulk of the information coming from chemical engineering mass transfer texts [2,8,9].

Solvent Transport

Solvent transport is related to the process conditions and the membrane characteristics. The fundamental filtration

TABLE 6.1. Conventional Membrane Separation
Classification (adapted from Reference [9]).

Membrane Type	Particle Size, Micrometer	Molecular Weight
Conventional filtration	>20	Not applicable
Microfiltration	0.02 to 10	>300,000
Ultrafiltration	0.001 to 0.02	300 to 300,000
Reverse osmosis	0.0001 to 0.001	<300

flux equation relates transport flux to the pressure difference driving force, membrane resistance, and collected cake resistance [8].

$$Nw = dP/(Rc + Rm) \qquad (6.1)$$

where Nw is the permeation flux through the membrane in moles/time/area, dP is the applied pressure difference, Rm is the membrane resistance, and Rc is the cake resistance.

Definitive units are not used throughout this discussion since the equations cited are mentioned to advance qualitative discussions. They are not presented as design methods.

In *conventional* filtration the cake resistance is time (flow) dependent and approximated as follows [8,9]:

$$Rc = \frac{\sigma w \, Vt(dP)^s \mu}{A} \qquad (6.2)$$

where σ is a constant dependent on the cake properties, w is the weight particles per unit volume of filtrate, Vt is the volume filtrate throughput, s is the cake compressibility exponent ($s = 1$ for completely compressible cake and 0 for noncompressible cake), μ is the filtrate viscosity, and A is the filter surface area.

Osmotic pressure is a membrane property resulting from low molecular species retention, which must be overcome by the applied system pressure. Osmotic pressure at a specific point can be approximated using the van't Hoff equation [9].

$$\pi = iCRT \qquad (6.3)$$

where C is the solute molar concentration, i is the number of ions formed from the solute molecule, R is the gas constant, and T is the absolute temperature.

Osmotic pressure increases with (1) decreasing species molecular size and weight, (2) increasing valence of the ions formed, (3) increasing concentration, and (4) increasing temperature. Since the applied pressure must be greater than the osmotic pressure, the pressure difference is the driving

force, which when applied to the *general* filtration model, Equation (6.1), results in revised Equation (6.4) [9].

$$Nw = (dP - d\pi)/(Rg + Rm) \qquad (6.4)$$

where $d\pi$ is the osmotic pressure difference.

Solute Transport

Solute can migrate from the bulk stream through the membrane by either *pore* or *diffusion* mechanisms. Solute transported with solvent is primarily related to membrane porosity and pore size distribution, and is the common transport mechanism for *micro-* and *ultrafiltration* systems.

Pore flow process can be defined in terms of solvent transport through the pores, expressed as [9]:

$$Ns = \epsilon Cs \, Nw = \epsilon Cs \frac{dP}{Rm + Rg} \qquad (6.5)$$

where Ns is the solute flux, moles/time/area; ϵ is the fraction of liquid feed passing through membrane pores accommodating solute; and Cs is the membrane upstream solute molar concentration.

Diffusion transfer is common for RO systems. The *diffusion process* depends on the species dissolving in the membrane liquid, and the resulting concentration gradient is the diffusion driving force. Solute diffusivity can be defined by the Stokes-Einstein relationship, expressed as Equation (6.6) [8,9]:

$$Dm = \frac{KT}{6\pi\mu r} \qquad (6.6)$$

where K is the Boltzman constant, T is the absolute temperature, μ is the viscosity, and r is the particle (diffusing) radius.

Diffusion solids transport is related to the selected species solubility and diffusivity properties, and the concentration gradient driving force, as indicated by Equation (6.7) [9].

$$Ns = H\frac{Dm}{t}(Cs - Cf) \qquad (6.7)$$

where H is the solute distribution coefficient, Dm is the solute diffusivity in the membrane liquid, t is the membrane thickness, and Cf is the bulk feed solute concentration.

Based on the Stokes-Einstein diffusivity equation, the solute transport by diffusion mechanism is primarily affected by its molecular weight and size of the species, and the solvent temperature and viscosity. Solute rejection is dominated by a large concentration gradient, so that pressure has a minor effect.

Rejection or Retention

Membrane effectiveness can be expressed by the rejection (or retention) efficiency, Rj, in terms of the bulk solute concentration Cb and the filtrate (effluent) concentration Cf [9]:

$$Rj = (Cb - Cf)/Cb \qquad (6.8)$$

Membrane performance is measured by the volume concentration ratio (VCR) and the probability of rejection (P). They relate the influent (Vo) and reject (Vr) volumes, as expressed by Equation (6.9a), and the fraction of solids passing through the membrane (σ), as expressed by Equation (6.9b) [2].

$$VCR = \frac{Vo}{Vr} \qquad (6.9a)$$

$$\sigma = 1 - P \qquad (6.9b)$$

In turn, the influent (Co) and effluent (C) concentrations can be expressed as indicated by Equation (6.10). Membrane filtration is not viable when σ is zero, the effectiveness increasing as σ increases.

$$C = Co \cdot VCR^\sigma \qquad (6.10)$$

Concentration Polarization

The primary driving force in a membrane system is pressure, with increasing flux resulting from increasing pressure. Initially, the gel resistance is negligible and the flux is proportional to the applied pressure difference. Wastewater flowing to the membrane surface transports solute, which when deposited forms a secondary flow resistance. As the secondary resistance increases, the flux decreases, and deviates from a straight line proportional to the pressure difference. Solute accumulation continues until cake resistance impedes liquid flow through the membrane. At this point pressure is not a significant process driving force. In fact, pressure could compact and further increase film resistance. This process, referred to as concentration polarization, produces a second resistance to wastewater transfer. This secondary resistance affects flux as indicated by the basic filtration Equation (6.1), where Rc, the cake resistance, is the gel resistance (Rg).

At some point the gel resistance exceeds that of the membrane, and an equilibrium is reached between convective transfer of solute to the membrane and diffusive transfer back to liquid, as expressed by Equation (6.11).

$$N \cdot C' - D \cdot dC/dx = 0 \qquad (6.11)$$

Forward convective transport − diffusive transport back = 0

where C' is the bulk concentration of solute, dC/dx is the concentration gradient of solute with distance x, and D is the diffusivity of solute.

Equation (6.12) is the integrated form of Equation (6.11); indicating flux (Nw) is independent of pressure, but dependent on concentration gradient and gel characteristic.

$$Nw = D/t \ln (Cs/Cb) \qquad (6.12)$$

where Cs is the solute gel concentration, Cb is the bulk solution solute concentration, D is the diffusivity of solute, t is the stagnant boundary-layer thickness, and Nw is the flux, moles/time/area.

This equation can be rearranged to what is commonly referred to as the polarization modulus (Pm).

$$Pm = Cs/Cb = \exp (Nwt)/D \qquad (6.13)$$

The polarization modulus (Pm) is defined by the ratio of membrane surface to bulk concentrations. An increasing polarization modulus results from increasing flux (Nw), decreasing diffusivity (D), and increased gel thickness. The gel thickness is affected by particle size, membrane pore size, solvent viscosity, fluid temperature, and fluid velocity. The predominance of this phenomena is dependent on the specific membrane process (as discussed below).

MEMBRANE SYSTEMS

MICROFILTRATION (MF)

Microfilters can restrain particles in the 0.02- to 10-μm range from passing through smaller membrane pores, generally involving solutes with molecular weights higher than 300,000 [9]. Membranes employed are constructed of highly porous, open-pore cellulose materials or synthetic polymers. The open-pore characteristic subjects the membrane to high plugging potential, requiring off-line membrane cleaning. Typically, microfiltration processes operate at flux rates ranging from 41 to 407 liters/minute/sq m (1 to 10 gpm/ft²), and pressures ranging from 69 to 690 kPa (10 to 100 psig) [6].

Process Models

The basic membrane equation can be modified to reflect specific MF characteristics; such as

(1) Separation is by size, with membrane pores relatively large compared to the other processes.

(2) Retention of liquid, and therefore osmotic pressure, is not significant.

(3) The process is pressure driven.

Neglecting the osmotic pressure, the solvent transfer Equation (6.4), reduces to Equation (6.14):

$$Nw = dP/(Rg + Rm) \qquad (6.14)$$

The gel (membrane surface cake) resistance (Rg) can be related to the system properties; increasing with pore plugging, increased thickness, and compacting under pressure. When the gel resistance is dominant relative to the membrane resistance, the MF process flux Equation (6.14) reduces to the Equation (6.15), substituting Equation (6.2) for Rg.

$$Nw = A \cdot dP^{1-s}/\sigma \cdot W \cdot Vt \cdot \mu \qquad (6.15)$$

where

σ is a constant dependent on the cake properties
W is the weight particles per unit volume of filtrate
Vt is the volume filtrate throughput
s is the cake compressibility exponent; $s = 1$ for completely compressible cake, and 0 for noncompressible cake
μ is the filtrate viscosity
A is the filter surface area
dP is the pressure difference driving force
Nw is the permeation flux expressed as moles/time/area

Solute transfer is basically driven by pore flow, as described by Equation (6.5).

Fouling Control

Pore fouling is the primary concern, resulting from "straining" mechanisms pushing near pore size particles into, but not out of, the membrane. Although gel formation is secondary to fouling, it can be accelerated by stagnant surface conditions and precipitation of components at the solubility limits. Fouling and gel control methods include (1) pretreatment to minimize solids loading to the membrane, (2) ion exchange, and similar pretreatment methods, to minimize components that could salt-out, (3) careful selection of anisotropic membranes with funnel structure, minimizing solid entrance but allowing clear passage of those that do enter, (4) water or chemical membrane cleaning, (5) utilizing a recirculation loop to develop a high velocity cross-flow to sweep the membrane surface, (6) ultrasonic cleaning methods, and (7) controlled flow rate.

Rearrangement of Equation (6.15) illustrates the effects of controlled flow rate on filtrate flux; filtrate flux is defined as total volume throughput per area, being a measure of polarization effects.

$$Vt/A = dP^{(1-s)}/\sigma w Nwu \qquad (6.16)$$

Reduction of the feed flow or flux (Nw) results in a corresponding increase in the total filtrate collected per unit area of membrane; ideally directly proportional, although in some cases tapering at higher feed rates. This is similar to increasing the net production rate as discussed in Chapter III-5, Filtration. This control method increases required filtration area, and resulting capital costs. However, where gel formation is an expected problem, a conservative design flux is an important consideration. This could result in operating existing units at reduced feed rate; or designing new units at a conservative flux, and additional or larger operating units installed.

ULTRAFILTRATION (UF)

Ultrafiltration can be used to separate large dissolved macromolecules or colloids from water. Separable solute diameters range from 0.0010 to 0.02 μm, generally larger than the solvent, and molecular weights from 300 to 300,000 range [9]. The process is pressure driven. However, because of the high molecular weights involved osmotic pressure is insignificant. Defining ultrafiltration limits by size is convenient, but not absolute. Retention is affected by solute shape, charge, and deformability characteristics. Highly structured molecules are more effectively removed than single flexible molecules. With proper surface management, solute accumulation is not a significant problem since macromolecules separation occurs away from the membrane, and surface velocities are generally maintained at 0.6 to 3 m/s (2 to 10 fps). Flux rates range from 204 to 815 L/day/sq m (5 to 20 gal/day/ft^2), at operating pressures from 69 to 345 kPa (10 to 50 psig), with some units at pressures up to 690 kPa (100 psig) [6].

Process Model

The basic membrane equation can be modified to reflect the specific characteristics of the UF process, such as:

(1) Process separation by size is not absolute, and membrane pores are not uniform. Rejection rate is influenced by defect holes and solute characteristics such as charge, shape, and deformability.

(2) Retention of liquid, and therefore osmotic pressure, is not significant.

(3) The process is pressure driven.

The solvent transfer Equation (6.4) reduces to the ultrafiltration Equation (6.17).

$$Nw = dP/(Rg + Rm) \qquad (6.17)$$

Because UF systems are susceptible to polarization concentration, the concentration polarization model and related principles are relevant to UF operation. Solute transfer is basically by pore flow as defined by Equation (6.5).

Polarization Control

Boundary layer control is critical to maintaining a viable ultrafilter. Because of the UF pore size it is extremely susceptible to gel formation discussed under *concentration.*

In the UF process the relative solute and membrane sizes result in the potential for maximum surface accumulation. Surface rather than penetration fouling is predominant [9]. The polarization modulus (Pm) is higher than in the MF and reverse osmosis process; directly related to the liquid flux and stagnant layer thickness, and inversely to the solute diffusivity. The resulting surface concentration (Cs) can become significant if polarization concentration control is not considered in the design and implemented in the operation. The primary control considerations are preventive measures, as well as membrane maintenance. These include (1) pretreatment, (2) closed-loop, high velocity cross-flow at the membrane surface, (3) backwashing to remove surface particles, (4) mechanical, chemical, or ultrasonic cleaning, and (5) using a conservative design flux.

Effect of Concentration Polarization on Retention

Solids retention in the concentrate can be represented by the Rejection Efficiency Equation (6.8):

$$Rj = (Cb - Cf)/Cb.$$

so that if $\sigma = Cf/Cb$, then $Rj = 1 - \sigma$; and where σ is the fraction of total membrane liquid discharging through pores large enough to pass solute molecules; a function of the bulk solute concentration Cb, the filtrate (effluent) concentration Cf, and the pore structure.

As the membrane surface concentration increases above the bulk concentration the filtrate concentration through the pores increases. At some point the filtrate concentration approaches the gel concentration, or $\sigma = Cf/Cb = Cs/Cb$; where Cs is the gel concentration, which is approximately equal to the filtrate concentration. Because polarization modulus (Pm) as defined by Equation (6.13) equals Cs/Cb, then the solids retention equation can be rewritten as

$$Rj = 1 - Pm \qquad (6.18)$$

so that as polarization modulus increases, Cs increases, and the system performance (Rj) decreases.

REVERSE OSMOSIS (RO)

Reverse osmosis is used to remove dissolved materials, commonly salts, under pressures ranging from 1400 to 10,300 kPa (200 to 1500 psig) and flux rates from 204 to 611 L/day/sq m (5 to 15 gpd/ft^2) [6]. RO separation is generally effective at solute molecular weights below 300, and related solute sizes from 0.0001 to 0.001 μm [9]. Solvent and solute size are generally of the same magnitude. In this service range, water transport is by diffusion, with the water becoming part of the membrane structure and exerting osmotic pressure. The osmotic pressure must be overcome by applied pressure. Solute rejection is not only by size, but also by those variables affecting diffusion.

Salt removal by cellulose acetate membranes generally follows the following criteria [5].

(1) Multivalent ions are more highly rejected than univalent ions.
(2) Undissociated or poorly dissociated substances have low rejections.
(3) Acids and bases are rejected to a lesser extent than their salts.
(4) Co-ions affect the rejection of a particular ion.
(5) Low-molecular-weight organic acids are poorly rejected.
(6) Undissociated low-molecular-weight organic acids are poorly rejected, and their salts well rejected.
(7) Trace quantities of univalent ions are generally poorly rejected.

Process Model

The basic membrane equation can be reviewed to reflect specific RO characteristics:

(1) The process involves separation by diffusion, the primary transport mechanism being water passing through the membrane and rejecting the solute.
(2) Liquid is retained in the membrane pores, osmotic pressure is significant.
(3) The process is pressure driven, with the applied pressure high enough to exceed the osmotic pressure.

Equation (6.4) is directly applicable to RO. The higher diffusivity values of the ionic species encountered in RO, along with the lower stagnant thickness, make the polarization modulus (Pm) inherently lower than in the UF processes.

A gel is seldom formed at the membrane because the concentration there rarely exceeds the solute solubility limits for most waste components. Any solids formed are generally a result of biological fouling by secondary treatment plant effluent, or water chemistry conditions resulting in precipitation of inorganics such as calcium carbonate, sulfates, iron oxide, manganese oxide, etc. The polarization modulus is inherently low because of high diffusivity. Low solvent flux results from a high flow resistant, low permeable membrane, and not the surface gel concentrations. As a result, the basic transport equation accurately represents the RO process, excluding the gel resistance.

$$Nw = (dP - d\pi)/Rm \qquad (6.19)$$

The above equation can be further simplified by relating the membrane resistance to its characteristics [8].

$$Rm = l/Pw$$

Pw is the specific permeability of the membrane, and l is the membrane thickness. So that if no concentration polarization is assumed, Equation (6.19) reduces to:

$$Nw = Pw(dP - d\pi)/l \qquad (6.20)$$

The specific permeability of the solvent Pw is defined by the liquid characteristics in the membrane, as indicated by Equation (6.21) [8].

$$Pw = (Cwm \cdot Dwm \cdot Vwm) \cdot R \cdot T \qquad (6.21)$$

Cwm is the membrane solvent mean concentration
Dwm is the membrane solvent mean diffusivity
Vwm is the membrane solvent molecular volume
R is the gas constant
T is the absolute temperature

The solute flux (Ns) is defined by the solid flux Equation (6.7) for diffusion transfer. Based on this equation, the quantity of dissolved solute transported with water is related to the distribution coefficient (H), the solute concentrations Cs and Cf, the solute diffusivity (Dm), and the membrane thickness (t).

Surface Fouling Control

As with the other processes, membrane gel control and process flux stability are important operating considerations, making surface management important. Surface fouling can occur as a result of

(1) Precipitation of iron or manganese oxides, calcium sulfate, or calcium carbonate
(2) Concentrations at the membrane film exceeding solubility limits resulting in solids deposit (in some rare instances)
(3) Solute molecule diameter close to membrane pore sizes resulting in possible plugging

In RO processes the osmotic pressure is significant, theoretically defined by Equation (6.3) [9], and is related to the species retained by the membrane liquid, its concentration and degree of ionization. As indicated by Equation (6.20), the osmotic pressure affects the process efficiency. A high osmotic pressure could result from poor membrane surface control if high concentrations are allowed to develop, and concentration polarization results as the fluid passes the membrane. The increased osmotic pressure decreases the available (net) pressure for solvent transport. However, relative to ultrafiltration, polarization is low and an insignificant process variable.

Although concentration polarization is low, fouling control techniques are still applicable to maintain flux stability. These include

(1) Pretreat to remove colloidals and prevent precipitation. Pretreatment includes prefiltration, ion exchange, adding reducing agents, addition of agents prohibiting precipitation, and pH control.
(2) Select membranes to minimize plugging.
(3) Use a closed recirculation feed loop, applying a cross flow design with a high velocity maintained over the membrane surface.
(4) Apply a conservative design flux.
(5) Reduce on-line treatment time to allow downtime for cleaning and membrane maintenance.
(6) Avoid excessive high pressure conditions which could cause compaction of any gel formed, and resulting flux deterioration.
(7) Clean the membrane.
(8) Avoid irreversible attachment of macromolecules and colloids to the membrane by employing the correction methods discussed.

MODERN INNOVATIVE APPLICATION—NANOFILTRATION

The most critical criteria governing membrane selection are controlling fouling, minimizing regeneration time, and extending membrane life and effectiveness. Complex membranes are often selected because of the waste characteristics, resulting in high maintenance and moderate performance. Researchers are constantly investigating improved membrane processes, and some modifications have been proposed allowing both high membrane flux rate and minimizing fouling [3,4,10–12]. In all of these processes the basic considerations involve waste preparation to adapt to the membrane properties, using a specially constructed membrane, and maintaining adequate scouring velocities using a concentrated solids recycle stream. A nanofilter is commonly employed in these modified systems; having a pore size of about 0.1 micron, falling between a micro and an ultrafilter. In addition to large pore size, nanofilter membranes are constructed with funnel type pores (small on top); the pore channels allow solids entering to be "pushed" through and not plug. The larger pore size and pore configuration make cleaning easier.

Such a system has been proposed as an advanced membrane technology for an industrial wastewater and groundwater treatment system [11]. The recommended filter consists of an inert, fluorocarbon based, tubular configurated membrane. It is suitable for severe services; being abrasive resistent, applicable over a wide pH range, and resistant to severe oxidizing or reducing conditions. All of these properties make the membrane appropriate to resist aggressive chemical and mechanical cleaning.

In addition to using membranes constructed for aggressive cleaning, waste management can be employed to minimize polarization; by applying waste pretreatment and a recirculation loop around the filter [12]. Pretreatment includes chemical precipitation and addition of coprecipitation scavenger metal cations or adsorbents. The precipitated wastes (resulting slurry) are injected into a loop recirculation around a membrane system, with a discharge flow equivalent to the influent. The recirculation loop is maintained at a high circulation rate, and a solids concentration of 3 to 5%, to assure a high scouring velocity.

A patented process known as seeded reverse osmosis has been proposed as an alternative to pretreatment [4]. Seed crystals introduced into the brine causes preferential precipitation of scaling components to occur on crystal surfaces rather than the membrane. Others have investigated a combined carbon adsorption and a membrane filter to treat industrial wastes containing priority pollutants, suspended solids, organics, and metals [14]. Powdered activated carbon is mixed with the wastewater in a pretreatment step, forming a one weight percent mixture, which is injected into a recirculation loop. The recirculating slurry stream enhances the system performance.

PROCESS ENGINEERING DESIGN

As indicated in Table 6.2, membrane filter effectiveness depends on the raw waste or conditioned feed characteristics. These affect both effluent quality and total run time. Consistent treatment performance depends on controlling concentration polarization, a result of membrane selection, pretreatment, and polarization control techniques. Membrane selection is a primary consideration in assuring an operable

TABLE 6.2. Operating Characteristics.

Variable	Operator Controllable	Critical
Waste Characteristics		
Waste generated	No	Yes
Composition	No	Yes
Concentration	No	Yes
Solid properties	No	Yes
Filtering properties	No	Yes
Operating Characteristics		
Flow rate	Minimal	Yes
Feed flux	Yes	Yes
Pretreatment	Yes	Yes
Polarization control	Yes	Yes
Recirculation	Yes	Yes
Pressure	Yes	Yes
Head loss	No	Yes
Cycle time	No	Yes
Backwash program	Yes	Yes
Backwash volume	No	Yes
Net production rate	No	Yes
Membrane life	No	Yes

system. Since membrane selection is a design function, the operator is left with pretreatment or polarization control techniques to control the treatment cycle and net production rate.

The critical filtration operating variable is hydraulic loading (flux), which is directly controlled by adjusting feed rate. *Feed rate* is a limited or nonexistent variable if upstream waste storage is not provided. In addition, cross-flow velocity can be a significant control variable, if recirculation is included in the design. The Process Engineer can increase operating flexibility by providing waste storage and equalization, allowing "batch" storage, testing, and pretreatment. In addition, maintaining upstream storage capacity allows the operator some flexibility in adjusting feed rate to optimize filter performance.

Pressure drop depletion, cycle time, and net production rate are all measures of process effectiveness, none of which are directly controlled by the operator. Cycle time is directly related to the pressure drop depletion rate, which is affected by the waste characteristics, polarization effects, and regeneration effectiveness. If the wastes contain high dissolved or colloidal solids capable of being transported and salting out within the membrane pores, cycle time will be greatly reduced and membrane cleaning time (down-time) greatly increased. Under severe membrane fouling conditions, there is little that the operator can do to improve regeneration without increasing down-time, which in some cases provides minimal process improvement. At some point, high regeneration frequency makes membrane filtration impractical as a treatment system.

Economically, process viability depends on membrane life, which is affected by all the factors discussed. The major contributors to membrane deterioration are polarization and excess pore plugging, and resulting excess membrane cleaning using strong chemicals and highly abrasive techniques.

REPORTED OPERATING DATA

Filtration systems are commonly designed using available water treatment or tertiary municipal plant criteria, assuming that equivalent effluent quality can be achieved by "proper" operation. In reality, treatment performance is significantly dependent on the specific waste characteristics, influenced by the influent and operating variations encountered, and defined by an achievable range rather than a specific value. As a frame of reference, results from approximately 18 industrial *reverse osmosis* facilities reported in the EPA *Treatability Manual* are summarized in Table 6.3 [13].

The data reported are for a variety of industries: paper mills, petrochemicals, iron and steel, hospitals, leather, textile, organic chemicals, pharmaceutical, rubber, and resin producing plants. These data are cited to demonstrate the range of performance possible, with no assurance that the facilities reported were properly designed or operated, that these results are the best that can be achieved, or that these

TABLE 6.4. **Required Design Data.**

Critical pilot plant treatability data specific to the waste
1. Design temperature
2. Feed flux
3. Media type
4. Required waste management
5. Pressure drop requirements
6. Reject volume and concentration
7. Cycle time before washing
8. Backwashing requirements
9. Effluent quality
Waste characteristics that should be obtained from laboratory studies
10. Concentration
11. Type solids
12. Constituent detrimental to the membrane
Selected operating characteristics
13. Flux rate
14. Required pretreatment
15. Recirculation
16. Operating cycle
17. Media life

results can be achieved with other facilities where influent characteristics are significantly different.

REQUIRED PROCESS DESIGN DATA

Membrane technology can be screened with laboratory scale columns to test *filtering* capabilities using laboratory methods discussed in Chapter III-5, Filtration. Such information is general in nature and provides no membrane filtration operating or design information. In all cases, final judgement can only be made with field studies utilizing full scale prototypes. Regardless of the manner of obtaining data, the required design criteria are listed in Table 6.4.

Depending on the length of the study, information obtained from field prototype units can include process suitability, acceptable membranes, cycle time, and required backwashing. Unfortunately, *membrane life* and durability information will not be obtained, unless there is an immediate failure. Therefore, for any new employment, economic feasibility is at best an "educated" guess.

In any prototype testing, upset conditions should be evaluated to determine whether the data developed are effective through the wide range of influent conditions encountered, especially varying waste composition and characteristics. A word of caution regarding pilot evaluation using single waste volumes. Waste characteristics can significantly change during manufacturing cycles, so that any one sample may not be representative of daily wastes. In addition, stored waste characteristics may change, altering test results. The best pilot studies are conducted in the field, utilizing continuous pilot units for extended periods.

TABLE 6.3. **Reported Operating Data (adapted from Reference [13]).**

	BOD$_5$	COD	TOC	TSS	O/G
Reverse Osmosis					
Effluent, mg/L					
Minimum	1	6	5	<4	<4
Maximum	429	736	50	<5	17
Median	2.7	25.5	8		7
Mean	43	73	<11	<4.5	<7
Removal, %					
Minimum	64	0	>5	>85	>20
Maximum	92	>99	96	>90	>72
Median	87	91.5	90		>50
Mean	83	87	84	>88	>40
Ultrafiltration					
Effluent, mg/L					
Minimum	12	444	66	2	5
Maximum	8890	36,600	939	539	195
Median	457	830	246	54	38
Mean	2850	9130	366	105	72
Removal, %					
Minimum	0	12	15	60	23
Maximum	88	99	97	>99	>99
Median	66	72	79	99	96
Mean	53	57	63	>92	>78

WASTE EVALUATION

The *primary* process consideration is that the wastes are compatible with the membrane selected, based on three basic criteria:

(1) Can a membrane be selected that can selectively concentrate or recovery specific waste constituents?
(2) Are the waste components compatible with the membrane operating limitations, and can they be effectively removed? Waste characteristics should be checked for components that could destroy or foul the membrane.
(3) How are the process rejects discarded? Membrane technology is basically a water reclamation process, effective as a final processing step to achieve water quality standards, or water reuse. The influent contaminants are concentrated, not destroyed, and must be reclaimed or the concentrated contaminants suitably disposed.

After membrane technology is proven acceptable to remove selected waste components, the effects of the other waste constituents must be evaluated, and pretreatment applied.

pH Considerations

The pH range affecting media deterioration depends on the membrane material. As an example, cellulose acetate membranes are limited to a maximum pH range of 4.5 to 9, hydrolyzing at lower values, and deacetylation resulting at higher values [2,8]. Cellulose triacetate and blended diacetate-triacetate membranes are somewhat more stable at the high and low ranges, but still a matter of concern when outside the recommended limits. Polyamide membranes are effective at a range of 3 to 12 because they do not hydrolyze in water [9,11]. Other membranes that are reported to be applicable at a wide pH range include polysulfone, polyacrylonitrile, and polyfuran [8].

Precipitation or "salting-out" resulting from an adverse pH could promote fouling, enhance membrane deterioration, or both. These effects are directly related to the waste chemistry, and generally follow the basic chemical principles governing solubility and solubility product, as discussed in Chapter I-6, Chemical Precipitation. Because waste composition changes cannot be predicted, protection of the membranes can be improved by removing active dissolved salts using an upstream ion exchange process.

Temperature

Waste temperature affects performance in several ways. First, membranes have a limited temperature operating range, that at a pH of 7 ranges from 55 to 90°C, 55°C for cellulose acetate, 60°C for polyacrylonitrile, 80°C for polyamide or polysulfide, and 90°C for polyfuran [8]. In addition, temperature can affect waste characteristics such as solubility, osmotic pressure, and viscosity, all of which influence either process efficiency or gel formation. Generally, membrane filters are applied at temperatures considerably lower than the membrane or waste controlling limits.

PRETREATMENT REQUIREMENTS

Membrane systems are extremely sensitive devices and considerable care must be taken to protect the membrane and optimize its life. Pretreatment is a waste management process to condition the waste to minimize membrane fouling by identifiable constituents by removing, dissolving, precipitating, or altering them. Waste parameters must be tailored to the specific membranes, some of which include limiting the following influent components or conditions:

(1) Suspended solids to prevent plugging
(2) Oil, grease, or floatables to prevent fouling
(3) Erosive substances to minimize membrane wear
(4) Temperature to the membrane operating limits
(5) pH to prevent membrane degradation
(6) Organics that promote membrane degradation or destruction

Waste components detrimental to membrane life, effective regeneration, or rejection rate must be conditioned to an inactive state. Certain compounds directly increase membrane fouling because they are at, or near, their solubility limits. They will "salt-out" either at the surface to form a gel, or as in the case of RO in the pores. In such cases, pretreatment to remove these inorganics is extremely important. In fact, activated carbon, conventional filtration, or ion exchange pretreatment may be required to reduce potentially high fouling conditions. Pretreatment can also include removing targeted contaminants by precipitation, adding seeding materials to promote precipitation and large floc sizes, adding adsorbent materials to remove constituents, or changing the contaminant characteristics by chemical oxidation or reduction. Waste preparation is an extremely important process step in managing the system performance.

Effluents from biological systems present a special fouling problem resulting from their biological activity. This becomes a major problem when the membrane process is run intermittently, with long stagnation periods. Under those circumstances, organic constituents in filter pores are susceptible to biological growth. Such systems should be continuously run, utilizing waste storage to assure continuous flow and down-time allowed for occasional chlorination to retard biological activity.

The quantity of waste solutes will directly influence membrane concentration polarization. High solute concentrations should be reduced using more economical treatment methods such as clarification, conventional filtration, ion exchange, or chemical treatment. In addition, wastes sent to membrane processes should be carefully selected and restricted to isolated streams which can be effectively treated. All plant wastes should not be processed because some will only increase filter size or decrease the on-line treatment time.

PROCESS DESIGN VARIABLES

Application of membrane technology requires a careful evaluation of these process considerations:

(1) System configuration
(2) Membrane characteristics
(3) Membrane selection
(4) Design flux rate
(5) Concentration polarization control
(6) System pressure
(7) System economics

System Configuration

Process configuration involves some important process decisions governing process flexibility, cycle time, net production rate, regeneration, and membrane life. Foremost, an evaluation must be made determining whether a batch, continuous, or multistage process best suits the waste generation volume. Significantly, whatever mode is selected, the basic configuration should incorporate a circulating loop into which is injected the feed stream, designed to maintain a 0.6 to 3 meter per second (2 to 10 fps) velocity across the membrane surface. A once through system is not only ineffective but limits the operating flexibility to maintain the membrane surfaces.

A batch system with the configuration illustrated in Figure 6.1 is viable if the waste volume is small. In this system the holding tank contents are recycled through the membrane, and returned; the tank volume being depleted in pro-

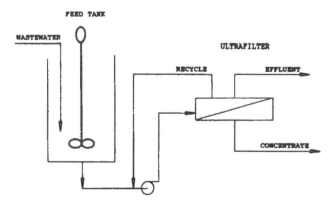

Figure 6.2 Ultrafilter feed and bleed system (adapted from Reference [2]).

portion to the concentrate discharge rate, and the process terminated when the effluent is at the discharge limits. If the daily waste volume is large, a continuous system is employed using a feed-and-bleed or multistage system.

The feed-and-bleed system illustrated in Figure 6.2 combines batch storage with a once-through continuous system; using a recirculation loop maintained at the required effluent quality, continuously discharging effluent (and concentrate) at a rate equal to the influent injection. The allowable flux is low because it is based on the final effluent (lowest) concentration. The membrane area can be optimized by utilizing a multistage system, which is basically a series of feed-and-bleed systems, each operating at an optimum flux rate based on successive effluent concentrations, as illustrated in Figure 6.3. Multistage units are frequently sized to equalize the area requirements, by adjusting the permeate flux through each unit, while still achieving the total required production rate.

Process selection must be based on minimizing capital costs, optimizing the equipment size and number of units, by carefully staging and selecting operable flux rates. Filtration systems are installed as series or parallel modular units to provide the required area at the required flux, allowable surface velocity, and allowable head loss. The units are commonly installed in a compact configuration because a relatively large filtering area is required to accommodate process capacity and expected downtime. However, the system must be installed with adequate space to be operable and serviceable. The amount of installed spare capacity should be carefully explored. Although over design may insure performance, excess capacity is extremely expensive, more so than most other treatment systems! The fact that membrane units are easily expanded should factor into the decision to minimize the initial capital expenditure by designing an expandable system.

Membranes

Besides selecting a specific membrane technology, membrane characteristics must be understood so that suitable and

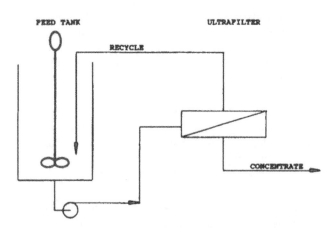

Figure 6.1 Ultrafilter batch configuration (adapted from Reference [2]).

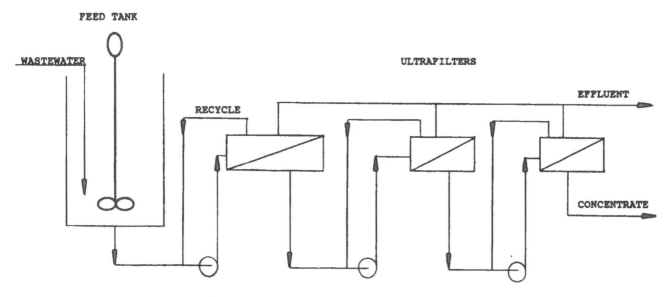

Figure 6.3 Multiple ultrafilters units in series (adapted from Reference [2]).

applicable media can be evaluated. Membranes are defined by their pore characteristics, specific separation limits, materials of construction, and available forms.

Membrane Type

Membranes consist of an integral and continuous structure constructed as a dense thin skin supported by a thick porous substructure, containing a spectrum of pore sizes. They are constructed in two forms, anisotropic or nuclepore structures. Those constructed as "sponge" or "track-like" pores are commonly referred to as anisotropic or Loeb-Sourirajan membranes. The skin retains the solute, passing the filtrate with a minimum of resistance. The structure is somewhat resistant to internal fouling since the solute that can penetrate the thin skin pores encounters increasingly larger pore sub-surface openings, making solute removal easier. Modern developments have allowed production of anisotropic membranes as composite membranes, where the dense thin skin is cast on a separate porous base to better tailor the membrane characteristics.

Track-like membranes, called nuclepores, are constructed with straight through cylindrical channels (pores), with well defined pore density and size. These membranes have porosities generally lower than the anisotropic, and their construction is prone to internal fouling since any solute that passes into the channel is subjected to the same orifice restriction throughout the unit.

Membrane Properties

Porosity and thickness are the two significant membrane properties defining its characteristics. Porosity is related to retention and penetration, thickness to allowable flux rate and physical strength. Where large particle sizes are to be retained and penetration limited, low porosity intensifies surface concentration making flux stabilization possible with high surface velocities and a simple backwash. Conversely, where particle penetration is likely, high porosity membranes increase solute storage capacity, increasing the potential for internal fouling. Anisotropic membranes are considerably more porous than a nuclepore, allowing more solute to enter. However, fouling potential is reduced because of their funnel-like construction.

Flux rate is inversely proportional to the membrane resistance, which is directly proportional to its thickness. Therefore, larger flux rates are possible with thinner and less resistant nuclepore membranes, which contain definitive routes for the filtrate to pass. The thin nuclepore membrane allows high filtrate flux due to lower resistance, even though its porosity is considerable less than the anisotropic. In addition to resistance, thickness must be related to the membrane's mechanical strength. If the unit must be utilized at very high pressures, or the waste has abrasive characteristics, physical endurance could be a primary consideration.

Membrane Materials

Membranes can be constructed of cellulose acetate, aromatic polyamide, polysulfone, or special materials specific to the filter type, as indicated in Table 6.5 [2]. Cellulose acetate membranes have the advantage of high performance and ease of manufacture and, therefore, are relatively inexpensive. However, in many services their disadvantages restrict their use. These include [2]

(1) Their relatively low operating temperature range (maximum 30°C) limits the allowable flux by restricting the

TABLE 6.5. Membrane Materials
(adapted from Reference [2]).

Material	Applied to		
	MF	UF	RO
Cellulose acetate	X	X	X
Cellulose triacetate	X	X	X
CA/triacetate blend			X
Cellulose esters (mixed)	X		
Cellulose nitrate	X		
Cellulose (regenerated)	X	X	
Gelatin	X		
Polyacrylonitrile		X	
Polyvinylchloride	X		
Polyvinylchloride copolymer	X	X	
Polyamide (aromatic)	X	X	X
Polysulfone	X	X	
Polybenzimidazole			X
Polybenzimidiazolone			X
Polycarbonate (track etch)	X		
Polyester (track etch)	X		
Polyimide		X	X
Polypropylene	X		
Polyelectrolyte complex		X	
Polytetrafluoroethylene	X		
Polyvinylidenefluoride	X	X	
Polyacrylic acid zirconium oxide skin		X	X
Polyethleneimine + toluene diisocyanate skin			X

high diffusivity and lower viscosity fluid properties achievable at elevated temperatures.

(2) They can be applied over the relatively narrow pH range (3 to 6) in which the material is stable.

(3) They are prone to chemical attack; some field studies indicating that some metal ions such as cobalt, ruthenium, iron, nickel, rhenium, palladium, and silver (in that order) increase membrane deacetylation rates [7].

(4) They exhibit a poor resistance to chlorine, which is applied to some secondary effluents from combined treatment systems.

(5) The potential for biological activity in the applied temperature range, and the fact that cellulose acetate is highly biodegradable, further restrict its use for tertiary treatment of biological system effluents.

(6) Its inherent property to compact in proportion to the applied pressure results in flux rate loss.

Aromatic polyamide membranes are a slight improvement over cellulose acetate but display some of the same limiting characteristics, especially to chemical stability in the presence of chlorine and other compounds mentioned. Polysulfone has characteristics allowing its application over a wider processing range, at temperatures up to 75°C, a pH range from 1 to 13, and improved chlorine resistance [2]. Fluorocarbon based materials tested under various conditions have

displayed results indicating the membrane is inert at any pH, and resistant to severe oxidizing and reducing conditions [11].

Membrane Form

Membranes are used in industrial applications in four common module designs—spiral wound, hollow fine fiber, tubular, and plate-and-frame—with spiral wound and hollow fine fiber units most commonly employed for water and wastewater treatment.

Spiral wound designs use sheets of flat membrane wrapped around a porous tube. Each sheet is separated with an impervious lining acting as a "spiral collecting" channel, directing flow to the center porous tube, a sandwich type construction. Influent passes parallel and into the sheets, rejecting the solute as it passes through. Concentrate reject flows through the sheets, while the filtrate passes through the membrane, perpendicular to the sheets to the "collecting channels," to the center tube, and out the system. The continuous wrapping around the tube allows a high membrane area in a single unit. The major advantage of this system is low capital cost, with high plugging potential (when poorly applied) and resulting high replacement costs being their major disadvantages [2].

Tubular membranes, as the name implies, are tubes with influent passing through the center carrying concentrated reject to the other end, whereas filtrate passes perpendicular to the tube to an outer shell. The design is similar to a shell-and-tube heat exchanger. They can be classified as narrow (2 to 10 mm) or wide (10 to 25 mm) bore [1]. Hollow fiber membranes are similar to tubes except that they are much smaller in diameter (1 to 2 mm), the influent entering into the shell side, with the filtrate passing into the tubes. The concentrated reject passes through the shell side. Plate and frame are similar to conventional filter presses, utilizing sheets of membrane as the separating media [1].

Membrane Evaluation

Membrane evaluation involves matching waste characteristics and required effluent quality with specific membrane properties. However, capital and operating costs are predominant factors impacting the final choice, with the major criteria being:

Membrane life
Membrane performance
Operating efficiency
Flux maintenance

For membrane separation to be viable, it must be economically feasible, which is directly related to *membrane life*. Less than 2 or 3 years membrane life could eliminate filtration as a viable treatment alternative because of the operating costs associated with repair, replacement, and start-up and lost operating time associated with frequent replacement.

Membrane performance must be carefully established prior to accepting membrane treatment as a viable process. This includes membrane physical and chemical stability, achievable effluent quality, low reject volume, and a high on-line operating efficiency. Performance must be based on a continuously operated system, over a significant period, and at varying operating conditions. Short-term batch tests are inexpensive, but do not correlate well to full scale plant performance. Any savings in piloting may not realistically reflect the installed equipment costs, and are insignificant if the full scale system is ineffective. Once the system is selected, process upgrading is limited and costly. Post-construction improvements are costly, both in studying the problem and in additional capital expenditures. Poor performance is almost always associated with failure to accurately appraise the considerations cited in *waste evaluation*.

Operating efficiency is a result of membrane life and stability, the essence of membrane system economics and reliability. If the process does not have a minimum of 80% on-line operating efficiency, including planned maintenance, the system will probably not be viable, not reliable, and an economic liability. The greatest factors reducing operating performance are (1) consistent membrane fouling involving frequent and long regeneration times, (2) membrane failure and replacement, (3) failure of high pressure pumps, and (4) failure of pumps and membrane equipment resulting from abrasive waste components.

Flux maintenance affects all the factors discussed above. It is difficult to specify what is a reasonable maintenance program, since that will be defined by the specific waste characteristics—before or after the installation and operation. However, the system design must start with the obvious, but often neglected, fact that the simplest system is the most practical from an operating viewpoint. Maintaining the membrane process should not require more elaborate maintenance skills than is required for the rest of the manufacturing facilities. Using a high recirculation flow rate to flush the membrane surface clean is simple, but involves operating (energy) dollars which should be factored into the economic evaluation. The backflush cycle to regenerate the membrane further increases the operating costs by reducing the unit's on-line operating time. Chemical treatment further raises the operating cost of the unit and could, in combination with the other operating vulnerabilities, make the operation uneconomical, inefficient, or both.

System economics are directly related to *flux maintenance*, defined by *flux stability*, which is specific to the selected membrane. Flux stability is directly related to polarization control, as detailed in the individual filtration discussions, the pretreatment section, and the polarization control section.

Membrane Selection

Once membrane technology is established as a viable process, a proper selection is crucial. Foremost, the least complicated membrane should be selected; a microfilter, rather than an ultrafiltration, or an ultrafiltration membrane rather than a reverse osmosis system. All other variables being equal, selection is based on matching specific membrane properties to the waste characteristics. This includes solute characteristics such as solute size, molecular weight, ionic charge, and solubility. Any waste constituents that affect membrane materials or performance should be considered in the selection. Any selection should be with the understanding that unlike any other waste treatment unit operation, membrane systems are difficult, expensive, and often impossible to correct after installation. Based on process simplicity and solute characteristics, an initial evaluation can be based on the following criteria:

(1) Microfiltration is a basic refinement to conventional filtration, capable of removing suspended solids and solutes of molecular weights greater than 300,000, sizes ranging from 0.02 to 10 μm, at process pressures ranging from 69 to 690 kPa (10 to 100 psig).

(2) Ultrafiltration is a refinement to microfiltration; capable of removing suspended and colloidal solutes of molecular weights ranging from 300 to 300,000, sizes ranging from 0.0010 to 0.02 μm, at operating pressures from 69 to 690 kPa (10 to 100 psi).

(3) Reverse osmosis is the most complex of the separation processes, capable of removing dissolved and suspended matter with molecular rates below 300, sizes ranging from 0.0001 to 0.001 um, at operating pressures from 1400 to 10,300 kPa (200 to 1500 psig).

(4) Nanofiltration is a modern development, based on specific membrane characteristics, generally classified between macro- and ultrafiltration. When combined with pretreatment and recirculation, studies cited in "Modern Innovative Applications" indicate superior membrane performance for certain wastes.

Next the membrane form must be selected. Tubular, hollow fiber, and spiral wound modulars could be considered for industrial application, based on the following defining characteristics [2]:

(1) Tubular modulars are capable of handling waste effluents with relatively high particle sizes, from 1250 to 2500 μm depending on the tubular size. They are easy to clean in place, and to replace, and relatively inexpensive. Pressure drops of 207 to 276 kPa (30 to 40 psig) are required, making them the highest energy consumers of the three modular forms. In addition, they have the lowest surface-to-volume ratio of the three configurations, requiring large floor space, thereby limiting the practical installed area and achievable concentration.

(2) Hollow fiber modulars have the highest surface-to-volume ratio, making them economically attractive for industrial application. Their economical advantage is enhanced because of the relatively low operating pressure drop of 35 to 138 kPa (5 to 20 psig), making them attractive in terms of energy consumption. Their "back flushing" capabilities make them relatively easy to clean. However, the small tube diameters make the fibers prone to inlet plugging, requiring that the influent be prefiltered to 20 to 100 μm. Their applications are limited to maximum operating pressures of 172 kPa (25 psig) because of the fiber rating, in some cases 240 kPa (35 psig) at low operating temperatures. Replacement is costly because of high unit costs, and because all fibers in a bundle must be replaced when one is inactive.

(3) Spiral wound modules are inexpensive because of low modular cost and economical to operate because of low power consumption resulting from low operating pressure drops of 103 to 138 kPa (15 to 20 psig). However, their area-to-volume ratio is intermittent between tubular and hollow fiber modulars; requiring careful selection of the spacing to optimize installed area, while balancing resulting operating pressure drop or potential for channel plugging. Channel plugging can be minimized by filtering the influent to the 5–25 μm, or 25–50 μm range, depending on the spacer construction.

Table 6.6 summarizes the characteristics of common configurations [2,8].

Some general guidelines which may be applicable to waste treatment, resulting from water treatment application, can be summarized as follows [5,15]:

(1) Spiral wound and hollow fiber are the most common water treatment units.

(2) Spiral wound modules have a wide range of applicability because of the inlet ability to better control concentration polarization.

(3) Hollow fiber modules are prone to fouling, and therefore almost always used for wastes with little or no suspended or colloidal matter.

Design Flux Rate

The primary factor affecting membrane performance is *flux*, which when related to total influent flow establishes the filter size. The selected design flux sets the required membrane area, affects concentration polarization control, and membrane life; all of which affect system costs. The factors influencing flux rate and flux stability include applied pressure, fouling or concentration polarization, and waste characteristics. All of these parameters have been covered and reference is made to the appropriate sections.

Dead-end systems and low influent flows result in low membrane velocities, providing poor concentration control conditions. In such cases, a recirculation system may be required to maintain a 0.6 to 3 meter per second (2 to 10 fps) flow velocity. A recirculation loop includes injecting the feed into the loop, and a corresponding stream discharged from the system. In such cases, the effluent concentration is that of the recirculation loop, and it will define the allowable flux rate. The system can be optimized by staging the process, sequencing the effluent concentrations and flux rates.

Flux rate influences the capital costs of the system by (1) establishing filter area, (2) affecting polarization control, and (3) affecting membrane life. There are no models that can adequately estimate a design flux since it is specific to the waste and the system configuration. Applicable flux rates are membrane dependent, and discussed in the respective sections.

Membrane Area

The required area can be estimated once a design flux is established and the design waste flow defined. The required area can be estimated using Equation (6.23).

$$\text{Area} = \text{waste flow/flux rate} \qquad (6.23)$$

Where flow can vary widely from very low, normal, and peak flows, the selected design flow can significantly affect

TABLE 6.6. Characteristics of Common Configurations (adapted from References [2,8]).

	Hollow Fiber	Spiral Wound	Tubular	Plate Frame
Flow path complexity	Fair	Fair	Good	Poor
Resistance mechanical damage	Good	Poor	Good	Good
Plugging potential	High	High	Low	High
Mechanical cleaning	Poor	Poor	Good	Fair
Area-volume-ratio	High	(1)	Low	(1)
Power consumption	Good	Good	High	Good
Membrane replacement costs	High	Low	Low	Low

Notes
(1) Intermittent between hollow fiber and tubular.
(2) Can be high, dependent on size and material.

the required area, and more importantly the capital costs. For these reasons, upstream equalization should be evaluated to average waste characteristics and flow. Seldom is membrane area, for peak or varying flows, more economical than flow equalization. Estimating the required area is easy, the more difficult design consideration is establishing whether the required area is economically acceptable!

Control Concentration Polarization

Fouling or concentration polarization potential, with and without control techniques, can only be determined by field testing. Polarization control techniques were discussed for each of the specific membranes, and generally can be categorized as follows:

(1) Pretreatment filtration to remove suspended solids
(2) Pretreatment to remove components that could "salt-out"
(3) Selecting an anisotropic membrane structure to minimize internal plugging
(4) Water backwash to flush collected surface particles
(5) Chemical backwash to remove surface fouling
(6) Use of a recirculation feed loop to provide a high velocity cross-flow sweep across the membrane
(7) Periodic ultrasonic cleaning for membrane restoration
(8) Periodic mechanical cleaning for membrane restoration
(9) Conservative flux rate to reduce polarization effects
(10) Design for a reduced on-line operating efficiency to allow for membrane maintenance

A planned membrane maintenance program must be developed, tailored to the specific system. This planned maintenance program must be developed early in the design phase and incorporated into the equipment selection and design.

Pressure

Pressure is an important design consideration, selected to overcome membrane resistance, osmotic pressure (in reverse osmosis), and cyclical pressures resulting from membrane gel or fouling resistance. The pressure should be adequate, controllable, and within reasonable design limits. Lower pressures result in lower gel and membrane compaction forces, improving flux restoration, minimizing flux deterioration, and increasing membrane life. Applicable pressures are membrane dependent, and discussed in the respective sections.

Economics

As important as technical considerations are to successful membrane application, the overwhelming consideration is the system cost. Specifically, the installation costs, required redundancy for expected downtime, operating costs, and

maintenance costs; factoring any recoverable costs for the sale of pollutants removed (usually small in waste treatment) and reusable treated effluent.

Significant operating costs include membrane replacement and energy expended. Membrane replacement costs are dominant if media life is less than two or three years. In addition, factors that contribute to membrane wear can result in erosion of other major equipment components. Energy is the next major operating cost, increasing with increasing system complexity, required pressure, and related power requirements. High pressure systems require significant energy costs, high capital costs for high pressure pumps and equipment, and high maintenance costs associated with equipment wear. Finally, consideration must be given to disposal of the concentrate from these systems, and the cost associated with treating, stabilizing, shipping, regulatory control, and site disposal.

FATE OF CONTAMINANTS

Large disposable waste quantities are usually not a concern since membrane filters are used for product recovery, or for tertiary treatment to meet water quality standards. In fact, membrane sensitivity and associated pretreatment result in minimum generation of secondary pollutants, since only select contaminants can be fed to the system.

The major discharge from a membrane filter is treated effluent for reuse, or final discharge either to a municipal sewer or directly to a receiving water way. The concentrated reject will consist of either a recovered concentrated chemical for sale or reuse, or a slurry to be disposed in a separate treatment system or a "select" point in an existing treatment. Select is a significant criteria, since the reject must be treated and disposed, or nothing has been accomplished except transferring the waste constituents from one stream to another.

The backwash volume generated as a result of membrane maintenance could be a major problem, especially if chemical treatment is employed. This by-product stream may require separate treatment, discharge to a select point in an existing on-site treatment system, or in extreme cases off-site disposal.

Vapors generated should be minimal since most of the vaporization should occur in upstream manufacturing or waste treatment processes. In addition, the closed nature of ultrafilters and their temperature minimize emissions. The membrane temperature restrictions limit the applicable wastewater temperature, which along with the low concentrations, negates vaporization and evaporation.

GENERAL ENGINEERING CRITERIA

The process components specific to a membrane filter system include

Equalization (optional)   Chapter I-4
Waste feed pH or pretreatment   Chapters I-5, I-6, I-8
High pressure feed system
High pressure recycle system
Membrane system

Each of these elements, except the *filtration system,* are discussed elsewhere.

A membrane system is a highly specialized machine, in which a system purchase includes detailed engineering, fabrication, and proprietary design. The owner purchases the supplier's design and operating experience. Unlike other treatment systems, the Process Engineer is seldom in a position to alter the system components, only able to question the process and become confident of the suppliers experience. Regrettably, no other treatment system performance effectiveness is as difficult to remediate after installation. The final evaluation is seldom a question of process performance—although this can sometimes be a problem—but of prohibitive operating costs generally specific to on-line treatment time, regeneration complexity, and membrane life. Some general design considerations are provided as a checklist for the Process Engineer to review specific proposed designs.

GENERAL DESIGN CONSIDERATIONS

Membrane filters are constructed as modulars, configured in series or parallel modes to satisfy capacity (parallel) or effluent quality (series) criteria. The relatively large membrane area requirements normally involve many units being installed in a compact manner to minimize space. However, the installation must not restrict membrane replacement, routine maintenance, process monitoring, and general operations.

As mentioned above, these units are designed and installed as proprietary systems, with the Process Engineer many times relying on the supplier's experience, conditioned upon compliance with a performance guarantee. Many times performance is based on single sample analysis, or no sample but an agreed waste balance, which may or may not adequately define actual full-scale operations.

The following elements of a process design should be reviewed:

(1) The membrane life is the principal economic consideration governing the successful application of membrane technology.

(2) Performance guarantees must be based on extended testing involving process efficiency and mechanical stability, within the range of expected waste variations.

(3) Regardless of the specific filtration system selected, the effect of suspended solids must be carefully considered.

(4) Regardless of the waste flow, provision for recycle should be included in the design as an operating control

of membrane velocity, influent concentration, and equalizing influent variations.

(5) The high membrane surface velocities required, and the high recirculation rates commonly employed, must be carefully evaluated to establish energy requirements and to minimize wear due to erosion.

(6) A complete waste characterization is essential, identifying constituents that salt-out, promote septic conditions common with biological effluents, and produce a high potential for membrane fouling or polarization concentration.

(7) The effects of influent pH or temperature variations must be carefully considered in the design.

(8) In developing the process configuration, pretreatment must be carefully evaluated to extend membrane life, including

   a. Equalization
   b. pH control
   c. Chemical treatment to improve the solute size characteristics
   d. Inexpensive prefiltration to protect the membrane equipment
   e. Ion exchange to remove dissolved components that will foul the system

(9) Staging should be evaluated to improve applied flux and optimize total filter area.

(10) Provision should be made for membrane cleaning, support facilities, and disposal of by-product wastes.

(11) Uniform load distribution to multi-unit systems is important.

(12) Standard membrane lengths should be specified to allow competitive replacement or installation of upgraded technology.

(13) Piping should be designed for easy access and with ample clean out connections to allow for in place maintenance.

(14) The system capacity should be designed for ample downtime for normal process maintenance, equipment maintenance, and emergency conditions. If the units are conservatively designed, any potential capacity problems should not be compounded by not allowing space and provision for expansion. The design capacity, and expected cycle times, should be carefully evaluated in the process design.

(15) Provision should be made to measure (and control) pressure and flows in and out of each modular.

COMMON FILTER DESIGN DEFICIENCIES

(1) Membrane technology not suitable for the wastes to be treated.

(2) Membrane technology suitable but the wrong filter selected.

(3) Design operating flux too high.

(4) Excessive fouling or polarization concentration affecting applied flux.

(5) Recirculation not provided to maintain membrane surfaces.

(6) System designed for inadequate pressure, or pumps selected not capable of maintaining the design pressure.

(7) Poor effluent quality.

(8) Poor volume reduction, excessive concentrate produced.

(9) Concentrate too expensive to dispose.

(10) No market for recovered product.

(11) Excessive backwash is generated.

(12) Excessive downtime for cleaning reduces the operating efficiency.

(13) Aggressive cleaning required to regenerate membranes.

(14) Membrane life considerably less than one year; a result of excessive plugging, aggressive cleaning, and abrasive waste components.

(15) Membrane deterioration dominant as a result of waste constituents, poor pH control, or biological activity.

(16) Waste constituents "salt-out" in the membrane pores.

(17) No provision made to monitor or control flows or pressures.

(18) Excessive wear of pump parts and transfer lines due to high pressures and excessive abrasion. Wrong equipment selected and poor design of piping system.

(19) Standard membrane lengths not selected, competitive replacements not possible.

(20) Poor equipment arrangement; difficult to access, operate, and maintain the equipment.

(21) Excessive noise from high pressure equipment.

CASE STUDY NUMBER 32

Perform a preliminary evaluation to establish the membrane area requirements to treat 114,000 liters per day of waste containing 300 mg/L of submicron contaminants, producing 91,200 liters per hour of usable effluent containing less than 10 mg/L contaminants. Assume a 0.99 separation factor, an applied flux rate of 35 L/sq m/hr at the feed condition, and a flux of 15 L/sq m/hr at the final raffinate condition.

(1) SINGLE BATCH CONDITIONS

(2) Establish flow balance

$$Feed = 114,000 \; L/hr$$
$$Permeate \; required = \underline{91,200 \; L/hr}$$
$$Raffinate = 22,800 \; L/hr$$

(3) Determine volume concentration ratio

$$VCR = feed/raffinate = 114,000/22,800 = 5$$

(4) Determine raffinate concentration

$$Co \cdot VCR^a = 300 \cdot 5^{0.99} = 1476 \; mg/L$$

(5) Determine material balances

Feed: 114,000 L/hr · 300 mg/L/1000 = 34,200 g contaminants
Raffinate: 22,800 L/hr · 1476 mg/L/1000 = 33,653 g contaminants
Permeate: 91,200 L/hr
(Difference) = 547 g/hr
Permeate concentration: [547/91,200] · 1000 = 6.0 mg/L

(6) Establish applicable initial, final, and design fluxes

$$Design \; flux = Fe + 0.33 \cdot (Ff - Fe) \; (\text{Reference } [2])$$

Final flux (Fe) = 15 L/sq m/hr
Initial flux (Ff) = 35 L/sq m/hr
Design flux = $15 + 0.33 \cdot (35 - 15) = 21.6$ L/sq m/hr

(7) Estimate required area

$$91,200 \; L/hr/ \; 21.6 = 4222 \text{ square meters}$$

(8) EVALUATE FEED AND BLEED SYSTEM. All balances and concentrations are identical to items 2 to 5.

(9) Establish recycle conditions. Recycle will be set to maintain a 2 to 10 fps flow velocity.

(10) Establish applicable design flux

$$Design \; flux = 15 \; L/sq \; m/hr \; @ \; effluent \; concentration$$

(11) Estimate required area

$$91,200 \; L/hr/15 = 6080 \text{ square meters}$$

(12) EVALUATE MULTI-STAGE CONDITIONS

(13) Estimate characteristics for equal membrane area

Stage	I	II	III	IV	
Feed, L/hr	114,000	79,000	54,000	38,650	
Raffinate, L/hr	79,000	54,000	38,650	22,800	
VCR	1.443	1.463	1.397	1.695	Totals
Effluent, L/hr	35,000	25,000	15,350	15,580	91,200
Flux, L/sq m/hr	35	25	15	15	
Area, sq m	1000	1000	1023	1057	4080

Calculation based on

a. Effluent selected to balance area
b. Raffinate = feed − effluent
c. VCR = feed/raffinate
d. Flux from available data
e. Area = effluent L/hr/flux

(14) Establish first stage concentrations

(15) Determine raffinate concentration

$$Co \cdot VCR^\sigma = 300 \cdot 1.443^{0.99} = 431.3 \text{ mg/L}$$

(16) Determine material balances

Feed: 114,000 L/hr · 300 mg/L/1000 = 34,200 g contaminants
Raffinate: 79,000 L/hr · 431.3 mg/L/1000 = 34,073 g contaminants
Permeate: 35,000 L/hr
(Difference) = 127 g/hr
Permeate concentration: [127/35,000] · 1000 = 3.6 mg/L

(17) Establish second stage concentrations

(18) Determine raffinate concentration

$$Co \cdot VCR^\sigma = 431.3 \cdot 1.463^{0.99} = 628.6 \text{ mg/L}$$

(19) Determine material balances

Feed: 79,000 L/hr · 431.3 mg/L/1000 = 34,073 g contaminants
Raffinate: 54,000 L/hr · 628.6 mg/L/1000 = 33,944 g contaminants
Permeate: 25,000 L/hr
(Difference) = 129 g/hr
Permeate concentration: [129/25,000] · 1000 = 5.2 mg/L

(20) Establish third stage concentrations

(21) Determine raffinate concentration

$$Co \cdot VCR^\sigma = 628.6 \cdot 1.397^{0.99} = 875.2 \text{ mg/L}$$

(22) Determine material balances

Feed: 54,000 L/hr · 628.6 mg/L/1000 = 33,944 g contaminants
Raffinate: 38,650 L/hr · 875.2 mg/L/1000 = 33,827 g contaminants
Permeate: 15,850 L/hr
(Difference) = 117 g/hr
Permeate concentration: [117/15,850] · 1000 = 7.6 mg/L

(23) Establish fourth stage concentrations

(24) Determine raffinate concentration

$$Co \cdot VCR^\sigma = 875.2 \cdot 1.695^{.99} = 1,475.7 \text{ mg/L}$$

(25) Determine material balances

Feed: 38,650 L/hr · 875.2 mg/L/1000 = 33,827 g contaminant
Raffinate: 22,800 L/hr · 1475.7 mg/L/1000 = 33,646 g contaminant
Permeate: 15,850 L/hr
(Difference) = 181 g/hr
Permeate concentration: [181/15,850] · 1000 = 11.4 mg/L

(26) Establish composite concentration

$$(127 + 129 + 117 + 181) \cdot 1000/91,200 = 6.1 \text{ mg/L}$$

(27) Establish recycle conditions. Recycle will be set to maintain a 2 to 10 fps flow velocity.

(28) Evaluate the effect of the separation factor. The effect of the separation factor on effluent quality from a four stage continuous unit is as follows:

Separation factor	0.90	0.92	0.94	0.96	0.98	0.99	1.0
Effluent, mg/L	56	45	35	23	12	6	0

DISCUSSION

The effluent quality is dependent on the separation factor, which is a characteristic of the membrane system selected. Likewise, the required recirculation rate depends on the membrane size, and is adjusted to maintain a 2 to 10 fps velocity sweep across the filter surface.

REFERENCES

1. Bemberis, I. and Neely, K.: "Ultrafiltration as a Competitive Unit Process," *Chemical Engineering Progress*, Pg 29, November, 1986.
2. Cheryan, M.: *Ultrafiltration Handbook*, Technomic Publishing Co., Inc, Lancaster, PA, 1968.
3. Eriksson, P.: "Nanofiltration Extends Range of Membrane Filtration," *Environmental Progress*, V 7, No 1, Pg 58, February, 1988.
4. Hess, M.C., et al.: "Wastewater Concentration by Seeded Reverse Osmosis-A Field Demonstration in the Electric Power Industry," *Environmental Progress*, V 7, No 1, Pg 7, February, 1988.
5. James M. Mongomery Consulting Engineers: *Water Treatment Principles and Design*, John Wiley & Sons, 1985 (*Weber Physiochemical Processes for Water Quality Control*, Wiley, 1972).
6. Kirk-Othmer: *Encyclopedia of Chemical Technology*, 4th Edition, John Wiley & Sons, 1991.
7. Murphy, A.P.: "Accelerated Deacetylation of Cellulose Acetate by Metal Salts with Aqueous Chlorine," *Research Journal WPCF*, V 63, No 2, Pg 177, March/April, 1991.
8. Perry, R.H. and Green, D.: *Perry's Chemical Engineers' Handbook*, Sixth Edition, McGraw-Hill, 1984.
9. Schweitzer, P.A.: *Handbook of Separation Techniques for Chemical Engineer*, McGraw-Hill, 1988.
10. Steinberg, S.: "Membrane Utilization in Hazardous Metal Removal from Wastewater in the Electronic Industry," *Environmental Progress*, V 6, No 3, Pg 139, August, 1987.
11. Teipel, E.W.: "Applications of Advanced Membrane Filtration to Industrial Wastewater Treatment and Groundwater Clean-

Up," *Water Pollution Control Association of Pennsylvania Magazine,* Pg 41, September/October, 1990.

12. Tran, T.V.: "Advanced Membrane Filtration Process Treats Industrial Wastewater Efficiently," *Chemical Engineering Progress,* Pg 29, March, 1985.

13. U.S. Environmental Protection Agency: *Treatability Manual,* Four Volumes, EPA-600/8-80-042a, 1980.

14. Van Gils, G. and Pirbazari, M.: "Development of a Combined Ultrafiltration and Carbon Adsorption System for Industrial Wastewater Reuse and Priority Pollutant Removal," *Environmental Progress,* V 5, No 3, pg 167, August, 1986.

15. Sanks, R.L.: *Water Treatment Plant Design,* Ann Arbor Science, 1978.

Dewatering

Dewatering is employed to reduce sludge volume by removing slurry water.

BASIC CONCEPTS

DEWATERING is employed to reduce total sludge volume to decrease transportation and disposal costs. Sludge dewatering performance depends on waste composition, solids characteristics, and water content. Basically, a cake's void characteristics define its dewatering properties by establishing its available storage capacity, the forces retaining the water, and flow resistance. Easily compressed solids with large voids have the best dewatering characteristics. Solids properties can be altered to improve dewatering by *chemical treatment* and *sludge conditioning*. Conditioning is used to increase free flowing water removal, reduce the force required to remove bound water, and prevent formation of a fragile sludge. Besides the solids properties, the factors influencing dewatering include applied *energy* such as force or heat, and extended processing *time*. Most dewatering devices minimize retention time by applying mechanical force.

Free water can be drained from the sludge by gravity forces through clarification and thickening devices. *Bound water* requires an applied force to compress the solids and "squeeze" water from the solid voids. Devices applying increased gravitational force include a *belt filter press, centrifugation, vacuum filtration,* and *pressure filtration*. Gravitational forces can be maximized by increasing the treatment time, such as in drying beds. Equipment such as flash dryers, rotary dryers, fluidized bed dryers, multiple-hearth furnaces, or microwave devices can be employed for heat treatment, functioning as evaporators to remove sludge water.

The performance of a dewatering device is defined by its centrate solids concentration and cake dryness, as illustrated in Figure 7.1. Solids recovery (*R*) is defined as the percent of feed solids captured in the cake, as indicated by Equation (7.1).

$$R = \frac{(\text{lb solids fed} - \text{lb solids in centrate})}{\text{lb solids fed}} \times 100 \quad (7.1)$$

The centrate concentration is a measure of the machine clarifying capabilities, that are directly related to solids recovery. Generally, the higher the recovery the greater the fines captured and combined with excessive water, and the higher the cake wetness.

An integral part of a dewatering system is the supplementary chemical addition system(s) included to agglomerate the solids and form a more effective floc, or to retard the activity of biological solids. Selecting a dewatering device and developing an integrated system depend on the principles discussed in the Process Engineering Design section, and specific criteria detailed in each of the dewatering device sections.

PROCESS ENGINEERING DESIGN

Dewatering system design requires evaluating critical system components to achieve acceptable effluent quality and sludge concentration over a range of anticipated conditions. The effluent quality achieved must be suitable for direct discharge or recycle into an existing treatment system. Sludge concentration defines the final cake water content, the acceptable dryness level being driven by disposal or final treatment costs.

REQUIRED PROCESS DESIGN DATA

Sludge dewatering characteristics can be screened in a series of laboratory settling, buchner funnel, and leaf tests. In all cases sludge conditioning should be evaluated to establish compacting and dewatering characteristics with (1) no conditioning, (2) a variety of conditioning chemicals, and (3) varying chemical dosages and treatment conditions. Settling criteria are described in Chapter II-8, Sedimentation. Results from sedimentation evaluation indicate the settling

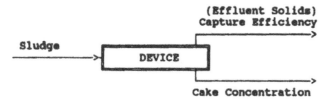

Figure 7.1 Dewatering device performance.

and thickening capabilities at gravity force. Theoretically, devices such as a vacuum filter, belt filter, pressure filter, and centrifuge should be successful at increased dewatering force if gravity settling is feasible. The buchner funnel tests should provide a measure of the water removal rate under an applied vacuum; indicating sludge filterability as measured by the specific resistance, and improvements that can be achieved by sludge conditioning. The leaf tests provide another measure of applicable filtration rates under vacuum conditions at varying applied chemical conditioning levels. These three tests give an indication of dewatering performance and cake compressibility with increased gravitational force, forces that could be applied by dewatering devices. Specific testing methods are detailed in basic waste treatment literature [10]. Poor gravity settling is a good indication that increased pressure may not be either beneficial or economical. In turn, buchner funnel and leaf tests indicate the benefits of vacuum for sludge compacting. If these tests prove unsuccessful there is an excellent chance that belt filters, centrifuges, or vacuum filters will not be practical.

Scale-up from bench scale is common, although prototype field testing is highly recommended. In fact, sludge tests using dewatering equipment prototypes are the best method of obtaining operating performance and design data, and of evaluating performance over a wide range of influent conditions. Equipment manufacturers should be contacted for available testing equipment. Regardless of how they are obtained, the data listed in Table 7.1 are required for process design.

PROCESS DESIGN VARIABLES

Specific process criteria are discussed in each of the dewatering device sections, general criteria applicable to all include the following:

(1) Waste evaluation
(2) Sludge conditioning
(3) Pretreatment
(4) Fate of contaminants

Waste Evaluation

Municipal sludge characteristics have been investigated by researchers in an attempt to identify specific sludge properties affecting dewatering [3,5,22]. Parameters investigated

TABLE 7.1. Required Design Data.

Critical laboratory data specific to the waste
1. Laboratory testing to indicated viability of mechanical dewatering
2. Design temperature
3. Applicable conditioning chemicals
4. Applicable chemical doses
5. Effective pH

Process criteria that should be obtained from full-scale laboratory studies, but can be estimated.
6. Hydraulic loading
7. Solids loading
8. Applied pressure or vacuum
9. Cake dryness
10. Solids recovery

Selected operating characteristics
11. Backwash requirements
12. Chemical injection points
13. Conditioning time
14. Conditioning shearing velocity (G)
15. Flocculating time
16. Flocculating shearing velocity (G)

included solids filament length, mean floc diameter, sludge protein, carbohydrate content, and bound water. Biological reactor process variables such as mean cell residence time, BOD:N ratio, DO concentration, and reactor configuration have also been investigated. A correlation between these variables and dewatering effectiveness has been difficult to establish; but it is generally concluded that decreased overall particle size, evidenced by the presence of large quantities of colloidal particles (less than 30 to 40 μm), is detrimental to dewatering efficiency. In addition, biological process conditions contributing to poor sludge settling resulted in poor dewatering characteristics. The reader is referred to Chapter II-2, Aerobic Biological Oxidation, for a complete discussion on this subject.

The EPA *Dewatering Design Manual* states that dewatering and conditioning processes are affected by the following sludge properties [14]:

(1) Particle surface charge and hydration influence solids repulsion and capillary action, increasing the resistance to dewatering mechanical forces.
(2) Decreased particle size is detrimental to the dewatering.
(3) Increased compressibility, allowing repulsion of trapped water, improves dewatering.
(4) Because viscosity decreases with increasing temperature, and gravitational forces increase with decreasing viscosity, dewatering will improve with increasing temperature.
(5) Sludge pH affects solid particle charge and the effectiveness of conditioning chemicals.
(6) The ratio of volatile to fixed solids could affect dewatering, with dewatering improving as the fixed solids content increases.

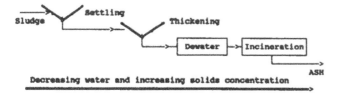

Figure 7.2 Sludge processing train.

(7) Factors affecting sludge stability could make dewatering more difficult. This includes septic conditions generating gases or reducing solids content.

Sludge Volume

It should be understood that solids or sludges contain a residual ash resulting in a disposal residue, regardless of the dewatering or treatment method. Sludge can be reduced to its ash content, which rarely (if ever) is zero. In fact, some sludge conditioning methods increase residual solid content. In most treatment facilities some residue will be shipped off-site, the quantity depends on the sludge processing train, as illustrated in Figure 7.2. The extent to which waste sludge is processed on-site prior to disposal depends on sludge quantity, transportation costs, the availability and capabilities of off-site sludge treatment facilities, and final off-site disposal costs.

Generally, if the sludge generated is less than one tank truck load a day off-site disposal should be investigated, especially if a municipal (or any centralized) facility with excess sludge processing capacity is available. Next, if gravity dewatering can reduce the sludge to less than one tank truck load a day, a simplified storage, gravity thickening, and decanting process should be contemplated.

Where large volumes are generated, complete on-site sludge treatment facilities must be economically evaluated, and must include adequate processing to minimize residue. In some cases incineration is justified to limit off-site disposal costs, since few industrial facilities will utilize potentially valuable on-site land for ash disposal. Chemical stabilization or fixation may be necessary if residue characteristics do not meet regulatory disposal criteria. In any overall, on-site facility evaluation treatment costs are more a factor than transporting costs, with incentives to reduce volume driven by reduced pumping and treatment equipment sizes, to decrease capital and energy costs.

Waste Properties

Sludge characteristics directly impacting dewatering effectiveness include solids settleability, free water drainage, pH, temperature, particle size, and solids compressibility. Solids settleability is a significant factor in establishing clarifier effectiveness, which along with thickening are the first concentration steps in a dewatering process. Next, the ability

of the *free water* to *drain* from the solids affects dewatering capabilities. Generally, water is easily drained from coarse solids, while sludges with small particle sizes, high particle charge, and high hydration capacity retain large quantities of water. Certain waste characteristics such as pH and temperature affect the dewatering process. Too high or low a pH can affect the solid particle charge, or reduce sludge conditioning effectiveness. Temperature affects the liquid viscosity, thereby affecting water drainage.

Particle *size* significantly affects void space and hydration capacity, with dewatering performance decreasing with decreasing particle size. This is because small solid particle size results in increased charge and decreased particle void space, all of which diminish draining capacity. In addition, small particle size increases solids hydration, increasing water retention.

Poor *solid compressibility* results in poor dewatering characteristics. For this reason, (compressible) organic solids are easier to dewater than inorganics. Solids should be light so that they can be easily compressed with applied mechanical force, easily rejecting water, and compacted into a condensed mass. But they should not be capable of being deformed and escape through the device restricting orifices.

Water Content

The most obvious sludge property is water content, the basis for implementing solids dewatering. The important properties affecting dewatering system performance are the raw sludge water content, the required water removal, and the solids hydration capacity. Needless to say, the dewatered sludge water content cannot be less than the solids dehydration capacity, since this level cannot be reached with mechanical dewatering devices. In fact, the required dewatering energy increases with decreasing cake dryness, becoming prohibitive as the hydration capacity is approached. At some point mechanical dewatering must be replaced with (heat) drying.

As illustrated in Figure 7.3, sludge water content consists of free and floc water, removable by gravity settling and dewatering [3]. Capillary water must be "squeezed" from the solids by extensive compaction, and bound water can best be removed with thermal energy. The important consideration is not how dry a cake can be obtained, but its final deposition. As an example, if the solids are to be disposed by land applica-

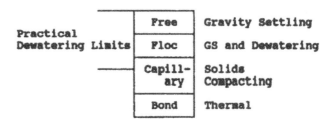

Figure 7.3 Sludge water removal limits.

tion, a high water content is favored for easy application. In any event, dry sludge becomes wet sludge after the first rainfall. At the other extreme, sludge that is to be incinerated should be dewatered to as low a water content as possible, taking into account that it takes about 4500 kcal/kg (8000 BTU/lb) of water retained to "incinerate" water.

Sludge Conditioning

Sludge conditioning variables affecting dewatering performance include the chemicals selected, chemical preparation method, mixing time, mixing intensity, and application point; the primary decision being the chemicals selected. Common conditioning chemicals used in dewatering include lime, ferric chloride, polymers, and flyash or readily available inert materials. The use of lime as a conditioning agent is primarily based on the reaction of calcium carbonate with the waste solids forming a higher porosity and better dewatering floc. Lime also helps stabilize organic sludge into a more inactive, less degradable, material. Ferric chloride acts similarly to lime in flocculating and forming a more stable and high porosity floc. Ferric chloride forms the inert iron oxide to assist in the process. Both lime and ferric chloride have the advantage of being relatively inexpensive, but have two significant disadvantages:

(1) Because of lime insolubility and ferric chloride corrosiveness these materials are difficult to handle, prepare, and transport, and work areas or machinery are difficult to keep clean.
(2) Both of these chemicals increase the cake volume, significantly increasing final processing and disposal costs.

Flyash, and similar materials, are sometimes used as inexpensive precoat materials in dewatering devices such as pressure and vacuum filters, but they too increase the final sludge volume. Table 7.2 summarizes the data reported in the EPA *Dewatering Design Manual* pertaining to the chemicals and chemical dosages commonly used for dewatering municipal sludges [14].

In theory, inorganic conditioning chemistry is relatively easy to understand. The chemicals promote a "new structure" more amenable to dewatering by interaction of sludge constituents with the inorganics. Preparation of inorganic conditioners may be cumbersome, but not complicated, and appropriate dosages easily determined within a relatively narrow range. Increased ash content resulting from inorganic conditioning makes final cake incineration expensive because of increased loading, increased potential for combustion chamber slag formation, and greater particulate loading to the air pollution control equipment. As a result, polymers are increasingly used in dewatering processes such as centrifuges and belt filters, where enhanced gravitation settling is promoted by larger and more settleable flocs.

In contrast, organic conditioning chemicals are more difficult to prepare, stability and effectiveness affected by preparation and storage time. In addition, polymer chemistry is complex and difficult to explain in absolute terms. Its application is highly dependent on the manufacturer's experience and recommendations, verified by extensive laboratory testing and on-going full scale evaluation. When a polymer application appears to be successful its effectiveness and dosage requirements can change with seemingly unexplainable cause. The major advantages of polymers are that they are usually effective at low dosages, and they produce a dewatered sludge that does not add a significant load to an incinerator, since the polymer is combustible. Their major disadvantages are relatively high cost, the complex preparation required to optimize solution reactivity, sensitivity to applied concentration and dosage, optimizing mixing with the sludge, and determining the best injection points.

A review of polymer chemistry and reported operating results will shed some light on sludge conditioning chemistry in general. Optimum polymer dosage has traditionally been presumed to be functionally related to the mean velocity gradient and mixing time product, $G \cdot t$, as illustrated in Figure 7.4.

As illustrated, a chemical dosage range can be assigned to a specified Gt product, the dosage increasing with increasing Gt product. This relation implies a definitive effective chemical dosage range for a specific set of conditioning parameters defined by the $G \cdot t$ product [6]. This requires a balance between providing sufficient polymer and solids contact (Gt) to assure good floc formation, and avoiding surplus energy which would shear the floc. On this basis, any additional polymer demand associated with increased Gt is probably a result of repairing increased floc disintegration. It therefore can be assumed that an optimum condition or a narrow range of conditions exists for a specific sludge which will minimize chemical addition and energy input.

This theoretical approach has been investigated at the laboratory scale using municipal sludge, the resulting data defined by Equation (7.2) [6,9]:

$$G^x \cdot t = k \qquad (7.2)$$

In Equation (7.2), the constant k is a function of polymer dosage. Exponent x is a function of the conditioned sludge characteristics, encompassing all the factors relating to the sludge properties and the conditioning process. The study results indicate that conditions defined by optimum polymer selection and dosage results in an x exponent value of 1. This implies that at optimum conditions shear and time are equal variables, the shearing intensity being a larger factor when the exponent is greater than 1, which is frequently the case. However, the data are insufficient to conclude that laboratory prototype conditions resulting in an exponent of 1 always result in optimum full-scale operating conditions.

TABLE 7.2. Common Municipal Sludge Chemical Conditioning (adapted from Reference [14]).

Dewatering Device	Chemical	Chemical Dosage, g/kg Dry Solids (lb/ton)			Potential Additional Dry Sludge, %
		Raw Primary	Mixed	Anaerobic Mixed	
Basket centrifuge	Polymer	0–2 (0–4)	0.5–2.5 (1–5)	1–3 (2–6)	0 to 0.30 Organic
Solid bowl centrifuge	Polymer	1–2.5 (2–5)	2–5 (4–10)	3–5 (6–10)	0.1 to 0.5 Organic
Belt filter press	Polymer	2–4 (4–8)	2–5 (4–10)	4–7.5 (8–15)	0.2 to 0.75 Organic
Vacuum filter	Polymer	2–5 (4–10)	3–6 (6–12)		0.2 to 0.6 Organic
Vacuum filter	Lime	80–100 (160–200)	90–160 (180–320)	150–210 (300–420)	10 to 27 Inorganic ash
	Ferric chloride	20–40 (40–80)	25–60 (50–120)	30–60 (60–120)	
Pressure filter	Lime	110–140 (220–280)	110–160 (220–320)	110–300 (220–600)	15 to 41 Inorganic ash
	Ferric chloride	40–60 (80–120)	40–70 (80–140)	40–100 (80–220)	

It should be pointed out that sludge shear includes stresses resulting from the conditioning process, sludge transport, and those resulting from the dewatering device. Therefore, optimum conditions at the conditioning tank may have to include "overdosing," to compensate for expected downstream shear stresses. However, excessive polymer "overdosing," besides not being economical, does not assure optimum dewatering conditions. Appropriate polymer dosage is influenced by various process conditions. They include polymer preparation and storage, the dosage range required for optimum dewatering, and any additional dosage to "repair" floc shearing or breakage in the processing. The additional chemical demand imposed by shears produced in the dewatering device or upstream transfer system may be significant, making location of additional feed points a critical operating parameter. Because of all the uncertainties discussed, the conditioning step must be designed allowing considerable operating flexibility in chemical preparation, dosage, and feedpoints. Supplementary polymer injec-

tion points should include the dewatering device, the transport line upstream from the device, or both.

The EPA *Suspended Solids Design Manual* cites flocculation criteria for water treatment sludge using alum or iron coagulants in flow-through flocculations as [12]

G up to 100 sec^{-1}
Gt ranging from 30,000 to 150,000

Corresponding criteria for activated sludge from an aeration basin are cited as

G 40 to 60 sec^{-1}
t of 20 to 30 minutes
Gt of 50,000 to 100,000

However, acceptable shear stresses minimizing sludge deterioration depend on the floc characteristics. Flocculation criteria at mixing times ranging from 20 to 30 minutes are generally defined in terms of the following range of applicable gradient velocities [19]:

Fragile flocs (biological): G of 10 to 30 sec^{-1}
Medium flocs (water treatment): G of 20 to 50 sec^{-1}
High-strength flocs (chemical ppt): G of 40 to 100 sec^{-1}

The following additional polymer dosages have been suggested to correct stresses imposed by specific dewatering devices [21]:

(1) Five mg/L of polymer for low stress conditions (Gt of 8000) similar to those imposed of drying beds
(2) Fifty mg/L of polymer for medium stress conditions (Gt of 90,000) similar to those imposed by a vacuum filter

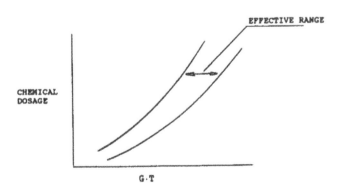

Figure 7.4 Effects $G \cdot t$ on chemical dosage (adapted from References [6,9]).

...ired twenty-five mg/L of polymer for high ...ditions (Gt of 800,00) similar to those imposed ...rifuge

...ing that inline stresses from both flocculating ...ng devices are important, it should also be ...at laboratory tests measure static conditions. ...sting includes only the mixing shear forces ...he test time and conditions, as affected by the ...ion. Repair of sludge damaged by transfer and ...ears are not included, and must be compen-...he process.

...s impossible to select chemicals or establish ...nical dosages and conditioning criteria based ...sludge sample, since in most processes sludge ...ficult to control, difficult to measure on an ...or continuous basis, and the variability is high. ...udge conditioning design must be based on a ...rating conditions, with specific consideration ...chemical alternatives, (2) chemical demand, ...strength, (4) chemical preparation conditions, ...eed points, and (6) varying conditioning times ...city.

...to sludge conditioning discussed above, im-...atment considerations include:

...ng of the feed to the dewatering device limits ...hydraulic overload and optimize performance ...the sludge pH for optimum chemical condi-...ffectiveness, especially for polymers ...large or abrasive materials from damaging ...tering system or device

...minants

...devices produce two primary products that ...fully evaluated—dewatered cake and effluent. ...d cake must be characterized to determine the ...tent, and the presence of RCRA listed constit-...which could restrict final disposal by landfill ...n.

...ischarges are a result of excess water removal, ...ondary streams created by backwashing opera-...which could contain significant conditioning ...emicals. They are generally discharged to the ...primary or secondary treatment system. In so

(2) The effluent composition, with special emphasis on difficult to remove solid "fines" or nonbiodegradable cleaning chemicals

Finally, volatile vapors emitted from open dewatering devices subjected to turbulence, such as belt filters and centrifuges, can cause either an air emission or an odor nuisance problem.

SELECTING THE DEWATERING DEVICE

Available municipal sludge dewatering experience can be readily adopted to industrial treatment systems. Performance of municipal dewatering devices is detailed in many design texts, and reviewed for industrial application in this section [7,14,20]. The primary consideration in selecting a dewatering device is cost effectiveness, as measured by the economical benefits of reduced sludge volume. In achieving that goal, process selection necessitates matching the sludge characteristics with the performance requirements, operating criteria, and site specific limitations.

PERFORMANCE REQUIREMENTS

Preliminary dewatering process development involves screening of dewatering, recovery, and conditioning design criteria to evaluate viable treatment alternatives. The required performance must conform with the dewatering and recovery capabilities of the device, employing minimum conditioning, and producing a readily disposable cake.

Dewatering Capabilities

Dewatering capabilities of specific devices are generally in the range indicated in Table 7.3. Final cake characteristics depend on the raw sludge characteristics, conditioning chemical dosages, operating conditions, and recovery limits. The required cake dryness must be carefully balanced with the required decant effluent quality because a higher recovery generally produces a wetter cake.

TABLE 7.3. General Dewatering Device Capabilities (adapted from Reference [14]).

Dewatering Device	% Water
Final settling	1
Thickening	2
Basket centrifuge	5–20

ake is required a pressure filter may
e need for a drier cake will be driven
ludge volume necessary to decrease
osts, disposal criteria, or the desire
loric value to at least 4500 kcal/kg
eration.

irements are specific to the decant
is generally recycled back to the
use primary clarification is not al-
trial treatment complexes, dewater-
ass through the secondary treatment
clarifier scum not easily removed.
icipal system design a 90% dewater-
if effluent suspended solids limits
gher, but a 95% or higher recovery
luent requirements are 20 mg/L or
strial applications recovery may be
riteria, driven by the necessity of
limits and avoiding a new separate
ubordinating cake dryness to a sec-
al capture efficiencies are tabulated

dewatering methods with final dis-
marized in Table 7.5 [14]. However,
t be offset with site specific realities
ns that include

osition
ntages and disadvantages
acilities
d by individual off-site facilities
ich could be imposed by landfilling
methods, primarily resulting from
ting of underground reservoirs

tering Device Recoveries
rom Reference [14]).

Common Capture, %
88–95
80–98
90–98

OPERATING CONSIDERATIONS

After performance conditions are met, *energy* and *labor* availability and costs will be major factors in measuring the viability of competitive dewatering devices.

Energy Requirements

Energy requirements are essentially dependent on required cake dryness, the specific device selected, and conditioning employed. Generally, energy requirements for mechanical devices are from moderate to high, and may not be the controlling factor in device selection. Selection may be significantly affected by total processing requirements, such as

(1) Long transportation costs can favor high concentrated, low volume sludges.
(2) On-site land disposal could accommodate higher cake wetness, with pumping costs a minor consideration.
(3) Incineration of the dewatered sludge would favor a dry, high caloric content sludge.

Energy considerations must take into account the available energy, the drive for energy conservation, requirements of competing devices, and the relation of energy costs to overall operating costs. Total energy costs include both direct and indirect costs. The direct energy costs are those required to generate the dewatering force for mechanical devices, or distribution pumping costs for sand beds. Direct costs are a result of process variables such as solids concentration, sludge throughput, number of machines, required recovery, and required dryness. Indirect costs are those associated with sludge transfer, storage, conditioning tanks, chemical preparation, etc. Typical energy requirements are indicated in Tables 7.6 and 7.7 [14].

Labor Requirements

The labor assigned to dewatering facilities is not usually crucial, because performance is not critical. First, unlike the treatment facility its performance is not motivated by regulatory limits. Therefore, poor performance may result in a plant full of sludge, but it will not shut down the facilities. In some cases labor is limited to what is available, and when it is available. In most industrial complexes dewatering operators are assigned on a part-time basis utilizing waste treatment or utility personnel. The support labor assigned depends on the treatment facility size, the size of the manufacturing plant labor force "pool," and the labor skills

TABLE 7.5. Comparison of Device with Disposal Method
(adapted from Reference [14]).

	Incineration	Composting	Agricultural Land Application	Landfill
Centrifuges				
Basket			X	X
Solid bowl	X	?	X	X
Belt filter	X	?	X	X
Vacuum filter	X	?	X	X
Filter press	X	X	X	X
Drying bed		X	X	X
Sludge lagoon			X	X

TABLE 7.6. Direct Dewatering Energy Requirements
(adapted from Reference [14]).

Dewatering Device	Direct Costs		Total Equivalent Electricity kwh/ton
	Fuel kcal/mton (BTU/ton)	Electricity kwh/mton (kwh/ton)	
Basket centrifuge		99–132 (90–120)	99–132 (90–120)
High-speed solid bowl centrifuge		66–99 (60–90)	66–99 (60–90)
Low-speed solid bowl centrifuge		33–66 (30–60)	33–66 (30–60)
Belt filter press		11–28 (10–25)	11–28 (10–25)
Vacuum filter		44–66 (40–60)	44–66 (40–60)
Fixed-volume filter press		44–66 (40–60)	44–66 (40–60)
Diaphragm filter press		39–61 (35–55)	39–61 (35–55)
Drying beds	5,600 (20,000)	1–2 (1–2)	3–4 (3–4)
Sludge lagoons	24,000–41,000 (88,000–146,000)	1–2 (1–2)	10–18 (9–16)

Basis: (1) 50 − 50% digested mix of primary and WAS at 3% feed. (2) 11,080 kj/kwh (10,500 BTU/kwh) and electricity efficiency at 32.5% of generation.

TABLE 7.7. Indirect (Conditioning) Dewatering Energy Requirements
(adapted from Reference [14]).

Dewatering Device	Conditioning Chemicals	Dosage g/kg (lb/ton)	Indirect Costs kwh/mton (kwh/ton)
Basket centrifuge	Polymer	3 (6)	0.7 (0.6)
Solid bowl centrifuge	Polymer	4 (8)	0.9 (0.8)
Belt filter press	Polymer	6 (12)	1.3 (1.2)

Industrial facilities are not operated as a public service, so that although any skill can be assigned to a treatment task, such an assignment in the long run must be economically justified. Economically justified means not only the costs assigned to the treatment facility, but as it affects the entire production facility. This could become an extreme burden on small plants, where many functions involved in operating and maintaining total waste treatment may have to be performed by one or two general operators, and routine maintenance often deferred. In this scenario the sludge management facilities are often shortchanged. The Process Engineer should keep this in mind and select an operating device consistent with the facility realities. This almost always means employing the simplest device that will work, and as maintenance free as possible. In any event, early inquiries should be made as to what labor is available for the facilities.

SITE-SPECIFIC, PHYSICAL CONSIDERATIONS

After performance requirements have been defined and evaluated, and labor and energy costs considered, final dewatering device selection will depend on site specific considerations such as (1) *plant size,* (2) *facility restraints,* and (3) total *environmental considerations.*

Plant Size

Plant size (total manufacturing, waste treatment, or both) is an important consideration in evaluating applicable dewatering devices. Smaller facilities should first consider off-site, contracted sludge disposal, or uncomplicated on-site systems. Complex equipment is ineffective in small plants because of the lack of available labor manpower and specific skills that can be totally devoted to operating and maintaining the devices. EPA published guidelines based on U.S. experience are summarized in Table 7.8, relating dewatering devices with (municipal) waste treatment plant size [14]. These guidelines are not absolute criteria but one of the factors to be considered.

Facility Restrictions

Dewatering facility location depends on its size and required area, and its relative distance from the sludge generating facilities. The allocated processing area includes that for the dewatering device plus associated accessories, as well as the potential need for future expansions. On-site facility restrictions could influence all aspects of the dewatering system design, starting with the dewatering technology selected, and including all ancillary systems or structures. Some constraints that may affect the design include:

(1) Existing dewatering facilities and conditioning facilities will strongly influence plant preference in any expansion or new proposed facilities. In such cases, upgraded technology can only be suggested on the basis of large economic advantages, especially if current applied technology is adequate, successfully applied, and the costs acceptable.

(2) The nature of the manufacturing facility may impose some restrictions. Concern over chemical handling and use would not be a problem with a chemical manufacturing facility, but may cause some concern at other type facilities, requiring selection of "safe" chemicals.

(3) At some facilities there may be an inherent tendency to consolidate all utility operations in a common treatment area, sharing available personnel. Compatibility with chemicals used in the waste treatment, water treatment, or manufacturing facility may influence the process selected.

(4) Where chemicals are not used in significant quantities, prepared bulk liquid chemicals may be a better consideration than dry chemicals, with necessary preparation and dilution facilities eliminated. In fact, the predominant consideration may be the required chemical quantities, their costs, and the cost of the associated facilities, subordinating all other considerations.

(5) Frequently, it is advantageous to house the dewatering equipment requiring a new building or selection of a suitable existing facility. Expansion or modification of

TABLE 7.8. Dewatering Compatibility with Plant Size (adapted from Reference [14]).

Dewatering Device	Municipal Plant Size, cu m/day (mgd)		
	<3785 (<1)	3785–37,850 (1–10)	>37,850 (>10)
Basket centrifuge		X	X
Solid bowl centrifuge		X	X
Belt filter press	low pressure	X	X
Vacuum filter		X	X
Filter press		X	X
Drying beds	X	X	
Sludge lagoons	X	X	

existing dewatering facilities could be a viable alternative. In some cases, there will be a tendency to utilize an existing (usually an old abandoned) building to shelter the dewatering operation. The use of any existing building must take into account the following:

a. The suitability of the building for the most effective dewatering system, and a device not selected to suit the building
b. The relative distance of the building to the sludge generating facilities, and a practical sludge transporting method
c. Its structural capacity for heavy weight equipment
d. The building height relative to the required depths for operating and removing proposed equipment
e. Available receiving, loading, and unloading equipment
f. Practical cost limits of upgrading an existing building compared to that of a new facility
g. Available utilities, drainage, and power
h. Structural condition of building concrete and steel

(6) Current practice may favor off-site contracted or owner-operated remote facilities. Initially these options may have been based on transferring ultimate liability, which under current regulations is not a viable consideration. Transport costs can only be justified on the basis of small volumes generated, the desire to centralize sludge disposal operations (more than one facility), or to locate such operations to a more suitable site relative to population density and possible nuisance complaints.

Environmental Considerations

Other environmental considerations which must be included in dewatering system selection include noise, vibration, odor potential, aesthetics, and groundwater contamination.

Noise

Any housed facility design will have to include noise control, which could be expensive because of heavy rotating equipment, large pumps, compressors, or vacuum equipment. Such facilities must meet OSHA requirements applicable to the operators, and at the same time eliminate any adverse effects to neighboring populated areas. This could result in equipment segregation or "sound-proofing," and impact where the facility is located.

Vibration

Structural design must take into account the vibration forces of devices such as centrifuges, which may peak during start-up or malfunctions. The equipment should be equipped with automatic devices to indicate misalignment, and in extreme cases terminate the unit operation.

Odor Potential

Odor is a variable seldom defined by the Process Engineer, but by the surrounding neighbors. Once identified, odors are seldom eliminated, even if the facilities are not operating. Most experts in this field believe that changes in odor concentration intensify the problem. Odor level changes can be a result of varying emission quantities, although even with a constant emission rate ground level concentrations will vary as a result of climatic (dispersion characteristics) changes. Many times the need to ventilate a working area (OSHA) results in strong ventilation and resulting uncontrolled emissions.

The Process Engineer must consider various emission control measures, the *first* being sludge stabilization to minimize potential release of odorous compounds such as sulfides, ammonia, and mercaptans. The next consideration is minimizing fugitive emissions, especially from outside facilities. This almost always requires enclosure or covering, which is effective but expensive, because any enclosed or covered facility requires controlled exhausts, and treatment of exhausted volumes.

Aesthetics (Visual Impact)

When all other factors have been considered and satisfactorily met, the Process Engineer must take into consideration that waste treatment facilities, and especially sludge treatment, many times are envisioned as "huge toilets." Visual impact will depend on the facility type, and could include the dewatering facility itself (lagoons or beds), transportation, or storage facilities. Therefore, the Process Engineer must consider compatibility of the dewatering equipment with the other plant equipment, isolating the system from public view, blending outdoor equipment with other facilities to reduce visual impact, or landscaping the area to downplay its existence. Visual impact should be evaluated in the early design stages.

Groundwater Contamination

This is a serious factor in evaluating drying beds or lagoons, which can percolate leachate to the groundwater. Such conditions are highly dependent on soil characteristics and the basin or bed construction. At a minimum they should be lined basins, although low permeable, concrete constructed basins equipped with leachate removal and treatment are preferred to protect against future liability.

PRACTICAL INDUSTRIAL APPLICATION

Generally, but not always, the size of an industrial wastewater facility simplifies the selection of dewatering equipment to belt filters, centrifuges, or similar devices. The rest of this chapter will emphasize these technologies. Operating convenience may sometimes override economic considerations, since the relative costs of competitive devices may not be appreciably significant in the capacities employed.

GENERAL ENGINEERING DESIGN CRITERIA

A complete sludge dewatering facility includes sludge storage, sludge transfer, pretreatment, sludge conditioning, and dewatering. Ancillary facilities are required for chemical storage, preparation and feed, backwashing, and sludge storage. Figure 7.5 illustrates a Belt Filter Dewatering Preliminary Concept Flowsheet.

FACILITY DESIGN CONSIDERATIONS

An EPA survey of dewatering facilities can be used as a process checklist of design considerations, and commonly encountered operating deficiencies. Some common considerations will be discussed, and the cited references should be reviewed for further details [13,14].

General Design Criteria

The following general design criteria are common to all dewatering systems:

(1) Scum removal should be included in all secondary treatment equipment upstream of the dewatering system. The scum should be collected, separately treated, and not mixed with the waste sludge.

(2) Dewatering units should be sized for planned or emergency downtime, based on an on-line operating time of about 80%. The system should be designed for maximum operating flexibility, minimum total shutdown, and optimum production. This could include using multiple dewatering devices, units sized so that remaining units can accommodate dewatering when one is down, provision for extra shift operation during an emergency, and sparing all critical pumps and equipment parts.

(3) Provide for peak sludge generation by supplying upstream sludge storage capacity.

(4) Evaluate the benefits of a small sludge day- or shift-tank to minimize process variations, provide constant feed to the conditioning tank, and minimize required operator attention.

(5) Check that plant utility capacities are adequate for peak sludge treatment demands, and that utility service lines to the equipment are sized for peak loads.

(6) Sludge conditioning systems should be designed so that operating conditions can be optimized to prevent floc shear. Provisions should be made to minimize floc deterioration during conditioning, transport, and dewatering. Multiple polymer addition points should be included to repair floc which could shear in the transport lines or the dewatering device.

(7) Conditioning systems will often require pH control to optimize the process. The pH probes in these systems should be selected for "dirty" conditions, with provisions for on-line cleaning and an in-line spare. The probe should be located to assure representative vessel content readings.

(8) Sample taps should be specified for all equipment inlets and outlets. They should be located with sufficient room to install sampling equipment, and collect daily samples.

(9) At a minimum, each dewatering device should have a feed flow measurement and control device. In addition, flow meters should be included to measure outlet flows and chemical additions. Chemical feed controls should allow monitoring and control of chemical feed rates, and the ability to proportion the rate to the sludge flow.

(10) Chemical lines should be piped to allow multi-point feed points into the contact tanks.

(11) Truck loading areas should be designed for freeze prevention and personnel protection during all weather conditions.

(12) Cake discharge systems should be designed for direct discharge into receiving containers, avoiding drop heights greater than 5 meters (15 feet), and avoiding messy and untidy work areas.

(13) Easy access should be provided to all equipment for process monitoring, inspection, and maintenance. Where multiple floor designs are considered, activities on sequential levels should include the "natural" flow of materials, with gravity flow included where possible.

(14) Provision should be included for extra truck or dumpster storage to accommodate any delays in transport to the final disposal site.

(15) Conditioning to curtail biological action will have to be included for any sludge or cake which will be stored for an extended period prior to processing or disposing.

(16) The corrosive nature of the sludge conditioning process, especially the chemicals, will require the use of coated steel or PVC for pipes, tanks, and mixing equipment.

(17) Sludge and lime pipe sizes should be at least 10 cm (4 inches), with provisions for flushing and "quick" disconnect couplings to allow access to plugged piping. These pipelines should be designed for velocities that prevent solids settling while minimizing erosion. The piping layout should not include any inline restrictions, use only wide sweep turns, and eliminate any configuration that will restrict solids flow. Plug prevention design considerations should include:
  a. Avoid dead-end piping by including a recirculation loop.

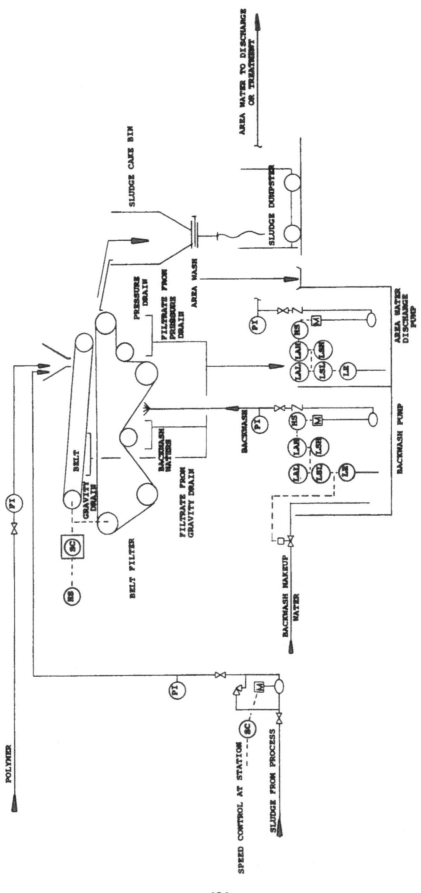

Figure 7.5 Belt filter preliminary concept flowsheet.

186

b. Install pipe takeoffs as close to the recirculation header as possible.

c. Use open channel gravity flow instead of pressure pumping wherever possible.

d. Presoften lime make-up water with hexametaphosphate, or use lime slurry supernatant.

e. Where feasible, use heavy duty flexible piping, or "quick" disconnect couplings if rigid piping must be used.

f. Design the piping to allow complete draining and flushing of all slurry lines when shut down.

(18) Piston and reciprocating equipment should be properly anchored to prevent excessive vibration, or transferring excessive vibration. Piping attached to vibrating equipment should be designed to avoid damage to the piping, or transferring vibration loading to sensitive equipment.

(19) All tanks should be provided with bottom drains for emptying and cleaning, and suitable facilities to accept tank residues or emergency dumps.

(20) Dried air should be provided for all pneumatically controlled instruments.

(21) All critical equipment should be housed for operator and equipment protection. The housing facilities should be designed so that the operating equipment is visible, can be easily monitored, and removed for maintenance. Floor drains should be strategically located to maintain a clean area and collect dewatering equipment discharges. Stairs and platforms should be provided to access all equipment.

(22) The sludge handling area should be well ventilated to prevent accumulation of generated gases such as hydrogen sulfide and ammonia. Vented gases must be scrubbed prior to exhausting.

(23) Provision should be made for dry areas to store bagged, humidity sensitive chemicals such as lime and polymer. Adequate area should be provided for storing the required inventory. Storage areas should be dry vacuum cleaned; washing stations must not be included in these water sensitive areas.

(24) Structural designs should include hoist supports for all large equipment, with adequate access for removal, repairing in place, or complete replacement.

(25) Motor control centers should be segregated from the dewatering equipment to prevent water damage. They should be installed in a separate room, and curbed from all possible water sources. Local control panels should be water-proof enclosures, curbed, and located to allow wash down of the processing area.

(26) Hydraulic or screw pumps should be used to transfer the dry cake from the collection bin to the final cake storage tank.

Belt Filters

Some specific design considerations applicable to belt filter dewatering systems include

(1) Belt selection is critical and should be based on providing adequate dewatering capacity, minimizing blinding and maximum belt wear.

(2) Because belt alignment and tension greatly influence performance, indicators should be provided to measure the tension level and alarmed to alert of any misalignment.

(3) Proper polymer addition is critical to controlling the process. Provision should be made for adequate sludge and polymer mixing, and feed point additions provided in the contact tank, transfer line, and dewatering device.

Common reported problems can be summarized as follows [13,14]:

(1) Improper belt selection

(2) Excessive belt tear resulting from applied tension and inadequate belt material

(3) Improper filter belt tracking causes the belt to constantly slide off the rollers.

(4) Safe access is not provided for the operator to adjust belt tension or tracking.

(5) Roller failures result from corrosion or damage from objects in the sludge.

(6) Frequent bearing failures result from improper roller alignment, inadequate bearing protection and seals, and poor lubrication.

(7) Inadequate control of the conditioning process results because of poor polymer and sludge mixing, not enough time is allowed for the mixing, and the inability to vary polymer addition points.

(8) Many systems lack interconnect controls to coordinate sludge feed, polymer feed, belt pressure, and sludge conveying during start-up, shutdown, or emergency conditions.

(9) Dewatering cannot be controlled because of the lack of sludge and filtrate sample points to measure performance and the inability to monitor flows.

(10) Poor conditioning performance results from improper polymer equipment sizes, constant equipment failure, the inability to control polymer to feed ratio, and the inability to control mixing.

(11) Poor sludge drainage results in excess overflow to the work area. The problem is a result of an insufficient drainage section, poor belt selection, excessive belt speed, poor belt cleaning, inadequate conditioning, or poor sludge distribution.

(12) Poor belt cleaning is a result of spray clogging and an inadequate water supply. No provision was made for

manual "emergency" stainless steel brush cleaning. Inlet filters are not provided when plant effluent is used for belt cleaning. No provision was made to use potable water as an alternative water supply.

(13) Poorly sealed spray wash unit results in a heavy mist to the area.

(14) Intermixing facilities are not provided to blend dewatered sludges, resulting in poor or highly variable cake properties.

(15) The process is difficult to operate because of major control deficiencies. A centralized control panel is not provided and local individual controls are poorly located or difficult to access. Controls are not interlocked to coordinate related control variables, and critical process flows are not controlled.

(16) No provision was made for automatic shut down in case of a belt drive failure, sludge conditioning tank failure, belt misalignment, insufficient belt tension, pneumatic or hydraulic system failure, low wash-water pressure, or a stop in the cake conveyance system. A general emergency trip wire to stop the system was not installed.

(17) Odor control is common in enclosed dewatering facilities with housed facilities that are inadequately ventilated. Point source ventilation hoods are not provided at critical process areas. Vented sources are not scrubbed prior to discharge. Drains are not hard piped and sealed to prevent fugitive emissions.

(18) Nonslip walks and floors were not provided.

(19) Noise levels were not adequately addressed.

Centrifuges

Some design considerations specific to centrifuge dewatering systems include

(1) Polymer addition is critical to assure dewatering performance. Provision should be made for adequate mixing of the sludge and polymer, and multiple feed addition points.

(2) Maintaining the solid cake integrity requires operating at lower speeds, which also reduces maintenance problems.

(3) The feed inlet should be designed to reduce impingement of the feed stream on the collected solids. Cocurrent machines are designed for this purpose.

(4) One feed pump should be provided per machine, an operating spare should be included if one machine is employed, and common operating spares installed for multiple machines.

(5) The sludge feed pumps should provide a continuous flow to the centrifuge, allowing local or remote variable feed control, and an interlock provided between the chemical and sludge feed pumps.

(6) Provision should be made to adequately flush the machine as recommended in the manufacturers operating manual.

(7) Allow for visual inspection of all critical operations.

(8) An automatically controlled backdrive should be provided to control speed differential between scroll and bowl.

(9) Centrifuges are commonly constructed of carbon steel, with stainless steel used for feed ports and other parts subject to erosion, and hard facing such as tungsten carbide material used for the conveyor and bowl. Bowls should be installed with wear strips so they do not erode.

(10) Rigid piping should be avoided for all connections to vibrating machines.

(11) Foundations must be designed for vibrating loads.

Common reported problems can be summarized as follows [13,14].

Basket Centrifuges

(1) Rigid piping is used for connections to the centrifuge.

(2) Structural support is inadequate.

(3) Inadequate solids capture results because of insufficient machine capacity, or no provision for polymer feed.

(4) Electrical control panels, located in the same room as the processing equipment, are exposed to a corrosive atmosphere.

(5) The process cannot be controlled because of inadequate sampling points to measure performance, and the inability to monitor critical flows.

Bowl Centrifuges

(1) Improper scroll tip materials result in excessive wear.

(2) Provisions are not made to access and remove the bowl assembly for maintenance.

(3) Rigid piping, used to connect the feed pipe to the centrifuge, as well as the connectors, crack and leak.

(4) Grit present in the sludge results in excessive centrifuge wear.

(5) Electronic controls, structural components, and fasteners are not designed to sustain or avoid machine vibrations.

(6) Electrical control panels located in the same room as the processing equipment are exposed to a corrosive atmosphere.

BELT FILTER PRESS: BASIC CONCEPTS

A belt filter employs a continuous moving belt to transport sludge through a sequence of stages to complete the dewatering tasks, as illustrated in Figure 7.6. Conditioned sludge is discharged onto a primary belt which accommodates free water draining. The partially dewatered sludge passes a two-belt zone, where a second top belt presses on the formed cake, and the belts and cake are subjected to increasing roller drum pressures. After shearing and compression, the belts separate and residual cake is scraped from the belt. The belts are then washed, returned to the press feed section, and the process repeated. Filtered secondary treatment plant effluent or recycled belt filter water is commonly used as belt wash. The filter components consist of a corrosion resistance frame, a belt system, rollers, bearings, a belt control system, cake discharge blades, and system controls. Ancillary equipment includes a coagulating agent preparation and feed system, a flocculating section, and a sludge feed system.

Reported characteristics of belt filters are that they (1) produce a dryer cake than most mechanical devices except a filter press, (2) require low power consumption, (3) are easy to operate, (4) allow the cake formation to be easily observed, (4) allow the operating variables to be easily changed, (5) allow operating at a low belt tension and thereby extend belt life, and (6) can be operated at low noise and machine vibration levels [14]. Cited disadvantages are (1) its sensitivity to feed characteristics and conditioning, (2) limited hydraulic throughput, (3) relatively short media life, (4) relatively high polymer requirements, (5) large quantities of wash water generated as a result of frequent belt and area washing, (6) required sludge prescreening and grinding because of machine sensitivity to large objects or fibrous material, (7) greater operator attention than a centrifuge because of critical effects of conditioning and scrapper adjustment, and (8) an open structure resulting in detectable odors.

As with all dewatering devices, belt filtration involves a balance between high solids capture, high effluent quality, and cake dryness. A high solids capture results in large recovery of solid fines and significant quantities of water with the fines, producing a wetter cake. Therefore, the Process Engineer must establish the primary criteria—a *dry cake* or *high solids capture*. If the effluent is directly discharged, high effluent quality associated with high solids capture is a significant criteria. If the effluent is returned to the head of the treatment plant, recycled water quality is still a criteria, but not a primary design concern.

PROCESS ENGINEERING DESIGN

As illustrated in Table 7.9, belt filter press performance is greatly dependent on the sludge characteristics, which are totally subject to the sludge generating processes, or the ability to condition the sludge to enhance dewatering qualities. Dewatering properties can deteriorate if storage conditions produce septic or degrading conditions. At the dewatering stage, the operator has little control of raw sludge properties. The significant operating variables are sludge conditioning and belt speed. Belt speed is a limited variable since it controls throughput, which must be equivalent to the average sludge generation rate. Belt speed also affects cake dryness, effluent quality, and the required polymer dosage. Therefore, the belt filter must be designed with considerable pretreatment and conditioning capabilities to prepare the sludge, producing a consistent sludge quality. This may require upstream storage and blending.

PRACTICAL LIMITS

Commonly achieved belt filter performances for raw, digested, secondary, conditioned, and combined municipal sludge are cited in Table 7.10. Details can be obtained from the cited references [14,15,16].

PROCESS DESIGN VARIABLES

In addition to the general design requirements discussed in the *Dewatering section*, specific process considerations

TABLE 7.9. Belt Filter Operating Characteristics.

Variable	Operator Controllable	Critical
Waste Characteristics		
Sludge generated	No	Yes
Composition	No	Yes
Concentration	No	Yes
Solids properties	No	Yes
Dewatering properties	No	Yes
Recovery properties	No	Yes
Operating Characteristics		
Flow rate	Limited	Yes
Loading rate	Limited	Yes
Chemicals	Yes	Yes
Chemical dosages	Yes	Yes
Conditioning	Yes	Yes
Speed	Yes	Yes
Belt tension	Yes	Yes
Belt washing	Yes	Yes

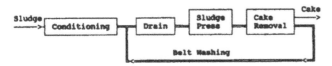

Figure 7.6 Belt filter system.

TABLE 7.10. Belt Filter Practical Limits
(adapted from References [14,16]).

Feed concentration	0.5–10%
Polymer addition	
Range	1–10 g/kg of dry sludge
Common	2–7.5
Cake dryness	
Range	10–35%
Common	20–25%
SS Recovery	
Typical	85–95%

include (1) sludge *preparation,* (2) *belt characteristics,* (3) belt filter *capacity* limits, and (4) *energy* requirements.

Sludge Preparation

Belt filter operating conditions are especially sensitive to the raw feed characteristics and sludge conditioning. Operator attention is required to avoid hydraulic overloading and belt blinding, maintaining the belt in an operable condition, and operating within the machine's capacity. Feed solids should possess bulk and fiber, requiring minimum conditioning. In addition, the sludge should be compressible, possess low water retention, and be easily agglomerated to achieve good sludge separation and a dry cake. Reference is made to the Waste Evaluation, Sludge Condition, and Pretreatment sections in the Process Engineering Design section for applicable criteria.

Belt filter performance is directly related to the *feed concentration.* Generally, higher feed solids concentration results in dryer cakes and lower capture efficiency, and therefore higher effluent suspended solids. Where cake dryness is a primary consideration, sludge thickening is frequently employed. The thickener tank or separate storage volume should be adequate to provide an extended period of reasonably consistent feed inventory. Upstream conditions affecting the sludge characteristics, producing large feed variations, could result in frequent adjustments to the dewatering operation.

Sludge conditioning is employed to correct raw waste properties resulting from poor sludge generating conditions. Organic polymers prepared at suitable dilute concentrations are commonly added at various feed points upstream from the conditioning step. Sludge conditioning is affected not only by polymer concentration and point of injection, but by polymer type, aging time, and mixing energy. Conditioning dosage is critical because under conditioned sludge will result in poor drainage and belt overflow, whereas over conditioning can result in rapid draining and poor distribution. In all cases, avoiding belt blinding is critical to successful filtering.

Chemical conditioning is complex because it is more an art than a definable science, requiring continuous jar testing and process optimizing, with performance quickly altered by un-

detectable changes in sludge quality. The best indicator of changing upstream process conditions is poor dewatering characteristics resulting in poor cake formation. This is easily observed since a belt filter is an open device. Investigation of municipal plant practices indicate that except for some installations using thermal conditioning, polymer addition is the most prevalent conditioning method [1]. A survey of municipal systems is summarized in Table 7.11. The significant process variables affecting polymer requirements and cake dryness are reported to be feed concentration, upstream operating conditions, and sludge quality. Most installations investigated used 3 to 8 g of dry powder per kg (3 to 8 lb/1000 lb) of dry sludge solids, depending on the sludge characteristics, with 4 to 6 g/kg (4 to 6 lb/1000 lb) common for most municipal systems surveyed [1]. The cited survey reported anaerobic treatment adversely affecting primary sludge dewatering, while aerobic and anaerobic digestion of secondary sludges resulted in reduced polymer requirements.

Belt Characteristics

Belt assessment begins with belt selection, followed by evaluating operating parameters such as speed and belt tension. Although in many cases these are inherent components of a proprietary manufacturer's equipment design, the Process Engineer should understand their significance when testing, rating, and selecting competitive units.

Belt Selection

Filter belts are generally 0.3 to 3.5 meters (1 to 12 feet) wide, with 2 meter (6.5 feet) widths common. Belt strength, life, and filtering characteristics depend on material and construction. Belts are commonly woven of polyester fiber for belt strength, and constructed in a coarse weave for effective dewatering. Belt strength must be adequate to withstand varying pressures and tensions, intensified by constant "stretching and relaxing" forces. A coarse weave minimizes draining resistance and blinding problems; but results in poor solids capture and greater polymer dosages, a costly operating system, and a system overly sensitive to pretreatment and sludge conditioning. A fine weave could theoretically eliminate conditioning, relying solely on the fine belt weave "straining" characteristics, but blinding and backwashing would be a major operating problem. The Process Engineer must balance these considerations in evaluating belt designs to optimize the overall filtering process. Belt life in municipal service varies from 400 to 12,000 hours, an average life of 2700 hours and 1000 to 2000 hours typical [1].

Belt Tension

Belt tension is a significant machine characteristic, specified by the manufacturer for an operating range compatible with the fixed machine variables; fine tuned to effectively balance flow rate, solids loading, and belt life. It is set at a

TABLE 7.11. Belt Filter Polymer Effects (adapted from Reference [16]).

Municipal Source	Sludge Type	Feed %		Dry Feed Loading		Dry Polymer, mg/kg dry sludge PPM (weight)	Cake Solids, %
				kg/hr/meter*	lb/hr/meter*		
Treatment plant	WAS	0.5–4		40–230	100–500	1000–10,000	20–35
Aerobic digested	P + WAS	1–3		90–230	200–500	2000–8000	12–20
Oxygen plant	WAS	1–3		90–180	200–400	4000–10,000	15–23
Anaerobic digested	WAS	3–4		40–136	100–300	2000–10,000	12–22
Treatment plant	P + WAS	3–6		180–590	400–1300	1000–10,000	20–35
Treatment plant	P + TF	3–6		180–590	400–1300	2000–8000	20–40
Anaerobic digested	P + WAS	3–9		180–680	400–1500	2000–8000	18–44
Anaerobic digested	P	3–10		360–590	800–1300	1000–5000	25–36
Raw	P	3–10		360–680	800–1500	1000–5000	28–44
Thickened and aerobic digested	P + WAS	4–8		40–230	300–500	2000–8000	12–30
Thermal conditioned	P + WAS	4–8	V	290–910	600–2000	0	25–50

*meter belt width

level to optimize solids capture, cake solids, and solids loading, based on a specified flow rate and belt speed. At any given flow rate and polymer dosage, higher tension results in maximum solids loading and a dryer cake. However, increased tension causes lower solids capture and excessive belt wear, and difficulties in containing the sludge and preventing side spillage. Belt tension is not used as an everyday operating parameter, rarely changed except in anticipation of extreme process changes. Flow rate, polymer dosage, and belt speed are the primary control variables.

Belt Speed

Belt speed is a critical operating variable affecting throughput, cake dryness, and solids recovery. Low belt speeds produce drier cakes, with polymer dosage used to control solids recovery. Increased belt speed allows higher hydraulic capacities (throughput); however, reduced drainage and press time results in decreased cake dryness and solids recovery, requiring increased polymer dosage to compensate for lower filtering effectiveness.

Belt Cleaning

An essential part of a belt filter operation is keeping the belt pores opening to allow dewatering. This is accomplished by scraping the belt surface of residual solids, and washing the solids from the belt pores. Washing is performed with on-line sprays recycling filtered effluent, or in some cases fresh water. Water conservation is critical because the filtrate and wash water can contain suspended solids levels ranging from 500 to 2000 mg/L, potentially causing both excessive hydraulic and solids loading when discharged to upstream treatment facilities [14].

Wash water must have both a low level suspended and dissolved solids content, low suspended solids to minimize potential pore plugging, and low dissolved solids because salting out could reduce belt life. In-line filters are usually employed to avoid spray nozzle plugging and to maintain an uninterrupted flow. In turn, a high dissolved solids recycle

may require some dilution (or complete replacement) with fresh spray water to increase belt life.

Large polymer additions alter the recycled wash water characteristics, potentially affecting the upstream secondary treatment facilities to which the water is discharged. As an example, large polymer concentrations could modify biological system aerator transfer characteristics, and if not biodegradable, the system performance. Recycled effluent containing large polymer quantities to a chemical treatment plant may alter the treatment chemistry and efficiency.

Belt washing flow rates range from 50 to 100% of the sludge feed rate to the machine. In terms of machine size washing rates as high as 5 L/s/m (25 gpm/ft) width may be required, with a typical rate in the range of 1 to 2 L/s/m (5 to 10 gpm/ft) belt width [1,14]. The water pressure is typically at 690 kPa (100 psi).

Capacity

A machine's capacity is limited by the applicable hydraulic and solids loadings, with solids loading generally being the more critical criteria. Based on reported operating practice, hydraulic loadings typically range from 1.5 to 3.5 L/s/m (7 to 17 gpm/ft) width [1]. Solids loadings reportedly ranged between 200 and 300 kg/m/hr (135 to 200 lb/hr/foot) and 500 and 700 kg/m/hr (340 to 470 lb/hr/foot), with the lower range probably a hydraulic loading limitation, and the higher range controlled by solids loading. An evaluation of operating data resulted in a belt filter capacity correlation based on feed concentration, expressed as (least square best fit) Equation (7.3) [1]:

$$L = 0.03256 + 0.0827X \tag{7.3}$$

where X is the feed solids expressed as percent of dry solids and L is the loading rate expressed as dry metric tons/hr/m.

Recognizing correlation error and possible limitations specific to the surveyed plants, this equation should be considered as a qualitative capacity measure, and not as definitive design data. What is significant is the apparent importance of solids feed concentration on allowable loading.

Belt filter capacities must be related to the corresponding energy requirements, relative to competitive devices. Based on the belt width, the average industry power requirement was reported as 5.7 kw/hour/meter, with a second reported higher range of 12.5 to 15.5 kw/hour/meter, probably indicative of the pumping power requirements included for some facilities [1]. The EPA *Dewatering Design Manual* [14] reports the energy requirements for belt filters as ranging from 0.011 to 0.029 kwh per kg of dry solids, or 10 to 25 kwh per short ton.

CENTRIFUGES: BASIC CONCEPTS

Theoretically, centrifugation follows the basic principles defined by Stokes' Law for particle settling, expressed by Equation (7.4) [4].

$$Vg = \frac{(Ss - Sl) \cdot d^2 \cdot g}{18 \cdot \mu} \qquad (7.4)$$

Centrifugation forces, defined by the speed (w) and radius (r) of rotation, increase the settling rate by increasing the gravitational force. Centrifugal force is defined by Equation (7.5).

$$G = w^2 \cdot r/g \qquad (7.5)$$

Combining Equations (7.4) and (7.5), Stokes' Law can be modified to include the increased gravitational force G, as indicated by Equation (7.6).

$$Vg = \frac{(Ss - Sl) \cdot d^2 \cdot G}{18 \cdot \mu} \qquad (7.6)$$

where Ss is the solids density, Sl is the density of water, d is the diameter, r its radius, w is the machine speed, μ is the water viscosity, and g is acceleration due to gravity.

As with all dewatering devices, cake dryness and recovery efficiency are the significant dewatering performance crite-

ria. Centrifuge performance can be correlated to the product of the number of gravity forces times the retention time. As an example, a clarifier operating at gravity force and one hour retention time operates at a 60 *g*-minute level, while a centrifuge operating at 3000 g and 0.5 minutes retention time has a 1500 *g*-minute effectiveness. In that way, machine variables can be explained in relation to variables affecting centrifugal force. Newton's Law defines centrifugal force (F) in terms of mass (m) times acceleration (a).

$$F = m \cdot a \qquad (7.7a)$$

When centrifugal force is applied, the acceleration (a) is equal to the distance between the particle and the axis of rotation (r) times the rotating speed (w) squared.

$$a = rw^2 \qquad (7.7b)$$

Expressed as multiples of gravity, the machine centrifugal force is proportional to the diameter (D) and the bowl speed (N):

$$F \equiv D \cdot N^2$$

Based on these definitions, it can be concluded that the machine variables affecting performance include applied force, retention time, bowl speed, and diameter.

TYPES OF CENTRIFUGES

Centrifuges are capable of producing three distinct functions involving cake formation, clarification of rejected water, and classification of separated solids according to size and density. Classification is not a very significant sludge treatment function, but producing a deposable cake and decant is important. Basic centrifuge configurations include basket, disk, and solid bowl. Table 7.12 [2] illustrates the general sludge dewatering, clarifying, and solids classifying characteristics of centrifuges based on the range of solids the units can treat. Perforated basket and solid bowl centrifuges are the basic dewatering machines, with solid bowls favored for continuous operation.

TABLE 7.12. Centrifuge Classifications (adapted from Reference [2]).

Type	Dewater	Clarify	Classify	Solids Processed
Basket				
Perforated	Very good	Poor	Good	Course
Imperforated	Fair	Good	Poor	Fine to course
Disc	Poor	Very good	Good	Fine
Solid Bowl				
Conical	Good	Poor		Fine to course
Cylindrical	Poor	Very good	Fair	
Combined	Good	Good		

Basket Centrifuge

Basket centrifuges operate semicontinuously, functioning as thickeners, and effective with difficult sludges containing high grit levels or varying solid characteristics. The batch nature of the centrifuge limits its use to relatively small facilities.

A basket centrifuge receives feed through the bottom of a rotating basket in a vertical axis, effluent discharging over a weir top, and cake formed on the bowl walls. Solids are initially skimmed through top nozzles. At a predetermined time feed is stopped, cake is scraped from a decelerating bowl and dropped from the machine bottom.

Basket centrifuges (Figure 7.7) have the advantages of (1) having both thickening and dewatering capabilities, (2) requiring little operator attention, especially if automated, (3) being very flexible in removing a wide range of solids, (4) capable of dewatering difficult sludges, although at reduced cake solids content, and (5) having a closed system which minimizes potential odor problems. Their disadvantages include (1) batch operation, with complex controls required to automate individual cycle phases, (2) high capital cost per rated capacity, (3) machine or process malfunction can produce a poor sludge which is not evident during operation, (4) mechanical requirements for structural support are greater than for a bowl unit, (5) and provision must be made for noise control.

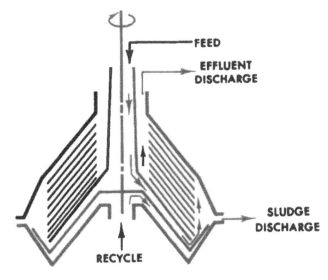

Figure 7.8 Disc centrifuge (copied from Reference [24]).

Disc Centrifuges

In principle, disc centrifuges (Figure 7.8) operate as vertical rotating baskets, except that the basket is divided into scores of narrow channels, with feed injected into each channel. In effect, the channels act as individual "baskets," or individual clarifiers, with applied centrifugal force. The short settling distance, unique to the disc channels, results in the feed solids being readily removed, and concentrated cake discharged through small openings. The disc centrifuge clarifying capabilities produce a very high quality effluent, much better than the basket centrifuge, and accordingly a wetter cake. The unit operates best at the same feed range as a clarifier, performing similarly to a clarifier, and is therefore considered a clarifier substitute rather than a dewatering device.

Solid Bowl Centrifuge

The solid bowl is the principal centrifuge used in waste sludge treatment because of its continuous nature and its combined basket and disc centrifuge characteristics. This combination results in a relatively dry cake and an acceptable effluent. Sludge is injected into a stationary feed pipe extending through the solids conveyor shaft, depositing the sludge in the rotating bowl pool. The solid bowl (Figure 7.9) is rotating at a velocity greater than the center conveyor, developing the centrifugal force separating the heavier solids from the liquid, and compacting the cake at the bowl wall. An annular ring separates the solid and liquid phases. The rotating shaft contains a screw conveyor (scroll) which transports the solids out of the system to the bowl incline (beach). At the beach, free water separates to a pool containing captured free fines, and the water is further clarified as it flows down the incline toward the discharge end. The rotating

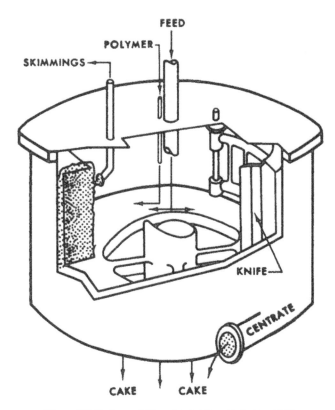

Figure 7.7 Basket centrifuge (copied from Reference [24]).

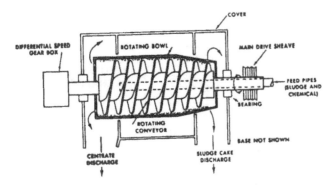

Figure 7.9 Solid bowl centrifuge (copied from Reference [24]).

conveyor moves the solids across the rotating bowl, up the incline, and to the exit port.

Bowl configuration can be conical, cylindrical, or a combined cylinder-cone, the specific design being critical in establishing the dewatering, clarifying, or thickening performance. A conical shape is effective for dewatering and solids classification, a cylindrical shape for clarification, and a cylinder-cone for a combined effect. Overall performance depends on machine characteristics such as bowl design, bowl speed, differential speed, and pool volume, as well as process variables such as feed rate, feed concentration, and conditioning. Although all machine operating parameters could be varied, the throughput rate and conditioning are usually adjusted to control operating performance. Pool volume and differential speed require machine downtime, and are not common operating variables.

Bowl centrifuges have the advantage of (1) not requiring continuous operator attention, (2) being able to vary the bowl design for thickening, dewatering, or combined performance, (3) being effective with highly variable feed solid concentrations at relatively low polymer dosages, (4) being effective at high feed rates by adjusting the polymer dosage, (5) being easily installed and requiring a relatively small area, (6) a closed continuous operation resulting in a clean appearance and clean area, and (7) being easy to start-up and shut-down. Their disadvantages include (1) scroll wear results in high maintenance requiring skilled personnel, (2) pretesting may be required to

design a machine to meet capacity, thickening, and/or dewatering criteria, (3) high power consumption for increased G capacity, (4) an enclosed machine that makes poor performance difficult to detect, (5) abrasive or large materials must be prescreened and ground, and (6) considerable mechanical engineering is required to minimize inherent noise levels and vibration at high G levels. Solid bowl performance resulting from changes in process and machine variables is summarized in Table 7.13 [2].

PROCESS ENGINEERING DESIGN

OPERATING CONTROL CHARACTERISTICS

As illustrated in Table 7.14, centrifuge effectiveness depends on the raw sludge properties; which are a result of the upstream process generating conditions, or the ability to improve the characteristics by conditioning. However, because many centrifuges are closed, machine cake formation is difficult to monitor, and performance is usually not established until the cake is analyzed. Machine performance can be fine tuned by adjusting differential speed, although this is a limited variable since capacity, cake dryness, and effluent clarity are all affected. Decreased differential speed results in reduced capacity, reduced cake dryness, and increased effluent clarity, the opposite being true with increased differential speed. Polymer dosage must be controlled to achieve an acceptable effluent clarity.

PRACTICAL LIMITS

Commonly achieved centrifuge performances for raw, digested, secondary, conditioned, and combined municipal sludge are cited in Table 7.15 [7,14,16,20]. Details can be obtained from the cited references.

PROCESS DESIGN VARIABLES

In addition to the general design requirements discussed in the Dewatering section, specific centrifuge process con-

TABLE 7.13. Effects of Machine Variable on Performance for Solid Bowl Centrifuges (adapted from Reference [2]).

Machine Variables to Improve:	Bowl speed	Pool volume	Conveyor speed	Differential speed
Recovery	inc	inc	dec	
Dryness	inc	dec	dec	dec
	Feed rate	Feed conc.	Temperature	Chemical condition
Recovery	dec	inc	inc	inc
Dryness	inc	dec	inc	dec

TABLE 7.14. Centrifuge Operating
Control Characteristics.

Variable	Operator Controllable	Critical
Waste Characteristics		
Sludge generated	No	Yes
Composition	No	Yes
Concentration	No	Yes
Solids properties	No	Yes
Dewatering properties	No	Yes
Recovery properties	No	Yes
Operating Characteristics		
Flow rate	Minimal	Yes
Loading rate	Minimal	Yes
Chemicals	Yes	Yes
Chemical dosages	Yes	Yes
Conditioning	Yes	Yes
Basket		
Speed	Yes	Yes
Volume	Yes	Yes
Cycle time	Yes	Yes
Solid bowl		
Differential speed	Yes	Yes
Pool depth	Yes	Yes

siderations include (1) sludge *preparation,* (2) *machine variables,* (3) *capacity,* (4) *configuration,* and (5) *energy* requirements.

Sludge Preparation

Conditioning can significantly impact floc dewatering characteristics such as shear resistance, compacting properties, and general dewatering parameters. Indeed, the primary goal of conditioning is to capture small particles and form a concentrated cake, maximizing the cake strength and shear resistance. The increased floc strength allows application of higher centrifugal forces, which increases recovery and produces a higher quality effluent. However, higher recovery usually results in poorer draining and a wetter cake.

TABLE 7.15. Centrifuge Practical Limits
(adapted from Reference [18]).

Feed concentration:	
Basket	0.5–5%
Scroll	0.5–15%
Polymer addition:	
Basket	500–3000 mg/kg
Scroll	500–7500 mg/kg
Cake dryness:	
Basket	5–20%
Scroll	15–30%
SS Recovery:	
Basket	80–98%
Scroll	90–98%

Conditioning must be adjusted to achieve the desired end result. Long contact time is advantageous to condition sludges containing considerable fines or heavy solids loadings. This can be achieved with a separate contact tank or by injecting the chemicals into a long feed line, upstream of the machine. Floc shear can be reduced by injecting additional conditioning chemicals directly into the machine. Typical centrifuge performance for municipal sludges, with and without conditioning, is cited in Tables 7.16 and 7.17 [7].

Machine Variables

The effects of machine and process variables are specific to the centrifuge type. Because the solid bowl centrifuge is the most adaptable for dewatering, its specific qualities will be discussed, although to some degree the basket and disc centrifuge have similar general characteristics. The principal bowl design objective is to balance required clarifying and cake thickening performance. The other machine variables include applied centrifugal force, bowl speed, pool volume, and conveyor speed. They are adjusted to achieve the required cake dryness or total solids recovery, at a range of throughputs, minimizing the chemical costs. Although the machine variables are discussed as independent entities, their effects are interdependent and controlled in a trade-off to achieve predominant performance requirements. Significantly, most of these machine controls are set at the factory, adjusted during start-up, and not manipulated on a day-to-day basis. Waste throughput and chemical conditioning are used as the primary controls.

Bowl Design

Bowl design is primarily the manufacturer's responsibility, tailoring the bowl geometry to meet required performance based on specified sludge properties, implementing proprietary machine design. A cylindrical configuration acts like a vertical clarifier operating at centrifugal forces many times that of gravity, resulting in a good effluent but a wet cake. A conical configuration operates as a thickener producing favorable dewatering conditions, but poor clarification.

The cylindrical-cone has the combined effect of balancing effluent quality with cake dryness, depending on the length to depth ratio and bowl angle. Achieving acceptable clarifying characteristics requires attention to the bowl angle and maintaining an adequate pool volume, the combination of which must provide adequate retention time. Bowl angle is directly related to the solids conveyability and recovery efficiency. High recovery requires a low bowl angle to collect finer and more fragile particle sizes prior to their being returned to the liquid pool. This produces a low slippage force and minimum solids return to the liquid pool. In addition, a low bowl angle increases the pool volume and retention time. The relationship between bowl angle, slippage, and gravitational

TABLE 7.16. Imperforate Basket Centrifuge Operating Data (adapted from Reference [18]).

Municipal Source	Sludge Type	Feed %	Dry Polymer, mg/kg dry sludge PPM (weight)	Cake Solids, %	Recovery, %
Raw	WAS	0.5–1.5	0	8–10	85–90
Raw	WAS	0.5–1.5	500–1,500	12–14	90–95
Anaerobic digested	50% P + 50% WAS	1–2	0	12–14	75–80
Anaerobic digested	50% P + 50% WAS	1–2	750–1,500	10–12	85–90
Anaerobic digested	50% P + 50% WAS	1–2	2,000–3,000	8–10	93–95
Aerobic digested	WAS	1–3	0	8–11	80–95
Aerobic digested	WAS	1–3	500–1,500	12–14	90–95
Treatment plant	TF	2–3	0	9–10	90–95
Treatment plant	TF	2–3	750–1,500	10–12	95–97
Treatment plant	70% P + 30% TF	2–3	0	9–11	95–97
Treatment plant	70% P + 30% TF	2–3	750–1,500	7–9	94–97
Treatment plant	50% P + 50% WAF	2–3	500–1,500	12–14	93–95
Treatment plant	60% P + 40% RBC	2–3	0	20–24	85–90
Treatment plant	60% P + 40% RBC	2–3	2,000–3,000	17–20	98 +
Raw	P	4–5	1,000–1,500	25–30	95–97

force is illustrated in Figure 7.10. Generally, a long bowl design with a length to diameter ratio of 2.5 to 4 improves clarification and is considered a better overall dewatering device than earlier configurations with a 1.5 to 1.8 ratio [2,20].

Centrifugal Force

A machine's centrifugal force limits are part of a manufacturer's proprietary design, dependent on bowl speed and internal machine diameters as represented by Equation (7.8) [2].

$$F = K \cdot (N^2)((D_p + D_b)/2) \qquad (7.8)$$

where

F is the multiple of gravity force, expressed in G's
N is the bowl speed, rpm
D_b is the bowl diameter, inches
D_p is the inner pool diameter, inches
K is a constant dependent on the units of g and the diameter, 0.0000142.

Bowl speed is a prime operating variable affecting its performance, since the centrifugal force generated in a machine is directly proportional to the square of the bowl speed. Commercial units have bowl speeds which develop forces equivalent to 500 to 3000 G. Increased speed generally results in higher fines capture and increased solids recovery, but not necessarily

TABLE 7.17. Solid Bowl Centrifuge Operating Data (adapted from Reference [18]).

Municipal Source	Sludge Type	Feed %	Dry Polymer, mg/kg dry sludge PPM (weight)	Cake Solids, %	Recovery, %
Treatment plant	WAS	0.5–3	5000–7500	8–12	85–90
Aerobic digested	WAS	1–3	1500–3000	8–10	90–95
Anaerobic digested	P	2–5	3000–5000	28–35	95+
Anaerobic digested	P	2–5	3000–5000	28–35	98+
Treatment plant	P + WAS	4–5	1500–3500	18–25	90–95
Raw	P	5–8	500–2500	25–36	90–95
Raw	P	5–8	0	28–36	70–90
Treatment plant	P + TF	7–10	0	35–40	60–70
Treatment plant	P + TF	7–10	1000–2000	30–35	98+
Raw	P	9–12	500–1500	25–35	82–92
Raw	P	9–12	0	30–35	65–80
Thermal conditioned	WAS	9–14	0	35–40	75–85
High lime		10–12	0	30–50	90–95
Treatment plant	P + WAS	13–15	500–2000	29–35	90–95

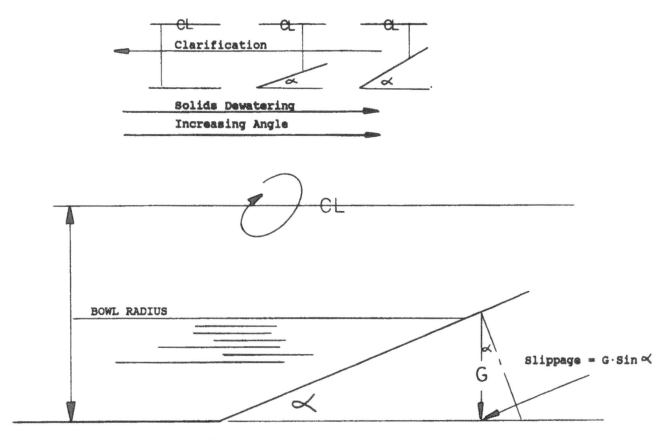

Figure 7.10 Bowl centrifuge angle and slippage (adapted from Reference [24]).

a drier cake [11]. Cake dryness depends on removal of water from the smaller particles, which in turn depends on the compacting and capillary properties of the smaller particles, all of which depend on bowl speed and particle size. Higher bowl speeds increase compacting, expelling water up to the cake draining limits, and usually improve solids recovery. However, increased solids recovery could mean increased cake fines, capable of retaining considerable water, increasing cake moisture content. Therefore, higher speed usually results in increased moisture in cakes with high fine content, and decreased moisture in coarse cakes. However, excessive speeds could produce decreased recovery, as a result of greater slippage and excessive filtrate (fine) solids content.

Pool Volume

Pool volume is set by adjusting the overflow weir, which in turn controls machine retention time and the centrifugal force generated. Commercial units maintain pool volumes equivalent to retention times ranging from 0.25 to 1.25 minutes [2].

Conveyor Speed

Conveyor speed controls the machine cake discharge rate. Speed differential, defined as the difference in the (higher) bowl speed and the helical scroll conveyor, is a critical control parameter. Its value establishes clarity efficiency, cake wetness, and capacity. Lower speed differential produce higher clarity efficiency, capturing a higher percent of poorer draining finer solids, resulting in a wetter cake. Conversely, higher speed differential results in higher (throughput) machine capacity, lower effluent quality and a dryer cake. Higher conveyor speeds and abrasive sludges increase conveyor blade wear. This wear reduces the ability of smaller particles to move to the beach, resulting in a lower solids recovery and a drier cake.

Capacity

Variables affecting bowl centrifuge capacity include differential speed, pool depth, feed rate, and conditioning. Although most of these machine variables affect centrifuge performance they are seldom adjusted on a routine basis. Instead the predominate operating variable is feed rate with volumetric capacities ranging from 0.25 to 16 liters per second (4 to 250 gpm), and dry solids throughputs ranging from 140 to 1600 kgs/hour (300 to 3500 lb/hour) [11]. Generally, increased feed rates (throughput) produces a dryer cake, but a poorer centrate quality as a result of decreased liquid (clarification) residence time. Lower feed rates improve clarification, and in some cases not significantly affecting dryness if the sludge characteristics inherently accommodate both high water drainage and solids capture, as is common with coarse solid particles [4].

A common basis for comparing operating characteristics is difficult to define because of the variety of machine models, varying operating conditions, and sludge differences. Unlike clarification, generalized centrifuge evaluations are not available. Instead, where appropriate, specific units are piloted for waste evaluation and significant performance characteristics are compared. However, centrifuge manufacturers have developed a comparison factor Q/Σ to theoretically compare machine performance based on the volumetric feed rate (Q), and a common capacity factor sigma (Σ), expressed in area units. Sigma is defined as the equivalent settler area, which theoretically produces results equivalent to the centrifuge. This comparison factor is defined by Equation (7.11) [23].

$$Q/\Sigma = Q/2\pi L(w^2/g)(3/4r_b^2 + 1/4r_p^2) \qquad (7.11)$$

where

Q is flow rate in cubic cm per sec
L is the clarification length of the bowl, cm
w is the bowl speed, radians per second
g is the acceleration of gravity, 980 cm per sec
r_b is the radius, from axis of rotation to bowl wall, cm
r_p is the radius, from axis of rotation to pond level, cm
Σ is sq cm, a machine property specific to the centrifuge

Manufacturer's Q/Σ ratio curves, correlating performance with recovery can be reviewed for relative effectiveness. These curves are generally prepared plotting the Q/Σ ratio with centrifuge recovery. This factor attempts to equate recovery and machine characteristics on a common basis.

Configuration

Centrifugal machines are often classified as low or high speed, and operated accordingly. High speed machines result in higher centrifugal forces, allowing smaller bowl diameters, resulting in greater solids capture and drier cakes. Their disadvantages include (1) more power is required, (2) more noise is generated, (3) excessive abrasion wear can result, and (4) more maintenance is required. Lower speed machines reduce these disadvantages but produce a wetter cake and a lower solids capture. One approach to offset the disadvantages of lower speeds is to operate the machine in a *countercurrent flow* instead of *cocurrent*, thereby, introducing the feed as far as possible from the cake discharge end and increasing the solids residence time.

Centrifuge capacity evaluation must include *energy requirements*. The EPA *Dewatering Design Manual* cites basket centrifuge energy demands as 0.105 to 0.140 kwh per kg of dry solids (90 to 120 kwh/ton), 0.035 to 0.070 khw per kg (30 to 60 kwh/ton) for low speed solid bowl centri-

fuges, and 0.070 to 0.105 kwh/kg dry solids (60 to 90 kwh/ short ton) for high speed bowl centrifuges [14].

CASE STUDY NUMBER 33

Investigate alternative dewatering systems for an industrial biological treatment system generating 100,000 gallons per day of sludge thickened to 20,000 mg/L concentration. Applicable design criteria are cited below.

PROCESS CALCULATIONS

(1) Quantity of solids processed per day

    0.10 mgd · 8.34 · 20,000 mg/L = 16,680 lb/day

BELT FILTER

(2) Select design criteria. Note: all loading are expressed on the basis of meter belt width.

Hydraulic loading, 1.5 liters/sec/meter belt width (24 gpm/ meter belt width)
Solids loading, 200 kg/hr/meter of belt width (445 lb/hr/meter of belt width)
Cake developed, 25% solids
Recovery, 90%
Chemicals required: 10 lb polymer per ton solids, as 1% solution
Washwater, 1.0 l/s/meter (16 gpm/meter)
Total energy (direct and indirect): 15 kwh/ton solids

(3) Calculate total chemicals required

    16,680 lb/day · 10/2000 lb/ton = 83 lb/day

    83 · 99/1 = 8217 lb water/day @ 1% solution

(4) Calculate total dry solids processed

    16,680 + 83 = 16,763 lb/day = 699 lb/hr

(5) Calculate the required belt size

(6) Size based on hydraulic loading

    100,000 gpd/1440/24 gpm/m belt ≈ 3 meter belt width

(7) Size based on solids loading

    699/445 lb/hr/meter width ≈ 1.6 meter belt width

    3 meter width belt selected

(8) Calculate cake characteristics

16,763 lb/day · 0.90 = 15,087 lb dry solids/day
15,087/0.25 = 60,348 lb wet sludge/day
Assume 100 lb/cubic foot of sludge
60,348/100 = 603 cubic feet sludge generated per day

(9) Establish effluent characteristics

(10) Effluent water

Total Influent: 100,000 · 8.34 =	834,000	lb/day
Influent solids	16,680	
Influent water (difference)	817,320	
Sludge water: 15,087 · (75/25)	45,261	(subtract)
Rejected influent water:	772,059	lb/day
Water with chemicals:	8217	lb/day

Backwash: 3 meter · 16 gpm/meter · 1440 · 8.34: 576,461 lb/day

Total effluent water: 1,356,737 lb/day

(11) Effluent solids

$$16,763 \cdot 0.1 = 1676 \text{ lb solids/day not recovered}$$

(12) Effluent solids concentration

$$1676/(1676 + 1,356,737) \cdot 1,000,000 = 1234 \text{ mg/L}$$

(13) Total effluent volume

$$(1676 + 1,356,737)/8.34 = 162,879 \text{ gallons/day}$$

(14) Energy required

$$15 \text{ kwh/ton} \cdot (16,763/2000) = 126 \text{ kwh}$$

Centrifuge

(15) Select design criteria

Cake developed, 25% solids
Recovery, 95%
Chemicals required: 5 lb polymer per ton solids, as 1% solution
Total energy (direct and indirect): 65 kwh/ton solids

(16) Calculate total chemicals required

$$16,680 \text{ lb/day} \cdot 5 \text{ lb/2000 lb} = 42 \text{ lb/day}$$

$$42 \cdot 99/1 = 4158 \text{ lb water/day @ 1\% solution}$$

(17) Calculate total dry solids processed

$$16,680 + 42 = 16,722 \text{ lb/day} = 697 \text{ lb/hr}$$

(18) Calculate the centrifuge size

100,000/1440 = 69 gpm
Select 75 − 100 gpm unit

(19) Calculate cake characteristics

16,722 lb/day · 0.95 = 15,886 lb dry solids/day
15,886/0.25 = 63,544 lb wet sludge/day
Assume 100 lb/cubic foot of sludge
63,554/100 = 635 cubic feet sludge generated per day

(20) Establish effluent characteristics

(21) Effluent water

Influent water:	817,320
Sludge water: 15,886 · (75/25)	47,658
Rejected influent water:	769,662
Water with chemicals:	4158
Total effluent water:	773,820 lbs/day

(22) Effluent solids

$$16,722 \cdot 0.05 = 836 \text{ lb solids/day not recovered}$$

(23) Effluent solids concentration

$$836/(836 + 773,820) \cdot 1,000,000 = 1079 \text{ mg/L}$$

(24) Total effluent volume

$$(836 + 773,850)/8.34 = 92,884 \text{ gallons/day}$$

(25) Energy required

$$65 \text{ kwh/ton} \cdot (16,722/2,000) = 543 \text{ kwh}$$

REFERENCES

1. ASCE Task Committee Belt Filter Presses: "Belt Filter Press Dewatering of Wastewater Sludge," *Journal of Environmental Engineering, Proceedings ASCE.* V 114, No 5, p 991, October, 1988.

2. Albertson, O.E., Guidi, E.E. Jr.: "Centrifugation of Waste Sludges," *Journal WPCF,* V 41, No 4, Pg 607, April, 1969.

3. Barber, J.B. and Veenstra, J.N.: "Evaluation of Biological Sludge Properties Influencing Volume Reduction," *Journal WPCF,* V 58, No 2, Pg 149, February, 1986.

4. Horenstein, B.K., et al.: "Fine Tuning Centrifuge Application," *Operations Forum,* Pg 12, March, 1991.

5. Knoche, W.R. and Zentkovich, T.L.: "Effects of Mean Cell Residence Time and Particle Size Distribution on Activated Sludge Vacuum Dewatering Characteristics," *Journal WPCF,* V 58, No 12, Pg 1118, December, 1986.

6. Lynch, D.P. and Novak, J.T.: "Mixing Intensity and Polymer Dosing in Filter Press Dewatering," *Research Journal WPCF,* V 63, No 2, Pg 160, March/April, 1991.

7. Medcalf & Eddy, Inc.: *Wastewater Engineering-Treatment, Disposal, Reuse,* Third Edition, McGraw-Hill, 1991.

8. Mininni, G., Spinosa, L. and Misiti, A.: "Evaluation of Filter Press Performance for Sludge Dewatering," *Journal WPCF,* V 56, No 4, Pg 331, April, 1984.

9. Novak, J.T. and Bandak, N.B.: "Chemical Conditioning and the Resistance of Sludge to Shear," *Journal WPCF,* V 61, No 3, Pg 327, March, 1989.

10. O'Connor, J.T. (Editor): "Environmental Engineering Unit Operations and Unit Processes Laboratory Manual," *Association of Environmental Engineering Professors,* July, 1972.

11. Sullivan, D.E. and Vesilind, P.A.: "Centrifuge Trade-Off, Operation Tips for Sludge Handling," *Operations Forum,* Pg 24, October, 1986.

12. U.S. Environmental Protection Agency: *Process Design Manual for Suspended Solids Removal*, EPA-625/1-75-003a, January, 1975.

13. U.S. Environmental Protection Agency: *Handbook for Identification and Correction of Typical Design Deficiencies at Municipal Wastewater Treatment Facilities*, EPA-625/6-82-007, 1982.

14. U.S. Environmental Protection Agency: *Design Manual for Dewatering Municipal Wastewater Sludges*, EPA-625/1-82-014, October, 1982.

15. U.S. Environmental Protection Agency: *Design and Operation of Belt Filter Presses*, EPA-600/D-85-192, August, 1985.

16. U.S. Environmental Protection Agency: *Design Information Report Belt Filter Presses*, EPA-600/M-86-011, May, 1986.

17. U.S. Environmental Protection Agency: *Design Information Report Recessed Filter Presses*, EPA-600/M-86-017, June, 1986.

18. U.S. Environmental Protection Agency: *Design Information Report Centrifuges*, EPA-600/M-86-023, September, 1986.

19. WPCF Manual of Practice No 8: *Wastewater Treatment Plant Design*, Water Pollution Control Federation, 1982.

20. WEF Manual of Practice: *Design of Municipal Wastewater Treatment Plants*, Water Environment Federation, 1992.

21. Werle, C.P. et al.: "Mixing Intensity and Polymer Sludge Conditioning," *Journal of Environmental Engineering, Proceedings ASCE*. 110, No 5, p 919, October, 1984.

22. Wu, Y.C., Hao, O.J. and Ou, K.C.: "Improvement of Activated Sludge Filterability," *Journal WPCF*, V 57, No 10, Pg 1019, October, 1985.

23. Hammond, D.S.: "Taking the Mystery out of Q/E in Configuration," Water and Sewage Works, August 1971, p 262.

24. US Environmental Protection Agency: Process Design Manual for Sludge Treatment and Disposal, EPA-625/1-79-007, 1979.

Index